THE CALORIC
THEORY OF GASES
FROM LAVOISIER
TO REGNAULT

THE CALORIC THEORY OF GASES
FROM LAVOISIER TO REGNAULT

BY

ROBERT FOX

OXFORD
AT THE CLARENDON PRESS
1971

Oxford University Press, Ely House, London W. 1

GLASGOW NEW YORK TORONTO MELBOURNE WELLINGTON
CAPE TOWN SALISBURY IBADAN NAIROBI DAR ES SALAAM LUSAKA ADDIS ABABA
BOMBAY CALCUTTA MADRAS KARACHI LAHORE DACCA
KUALA LUMPUR SINGAPORE HONG KONG TOKYO

© OXFORD UNIVERSITY PRESS 1971

PRINTED IN GREAT BRITAIN

PREFACE

THE caloric theory of gases has now been discredited for well over a century. Scarcely a mention of it appears in our modern textbooks of physics and even historians have paid little attention to it. Yet, as I hope to show, the theory was one of considerable distinction and historical interest. It was adopted and developed by some of the greatest men of eighteenth- and nineteenth-century science, men whose support for it manifestly did not arise from ignorance or blind, uncritical adherence to established doctrine, and it provided the theoretical basis for important innovations in the science of heat, of which the most notable is probably the famous Carnot cycle. In short, it was a theory with genuine merits and extensive influence, and as such it has long called for serious study without ever being given its due.

Since this book originated as a doctoral thesis presented in the University of Oxford in 1967, my first acknowledgement must be to my supervisor in Oxford, Dr A. C. Crombie, whose guidance and ready encouragement over a long period have been invaluable. Of my many debts to other scholars the greatest is undoubtedly that which I owe to Dr J. R. Ravetz. My discussions with him have been a constant source of inspiration and the benefits of his incisive criticisms and advice are apparent (at least to me) throughout this book. Special thanks are also due to Dr D. S. L. Cardwell, who has advised and helped me on many points since he first read the book in its thesis form some two years ago. I am also greatly indebted to Dr J. H. Brooke and Dr G. R. Talbot for their valuable criticisms of certain parts of the book which they kindly agreeed to read for me. Wherever possible other specific acknowledgements have been included in the footnotes but it seems more appropriate to use the preface to mention the correspondence I have had with Professor L. Pearce Williams (whose letter first aroused my interest in the historical problems concerning the caloric theory of gases), Mr A. M. Duncan, Mr P. M. Heimann, and Professor T. S. Kuhn.

PREFACE

My debt to the many British and continental libraries in which I have consulted books and manuscripts is very great, and the standard of service and courtesy has been so consistently high that it seems somewhat unfair to select any of the libraries for individual mention. However, no list of my acknowledgements would be complete without special reference to the help I have received in the Radcliffe Science Library, Oxford, the British Museum, and the university libraries at Lancaster and Leeds.

I am grateful to the Manchester University Press for permission to reproduce, in Chapter 3, the appendix to my paper on 'Dalton's caloric theory' which appeared in D. S. L. Cardwell (ed.), *John Dalton & the progress of science* (Manchester, 1968).

Finally I am happy to record my gratitude to the institutions and organizations that have supported me in my research. The Department of Scientific and Industrial Research (later the Science Research Council), which awarded me a research studentship in the period 1963–5, and The Queen's College, Oxford, where I was elected Clifford Norton Research Fellow in 1965, deserve special mention; but I have also received generous grants towards my research expenses from the Royal Society, the Genner Fund of Oriel College, Oxford, and the University of Lancaster.

ROBERT FOX

Lancaster
June 1969

CONTENTS

LIST OF PLATES	ix
REFERENCES AND ABBREVIATIONS	xi
NOTE ON TERMINOLOGY, UNITS, AND DATES	xv

1. INTRODUCTION

2. THE STUDY OF GASES AND HEAT TO 1800 6
    The caloric theory of gases: origins 6
    Gases and heat before 1783 20
    Specific heats of gases 32
    Adiabatic heating 39
    Expansion by heat 60

3. THE SPECIAL STATUS OF GASES 68
    Expansion by heat 69
    Adiabatic heating 79
    Rumford 99

4. THE RIVAL CALORIC THEORIES 104
    Theories of heat in Britain 105
    Theories of heat in France 121
    The 1812 competition 134
    After 1812 150

5. TWO GREAT CALORISTS: LAPLACE AND CARNOT 157
    c_p or c_v ? 157
    Laplace 165
    Sadi Carnot 177
    Other work 191

6. THE CALORIC THEORY OF AMEDEO AVOGADRO 196
    Affinity for light 196
    Affinity for heat 202
    Avogadro's new approach 218

CONTENTS

7. THE CALORIC THEORY IN DECLINE — 227
   The beginnings of revolt — 227
   Petit and Dulong on the nature of heat — 238
   The study of the thermal properties of gases, 1820–40 — 248
   Dulong and the positivists — 262
   The decline of Laplacian influence — 270

8. THE AGE OF VICTOR REGNAULT — 281
   Specific heats and the atomic theory — 282
   The physics of Victor Regnault — 295
   The kinetic theory of gases — 302

9. CONCLUSION: A SCIENCE IN DECLINE — 314

APPENDIX — 319

TABLES
   A. Specific heats of gases determined at constant pressure — 322
   B. Specific heats of gases determined at constant volume — 323
   C. The expansion of air by heat — 324
   D. The expansion by heat of gases other than air — 326
   E. Affinities for caloric and other quantities — 327

BIBLIOGRAPHY — 329

INDEX — 361

LIST OF PLATES

1. Cylinders of caloric, illustrating the Irvinist doctrines *facing p.* 26

2. Three European fire pistons 92

3. Illustration of Dalton's first theory of mixed gases 111

4. Atmospheres of caloric, illustrating Dalton's second theory of mixed gases 112

5. Diagram showing the diameters of the atoms of various gases, according to Dalton 113

6. Apparatus used by Delaroche and Bérard in the Institute's prize competition of 1812 138

7. Apparatus used by Clément and Desormes in the 1812 competition 143

8. Sketch of the thought experiment used by Clément and Desormes to calculate the output of an ideal heat engine operating expansively 180

REFERENCES AND ABBREVIATIONS

In all references the place of publication has been omitted for books with English titles which were published in London and for those with French titles which were published in Paris. Except where indicated, references are to the first editions of books.

Acta helvet.	*Acta helvetica, physico-mathematico-botanico-medica.* Basle.
Am. Scient.	*American scientist.* New Haven, Conn., etc.
Am. soc. Rev.	*American sociological review.* New York.
Ann. Phil.	*Annals of philosophy* (ed. T. Thomson). London.
Ann. Sci.	*Annals of science.* London.
Annali chim.	*Annali di chimica e storia naturale.* Pavia.
Annls Chim.	*Annales de chimie.* Paris.
Annls Chim. Phys.	*Annales de chimie et de physique.* Paris.
Annls scient. Éc. norm. sup., Paris.	*Annales scientifiques de l'École Normale Supérieure.* Paris.
Archs int. Hist. Sci.	*Archives internationales d'histoire des sciences.* Paris.
Atti Accad. Sci. Torino	*Atti della Reale Accademia delle Scienze di Torino.* Turin.
Bibliothèque universelle	*Bibliothèque universelle des sciences, belles-lettres, et arts. Sciences et arts* [later *Archives des sciences physiques et naturelles*]. Geneva, Lausanne, Paris.
Bibltca ital.	*Biblioteca italiana ossia giornale di letteratura, scienze, ed arti.* Milan.
Br. J. Hist. Sci.	*The British journal for the history of science.* London.
Br. J. Phil. Sci.	*The British journal for the philosophy of science.* Edinburgh, Cambridge.
Bull. Sci. math.	*Bulletin des sciences mathématiques, [astronomiques], physiques et chimiques. Première section du Bulletin universel des sciences . . . de M. le Baron de Férussac.* Paris.

REFERENCES AND ABBREVIATIONS

Bull. Soc. Encour. Ind. natn.	*Bulletin de la Société d'Encouragement pour l'Industrie Nationale.* Paris.
Bull. Soc. philomath. Paris	*Bulletin des sciences par la Société Philomathique de Paris.* Paris.
Bull. Soc. R. Sci. Liège	*Bulletin de la Société Royale des Sciences de Liège.* Liège.
C. r. hebd. Séanc. Acad. Sci., Paris	*Comptes rendus hebdomadaires des séances de l'Académie des Sciences.* Paris.
D.N.B.	*The dictionary of national biography* (eds. L. Stephen and S. Lee), 22 vols. London, 1885–1901.
Edinb. new phil. J.	*The Edinburgh new philosophical journal.* Edinburgh.
Edinb. Rev.	*The Edinburgh review, or critical journal.* Edinburgh.
G. Fis.	*Giornale di fisica, chimica e storia naturale* (eds. P. Configliachi and L. G. Brugnatelli). Pavia.
Gilb. Ann.	*Annalen der Physik* (ed. L. W. Gilbert). Halle.
Hist. Acad. Sci. Berlin	*Histoire de l'Académie Royale des Sciences et Belles-Lettres.* Berlin.
Introduction aux observations sur la physique	*Introduction aux observations sur la physique, sur l'histoire naturelle et sur les arts* (ed. J. Rozier). Paris.
Jber. Fortschr. phys. Wiss.	*Jahres-Bericht über die Fortschritte der physischen Wissenschaften von Jacob Berzelius.* Tübingen.
J. Éc. polytech.	*Journal de l'École Polytechnique.* Paris.
J. Mines	*Journal des mines.* Paris.
J. Phys.	*Journal de physique, de chimie, d'histoire naturelle et des arts.* Paris.
Mélang. Phil. Math. Soc. R. Turin	*Mélanges de philosophie et de mathématique de la Société Royale de Turin.* Turin.
Mém. Acad. Sci.	*Mémoires de mathématique et de physique, tirés des registres de l'Académie des Sciences.* Paris.
Mém. Acad. Sci. Inst. Fr.	*Mémoires de l'Académie [Royale] des Sciences de l'Institut de France.* Paris.
Mém. Docums. Soc. Hist. Archéol. Genève	*Mémoires et documents publiés par la Société d'Histoire et d'Archéologie de Genève.* Geneva.

REFERENCES AND ABBREVIATIONS

Mém. Hist. Sci.	Mémoires pour l'histoire des sciences & des beaux arts. Trévoux.
Mém. Phys. Chim. Soc. Arcueil	Mémoires de physique et de chimie de la Société d'Arcueil. Paris.
Mém. prés. div. Sav. Acad. Sci.	Mémoires de mathématique et de physique présentés à l'Académie Royale des Sciences par divers savans, & lûs dans ses assemblées. Paris.
Mém. prés. div. Sav. Inst. Fr.	Mémoires présentés à l'Institut des Sciences, Lettres et Arts, par divers savans, et lus dans ses assemblées. Sciences mathématiques et physiques. Paris.
Mem. Proc. Manch. lit. phil. Soc.	Memoirs [and proceedings] of the Literary and Philosophical Society of Manchester. Warrington, London, Manchester.
Mém. Sci. math. phys. Inst. Fr.	Mémoires de la classe des sciences mathématiques et physiques de l'Institut [National, Impérial] de France. Paris.
Mém. Soc. Phys. Hist. nat. Genève	Mémoires de la Société de Physique et d'Histoire Naturelle de Genève. Geneva.
Memorie Accad. Sci. Torino	Memorie della Reale Accademia delle Scienze di Torino. Turin.
Memorie Mat. Fis. Soc. ital. Sci.	Memorie di matematica e di fisica della Società Italiana delle Scienze residente in Modena. Modena.
Nicholson's Journal	A journal of natural philosophy, chemistry, and the arts (ed. W. Nicholson). London.
Notes Rec. R. Soc. Lond.	Notes and records of the Royal Society of London. London.
Nouv. Bull. Soc. philomath. Paris	Nouveau bulletin des sciences par la Société Philomathique de Paris. Paris.
Nova acta R. Soc. Scient. upsal.	Nova acta regiae societatis scientiarum Upsaliensis. Uppsala.
Observations sur la physique	Observations sur la physique, sur l'histoire naturelle et sur les arts (ed. J. Rozier). Paris.
Phil. Mag.	Philosophical magazine. London.
Phil. Trans. R. Soc.	Philosophical transactions of the Royal Society. London.

REFERENCES AND ABBREVIATIONS

Pogg. Ann.	*Annalen der Physik und Chemie* (ed. J. C. Poggendorf). Leipzig.
Proc. R. Soc.	*Proceedings of the Royal Society.* London.
Proc. R. Soc. Edinb.	*Proceedings of the Royal Society of Edinburgh.* Edinburgh.
Procès-verbaux	*Académie des Sciences. Procès-verbaux des séances de l'Académie tenues depuis la fondation de l'Institut jusqu'au mois d'août 1835*, 10 vols. Hendaye, 1910–22.
Q. Jl Sci.	*The quarterly journal of science, literature, and art.* London.
Revue Hist. Sci. Applic.	*Revue d'histoire des sciences et de leurs applications.* Paris.
Taylor's scientific memoirs	*Scientific memoirs, selected from the transactions of foreign academies of science and learned societies, and from foreign journals* (ed. R. Taylor), 5 vols. London, 1837–52.
Trans. R. Ir. Acad.	*Transactions of the Royal Irish Academy.* Dublin.
Trans. R. Soc. Edinb.	*Transactions of the Royal Society of Edinburgh.* Edinburgh.

NOTE ON TERMINOLOGY, UNITS, AND DATES

IT has generally been thought unnecessary to define or explain the elementary scientific terms used in this book. However, it should be noted that the quantity specific heat is taken throughout to indicate the heat required to raise the temperature of unit *mass* of a substance through one degree. It is represented by the symbol c_p (for the case in which the pressure remains constant during heating) or by c_v (for the case in which the volume is kept constant). Where volume specific heat is referred to, this is stated explicitly and the quantity is designated $c_p(v)$ or $c_v(v)$.

The terms 'capacity', 'heat capacity', or 'thermal capacity' indicate the heat required to raise the temperature of a body through one degree. In certain contexts they can be used interchangeably with specific heat, although it should be observed that in such cases the mass of the body is not necessarily unity.

The use of the adjective 'adiabatic' to describe conditions other than those in which there is no heat exchange at all between a gas and its surroundings has been thought unavoidable. In this book the term is frequently introduced in connection with experiments in which conditions were far from being truly adiabatic but in which thermal insulation was sufficient to allow temperature changes to occur when gases were rapidly compressed or expanded.

Celsius, Fahrenheit, and Réaumur scales of temperature have all been used but in most cases the conversion to the Celsius scale has been made for ease of comparison. The conversion from Rhenish and French feet to English feet has been made wherever necessary.

In most French primary sources of the period 1793–1805 dates are given according to the revolutionary calendar that was in use at that time in France, but these have normally been converted to the corresponding dates on the more familiar Gregorian calendar. In a few cases in which it seemed particularly instructive to retain the original form of the date the conversion has not been made.

1

INTRODUCTION

ONE of the most important events in the history of nineteenth-century physical science was the emergence, in the late 1850s, of a well-founded and generally acceptable kinetic theory of gases. According to this theory gases were thought to be composed of particles that moved about rapidly in straight lines, colliding frequently with the walls of the containing vessel and thereby giving rise to the phenomenon of gas pressure. This, of course, is basically the theory that we accept today and it is not surprising, therefore, that its origins and early development have long provided a popular and profitable subject for historical research.

The same cannot be said of the view of gas structure which the kinetic theory replaced. Admittedly it is well established that before the 1850s gases were normally thought to be composed of stationary, equally spaced particles between which repulsive forces were postulated in order to account for pressure and the other characteristic properties associated with the gaseous state. But the brief accounts of this 'static' theory that are at present available leave all too many questions unanswered. They do not tell us, for example, whether the theory had developed over the years and attained some real degree of sophistication or whether it had never progressed beyond the status of a vague and arbitrary guess. Was it a fruitful theory that had been at the centre of scientific debate, throwing up important theoretical issues, or was it rather an inhibiting factor? Why were the kineticists of the eighteenth and early nineteenth centuries so few in number and why was their influence so slight? Above all, why did so many of the great men of early nineteenth-century science—Dalton, Laplace, Gay-Lussac, and Avogadro, for example—adopt the static theory and make it the basis for some of the most elaborate experimental and theoretical work of the period? What could they achieve when their basic beliefs about gases were so seriously in error? Such questions as these, hitherto largely ignored by historians, are the ones that I have set out to answer. For the period covered in this book it would be quite impossible to separate the study of gases from that of heat, because from the late 1770s to the 1850s acceptance of the static theory generally implied belief in the imponderable, highly elastic fluid of heat, or caloric.† The reason for this is that it was the presence of caloric, either round or between the particles of gases, that was usually thought to account for the all-important repulsive forces between the particles, so that an interest in gases often arose from an interest in heat, and vice versa. Hence this book is closely concerned with the contemporary problems of the caloric theory in general and also with the problems of other theories involving imponderable fluids.

Inevitably the book touches on the still-unanswered question of why these theories lost their plausibility so rapidly during the second quarter of the nineteenth century. No complete solution is offered, although in Chapter 7 some attempt is made to associate the changing attitude towards the imponderables with a marked decline in the authority and acceptability of a style of science that is there termed Laplacian. This style of science, based on what were thought to be sound Newtonian principles and associated particularly with the names of Laplace, his close friend Berthollet, and his disciples Biot and Poisson, was dominant in France throughout the Napoleonic period. And in so far as the imponderable fluids were one of its essential ingredients, it did much to maintain the position of all theories based on the fluids, at least in France, where caloric and the corpuscular theory of light appear to have been more secure in the early years of the century than they were in Britain at that time. Indeed, the rejection of Laplacian science after 1815 is here put forward as a major cause of the discrediting of the caloric theory, although I do not suggest that this was the only cause—far from it. Any definitive study of the fortunes of caloric in general would have to take a much fuller account than I do here of the doubt that was cast on the materiality of

† This statement is, however, less true of the period from the 1820s to the 1850s, when the existence of caloric (though not the static theory itself) came increasingly to be doubted.

INTRODUCTION

heat by Berzelius's advocacy of his electrochemical theory in the decade 1810–20; this theory certainly did not disprove the existence of caloric but it went a long way towards making it an unnecessary assumption. Another development that served to shake confidence in caloric during this same decade was the rejection of the corpuscular theory of light as a result of Fresnel's successful advocacy of the wave theory after 1815. Again there was no question of actually disproving the caloric theory—this was not done until the advent of energy conservation†—but once waves in an all-pervading ether were accepted as the means for the transmission of light, then it became more natural to see radiant heat as a wave motion in the ether and hence to see the vibrations of the particles of ordinary ponderable matter as the obvious source of this motion. The electrochemical theory and the acceptance of the wave theory of light did no more than make the caloric theory less plausible, but this was enough to transform what had previously been seen as anomalies and imperfections in an otherwise satisfactory theory into major objections.‡ As it happened, the result in the 1820s was not a sudden turning towards our modern vibrational theory but a period of generally

---

† The great achievement of the energy conservationists in this respect was to discredit the prevailing view that heat was conserved in all processes, chemical and physical. This view was quite fundamental to work in the caloric theory, but it is important to note that until the 1840s it was also shared by nearly all those who opposed the theory. Sadi Carnot, who at some time after 1824 was converted to the vibrational theory (the view that heat consists in the vibrations of the particles of ordinary matter), was a notable exception.

‡ The effect of the acceptance of the wave theory of light is particularly clear in the posthumously published notes of Sadi Carnot; see *Sadi Carnot, biographie et manuscrit*, ed. É. Picard (1927), pp. 76–7, where Carnot uses the analogy between light and radiant heat as strong evidence in favour of the vibrational theory of heat. Ampère appears to have been influenced in the same way; see 'Idées de Mr. Ampère sur la chaleur et sur la lumière', *Bibliothèque universelle* **49** (1832), 225–35. Philip Kelland of Queens' College, Cambridge, who admired the work of Ampère, also saw the analogy as evidence for the vibrational theory; see his *Theory of heat* (1837), p. iii. The emergence of the electrical theory of chemical heat as a rival to Lavoisier's theory, and the changing attitude to the nature of light, seem to me the most important of the factors that served to undermine the position of caloric in the early years of the nineteenth century. I suspect that they had a considerably wider influence than, for example, the criticisms of caloric made by the followers of the Kantian tradition, which are discussed by L. P. Williams in *History of science* **1** (1962), 7–8, and by R. Kargon in his paper 'The decline of the caloric theory of heat: a case study', *Centaurus* **10** (1964), 35–9.

acknowledged agnosticism with regard to the nature of heat, a period that lasted until the caloric theory was finally abandoned about 1850. In view of this agnosticism it is hardly surprising that the caloric theory was such an easy target for the energy conservationists; at the middle of the century it was still technically the prevailing theory of heat but by then it carried very little conviction.

Comparatively little mention is made in this book of Count Rumford, a figure so beloved of the writers of textbooks on heat. The reason is that his celebrated cannon-boring experiments described to the Royal Society in 1798 by no means 'annihilated' the material theory of heat, as John Tyndall and so many others since Tyndall's day have maintained.† How could it be so when reputable eighteenth-century writers of the stature of Boerhaave and Lambert, to say nothing of Erasmus Darwin,‡ were able to explain the enormous heating effect associated with friction while maintaining their belief in the materiality of fire or heat and without adopting Rumford's view that heat consisted in the vibrations of the particles of ordinary ponderable matter? The reality of the situation is that Rumford's experiments raised at the most one new issue (discussed in Chapter 3) and, apart from this, did no more than keep alive the debate concerning the nature of heat. Rumford's work is, I believe, a red herring for the historian, and I hope that my placing of the great era of the caloric theory in the first quarter of the nineteenth century—after Rumford's famous 1798 paper—will serve to make the point.

I would stress that by concentrating attention on the caloric theory of gases, as this is defined at the beginning of the next chapter, I do not wish to imply that the theory had no rivals. In one form or another kinetic theories of gas structure had a following right through the eighteenth century and down to the middle of the nineteenth century, when, thanks chiefly to the

† Tyndall, *Heat considered as a mode of motion* (1863), p. 25 n.
‡ Boerhaave, *Elementa chemiae* (Leiden, 1732), vol. 1, pp. 176–81, and Lambert, 'Tentamen de vi caloris, qua corpora dilatat, eiusque dimensione', *Acta helvet.* 2 (1755), 175. In making this point it has to be stressed, however, that Boerhaave and Lambert both laid great emphasis on the *motion* of fire particles as the cause of heat. Thus they differed from the true calorists, for whom the motion of caloric was not a cause of heat. Darwin is a good example of a true calorist of Rumford's day who explained the heat of friction (at least to his own satisfaction); see p. 59 n. §.

work of Clausius and Maxwell, our modern kinetic theory gained the ascendancy which it has never since lost. Nevertheless, the fact remains that in the period considered here the caloric theory of gases was easily the most widely accepted theory of gas structure. As a theory, it had obvious deficiencies; many points of detail were never worked out as fully as they should have been and the experimental data that supported it left a good deal to be desired. But it also had consistency, elegance, and other genuine merits that must commend it to the historian and remove any feeling that its short period of success and popularity was at all surprising.

2

THE STUDY OF GASES AND HEAT TO 1800

The caloric theory of gases: origins

IN the form in which it was generally understood from the 1770s until it was finally rejected in the 1850s the caloric theory of gases had two distinguishing characteristics. First, the particles of gases were conceived to be stationary, being held in position by repulsive forces that were thought to exist between them. Secondly, these repulsive forces were themselves attributed to the presence, either round or between the gas particles, of the subtle, weightless, and highly elastic fluid of heat that from 1787 was known to most men of science as *caloric*.†

Although a theory embracing both of these characteristics, and so recognizable as the caloric theory of gases, does not appear to have been expounded before the 1770s, the sources for such a view of gas structure can be traced at a much earlier date. For example, the belief that the elastic properties of gases could be accounted for by supposing that gas particles were stationary and subject to mutually repulsive forces was wholly Newtonian and it was readily acknowledged as such by most of the eighteenth- and nineteenth-century writers who accepted it. In the *Principia* Newton had shown how Boyle's law could be predicted on such a basis if it was assumed that the repulsive force between any two adjacent particles of a gas expanding or

† There is evidence that the term *calorique* was current in Lavoisier's circle by 1784, for in his manuscript 'Discours d'introduction à mon cours de physique' (Library of the Institut de France, MS. 2104), stated to have been read on 2 December 1784, J. A. C. Charles refers to the use of the word by 'modern chemists'. *Calorique* was first used in print in L. B. Guyton de Morveau, A. L. Lavoisier, C. L. Berthollet, and A. F. de Fourcroy, *Méthode de nomenclature chimique* (1787), p. 31. The English form 'caloric' appeared in James St. John's translation of this work, published in London in 1788. After that date the term quickly came into common use, although a few authorities, notably the Scottish mathematician and natural philosopher John Leslie, rejected it. For Leslie's views see especially his preliminary dissertation to the seventh edition of the *Encyclopaedia Britannica* (Edinburgh, 1842), vol. 1, p. 645.

THE STUDY OF GASES AND HEAT TO 1800   7

contracting isothermally was inversely proportional to the distance between them.† At least in the *Principia* Newton was careful to couch his argument in terms appropriate to a mathematical hypothesis, maintaining that the question whether gases really were constructed in the way he had postulated was unsolved and a matter for future discussion. But in the *Opticks* he was somewhat less guarded, and in Queries 21 and 31 there are passages that certainly imply that he regarded his treatment of gases as somewhat closer to a physical truth than he cared to admit in the *Principia*.‡ In view of Newton's own vacillation on this point, it is hardly surprising that the cautious attitude of the *Principia* was not accepted by most of the eighteenth-century philosophers who are generally recognized to have worked in the Newtonian tradition. Willem Jakob Storm van 'sGravesande and Jean Théophile Desaguliers, for example, both stated without any reservations that the repulsive force between two gas particles did vary in inverse proportion to their distance apart,§ and the Revd. Stephen Hales's comment on the matter was only a little less dogmatic (and scarcely less influential).¶ Others, among them Herman Boerhaave and Pieter van Musschenbroek in Holland, made no mention of the relationship between force and inter-particle distance but even they gave sympathetic consideration to the view that a repulsive force of some sort did exist in the gaseous state.‖

Although the great textbook writers of the eighteenth century thus tended to lose sight of Newton's original intentions on the question of gas structure (at least as these intentions were

† Newton, *Philosophiae naturalis principia mathematica* (1687), pp. 301–3 (Book II, proposition xxiii). The numbering of this proposition was unchanged in all the later editions of the *Principia*.

‡ Newton, *Queries: or, a treatise of the reflections, refractions, inflections and colours of light* (4th edn, 1730), pp. 326 and 371. The numbering of the Queries used here is that found in the second and subsequent editions of the *Opticks*.

§ 'sGravesande, *Mathematical elements of natural philosophy*, trans. J. T. Desaguliers (5th edn, 1737), vol. 1, p. 216, and Desaguliers, *A course of experimental philosophy* (2 vols., 1734–44), vol. 2, p. 262. The similarity, even in wording, between the comments in these two books is largely accounted for by the fact that between 1720 and 1747 Desaguliers translated no fewer than six editions of 'sGravesande's book into English. In the earliest editions of the book, notably the original Latin version *Physices elementa mathematica* (2 vols., Leiden, 1720–1), 'sGravesande made no mention of the law of proportionality.

¶ Hales, *Vegetable staticks* (1727), p. 207.

‖ Boerhaave, *Elementa chemiae* (Leiden, 1732), vol. 1, p. 395, and Musschenbroek, *Introductio ad philosophiam naturalem* (Leiden, 1762), vol. 1, p. 856.

8   THE STUDY OF GASES AND HEAT TO 1800

expressed in the *Principia*), most of them were good Newtonians in one respect, namely in their reluctance to commit themselves on the cause of the repulsion between the particles. On this point Newton had made his own attitude quite clear in the *Opticks*. In Query 31 he had stated that it was sufficient merely to postulate the existence of a 'repulsive force', the cause of which was unknown, and he had specifically rejected mechanical explanations of the elastic properties of air based on the supposition that its particles were 'springy and ramous or rolled up like hoops'† (evidently a reference to the well-known comparison between the behaviour of air and that of a fleece of wool that had been made by Robert Boyle‡). Thus in declaring that 'we are entirely ignorant of the cause of this force [between the particles of air]', both 'sGravesande and Desaguliers were adopting an attitude very similar to Newton's own;§ and they were not alone in this. Stephen Hales, for example, considered that his experiments on the gaseous products of fermentation had quite discredited Boyle's simple mechanical picture¶ and he offered no alternative explanation. Similarly Musschenbroek, despite a tentative suggestion that electricity might in some way be the cause of repulsion, preferred not to commit himself.∥

When the true caloric theory of gases emerged, it did so against this background of widespread, though by no means universal, acceptance of the 'Newtonian' view of gas structure, allied to a generally acknowledged ignorance concerning the cause of the repulsive force between gas particles.†† The earliest

† Newton, *Opticks* (1730), p. 371.
‡ Boyle, *New experiments physico-mechanicall, touching the spring of the air, and its effects* (Oxford, 1660), pp. 23–4. Cf. the similar analogy, though in this case between air and cotton, made by Edme Mariotte in his 'Discours de la nature de l'air' of 1676; see *Œuvres de Mariotte* (Leiden, 1717), vol. 1, p. 173. Boyle also described an explanation of elasticity put by Descartes and based on the view that air was a mass of whirling particles. He considered the 'fleece of wool' analogy 'somewhat more easie' than this but declined to commit himself on the cause of the air's 'spring'. For Descartes's explanation see his *Principia philosophiae* (Amsterdam, 1644), pp. 217–18 (Book IV, paragraphs xlv–xlvii).
§ 'sGravesande, *Mathematical elements* (1737), vol. 1, p. 216, and Desaguliers, op. cit., vol. 2, p. 262.
¶ Hales, op. cit., pp. 207–8.   ∥ Musschenbroek, loc. cit.
†† For an excellent summary of theories of gas structure that deviated from the 'Newtonian' position see G. R. Talbot and A. J. Pacey, 'Some early kinetic theories of gases: Herapath and his predecessors', *Br. J. Hist. Sci.* **3** (1966–7), 133–49, where it is shown that various kinetic theories had a small

published exposition of the theory in something resembling its complete form appears to have been that given by William Cleghorn, a graduate of Trinity College, Dublin, and a nephew of George Cleghorn, a well-known professor of anatomy at the college.† In September 1779, when he was not yet 25 years old, Cleghorn presented a dissertation, with the abbreviated title *De igne*, for the degree of M.D. at Edinburgh University and in the course of this short work he described a fluid 'fire', having all the essential properties that very quickly were to become associated with caloric.‡ His fire, like caloric, was composed of particles (*ignis particulae* or *igneae particulae*) that were mutually repulsive but were attracted by ordinary ponderable matter with a force determined by the nature of the matter concerned. It was the mere accumulation of fire in a body that caused the sensation of hotness and it was because of the repulsion between its particles that fire was able to oppose and even overcome the natural attractive forces of cohesion that gave solids and liquids their characteristic structure. Fire, in fact, was responsible for the transition to the vapour state and for the great elasticity associated with that state.

It must be pointed out that by 1779 two men at least had already publicly described views on fire that bore some, though not a complete, resemblance to Cleghorn's. One of these men was none other than Lavoisier, who, in a paper presented to the Académie des Sciences in Paris on 5 September 1777, expounded the view that gases and vapours resulted from the combination of 'matter of fire' (*matière du feu* or *fluide igné*) with a 'base', which could be either a liquid or a volatile solid.§ The mechan-

---

but by no means inconsiderable following throughout the eighteenth century; also C. A. Truesdell, 'Early kinetic theories of gases', in Truesdell, *Essays in the history of mechanics* (Berlin, Heidelberg, and New York, 1968), pp. 272–82.

† For biographical details see D. McKie and N. H. de V. Heathcote, 'William Cleghorn's *De igne* (1779)', *Ann. Sci.* **14** (1958), 1–5, and the *D.N.B.* article on George Cleghorn.

‡ Cleghorn, *Disputatio physica inauguralis, theoriam ignis complectens* (Edinburgh, 1779), especially pp. 40–1. The work is conveniently reproduced, translated, and commented upon in McKie and Heathcote, op. cit., pp. 1–82, but all my references are to the original 1779 edition.

§ Lavoisier, 'De la combinaison de la matière du feu avec les fluides évaporables', *Mém. Acad. Sci.* (1777), pp. 420–32. Although this was the first occasion on which Lavoisier had made his views public, he had outlined them some five years earlier in an unpublished essay. The manuscript essay, dated August 1772, has been identified by Professor Henry Guerlac as Lavoisier's

ism of this combination, as he explained in a second paper (also dated 5 September 1777), resembled that of a normal chemical union.† Heating occurred in a process of combustion, for example, simply because the base of *air vital* (oxygen) had a greater affinity for the inflammable substance than for the matter of fire and so combined with it, allowing the fire to escape and become free. It was quite fundamental to this explanaton of chemical heat that fire could exist in two states, combined or free. When fire was combined with ordinary matter it was undetectable, and it was only when it was free that it affected the thermometer and produced the sensation of heat.‡ In fact, Lavoisier defined the true measure of hotness simply as the quantity of free, uncombined fire in a body. This distinction between the two states of fire was to be an important one in the theory of heat, since it was soon to provide one of the central issues in the debate between those calorists who saw themselves as the supporters of Lavoisier and Laplace and those who claimed to be following the chemists of the Scottish school, William Irvine and Adair Crawford.

The care with which Lavoisier made the distinction does suggest that his views on fire and heat were clearly formulated by 1777, but it seems that his conception of the nature of his matter of fire was somewhat vague at this time. Although he was in no doubt that its liberation accounted for the appearance of

'Système sur les élémens', which was handed to and initialled by the secretary of the Académie des Sciences on the 19th of that month; see Guerlac, *Lavoisier —the crucial year* (Ithaca, N.Y., 1961), pp. 96–7, and 'A last memoir of Lavoisier', *Isis* **50** (1959), 126–7. It is reproduced in R. Fric, 'Contribution à l'étude de l'évolution des idées de Lavoisier sur la nature de l'air et sur la calcination des métaux', *Archs int. Hist. Sci.* **12** (1959), 140–5, and in Guerlac, *Lavoisier*, pp. 215–23. In the essay Lavoisier states that evaporation is the result of a combination of the 'matter of fire' with another substance (*une matierre quelconque*). In a slightly later manuscript paper, dated 15 April 1773 (see Fric, op. cit., pp. 147–51) Lavoisier puts the view that air too is a combination of the matter of fire with *un fluide particulier*.

† Lavoisier, 'Mémoire sur la combustion en général', *Mém. Acad. Sci.* (1777), p. 595.

‡ Lavoisier had also made the distinction between combined fire and the fire which caused the sensation of heat (according to its 'intensity' in any body) in the manuscript paper of August 1772 cited in the previous note. Lavoisier seems to have gained the idea for this distinction from his teacher G. F. Rouelle; see Guerlac, *Lavoisier*, p. 33. But, as Dr G. R. Talbot has pointed out to me, a similar distinction was made by no less an authority than Macquer; see, for example, Macquer's *Élémens de chimie théorique* (1749), p. 12, and his *Dictionnaire de chymie* (1766), vol. 1, p. 498.

THE STUDY OF GASES AND HEAT TO 1800

both heat and light, he gave few details of its properties, referring to it merely as 'a subtle, rare, highly elastic fluid which everywhere surrounds the planet that we inhabit, entering all the bodies which make up the planet with varying degrees of facility and tending, when it is free, to attain a state of equilibrium within them'.†

The other authority whose views on fire in the 1770s bore some resemblance to Cleghorn's was the Irish physician and chemist Bryan Higgins, the uncle of the William Higgins who was later to dispute the priority for the discovery of the chemical atomic theory so vigorously with John Dalton. In a syllabus for a series of lectures by him that was to begin in London in November 1775 Higgins described fire as an 'elastic fluid' whose elasticity was 'the result of repulsion'.‡ Since he specifically invoked the analogy with 'other fluids', there can be no real doubt that he conceived fire to be made up of mutually repulsive particles, like Cleghorn's fire particles. Some years later, in 1786, he enlarged on his ideas in a detailed description of the matter of fire and its properties and in doing so gave what is immediately recognizable as the true caloric theory of gases.§ The repulsion between the particles of gases and vapours were all attributed to the 'charges of the repellant [fiery] matter' which formed 'distinct atmospheres' round the 'grosser parts' and so caused them 'to recede from each other contrary to their inherent and incessant attractive powers', and Higgins even went so far as to specify that the density of the atmospheres varied 'reciprocally as the distances from the central particles, in a duplicate or higher ratio'.¶

---

† Lavoisier, *Mém. Acad. Sci.* (1777), p. 595.

‡ Higgins, *A syllabus of the discourses and experiments, with which the meetings of the subscribers are to be opened, after the course of chemistry is concluded* (n.d. but almost certainly 1775), p. 56. The copy of this syllabus that I have used is in the library of the University of Manchester, but the syllabus was reproduced, with only minor alterations, as an introduction to Higgins's *A philosophical essay concerning light* (1776), where the relevant passage is on pp. xlii–xliii. It is interesting to note that fire was not one of Higgins's seven elements but a 'compound' of two of these, phlogiston and light; see Higgins, *Syllabus*, p. 51, and *Philosophical essay*, p. xl.

§ Higgins, *Experiments and observations relating to acetous acid* (1786), pp. 301–34, but especially p. 306. Higgins's theory of gases in this work is very similar to the one that Dalton was to adopt some years later; see pp. 109–15.

¶ Cf. the view of Gowin Knight (1748), the density of whose atmospheres of repelling matter round the particles of ordinary matter varied inversely with

The extent of Cleghorn's debt to his contemporaries and to earlier writers on heat is not easy to assess. His, after all, was far from being the first theory of heat to be based on the existence of a fluid fire, even if we except the theories adopted by Lavoisier and Higgins (of which Cleghorn was almost certainly ignorant). In fact, a number of very reputable theories that in various ways resembled the one expounded in the *De igne* (however tenuously) would have been well known at the time Cleghorn was writing.

By the 1770s the most celebrated of the theories based on the materiality of fire was the one that Boerhaave had taught for so many years at the University of Leiden earlier in the eighteenth century. This theory was fully described in the numerous editions of the great *Elements of chemistry*, a work that Cleghorn had certainly read.† Boerhaave's fire resembled Cleghorn's, and also caloric, in that it was composed of particles.‡ These were the smallest and most solid of all bodies yet known, and were also weightless, a property that was to be quite fundamental to the caloric theory, although Cleghorn himself did not comment on this particular point. Boerhaave's theory differed most markedly from that of the calorists in the very great importance that was attached by him to the motion of the particles of both fire and ordinary ponderable matter. In Boerhaave's opinion it was the motion of the particles of ordinary matter (conceived as a vibration) which was responsible for the phenomena of heat and it was the function of fire, by its own movement, simply to cause and sustain this motion.§ By contrast, of course, the movement of fire particles had no place in the true caloric theory, in which heat resulted from the mere accumulation of caloric. But theories of heat in which the ceaseless movement of fire, or some such fluid, played a crucial role did enjoy a considerable popularity

the distance from the centre; see p. 41 of the work by Knight cited on p. 18 n. ‡.

† On Cleghorn's reading of the book see p. 14. In his *Bibliographia Boerhaaviana* (Leiden, 1959), pp. 81–6, G. A. Lindeboom lists no fewer than thirty-nine editions and impressions of the *Elements* which appeared, in Latin, English, French, and German, between 1732 and 1760. Even this figure does not include abridgements (which were available in English, German, and French) and spurious editions.

‡ Boerhaave, *Elementa chemiae* (1732), vol. 1, pp. 389–93.

§ See, especially, Boerhaave, op. cit., vol. 1, pp. 142–3.

THE STUDY OF GASES AND HEAT TO 1800 13

throughout the eighteenth century, and such theories were still to be found in use even after the caloric theory had been formulated and widely adopted. The fact that they lost favour rapidly during the last quarter of the century means that comparatively little will be said of them in this book, but this rather scant treatment in no way reflects their interest and complexity, or their true importance in the early history of theories of heat. Hence I do not claim to be doing them justice in discussing them here as just one of the sources for the caloric theory proper.

It is important for the present purpose to emphasize that Boerhaave was only the most prominent and influential of the many eighteenth-century authorities who based their theories of heat on the existence of a material fire. Indeed, it seems that belief in the materiality of fire was part of standard doctrine, especially among chemists, long before Lavoisier gave his sanction to the caloric theory. Even before the earliest published version of Boerhaave's lectures had appeared the widely read 'sGravesande was already writing that the sensation of heat was caused by a motion (*agitatio*) both in the particles of bodies and in the fire that all bodies contained.† And still earlier, in the first few years of the century, the chemists Nicolas and Louis Lemery and Wilhelm Homberg were generally known to hold views on fire similar to those that only became associated with the name of Boerhaave from the 1720s, and it is possible to trace the origins of their views back still further to the Cartesian doctrine of the all-pervading subtle fluid or ether.‡ Moreover, Boerhaave's death in 1738 did not cause any noticeable decline in the popularity of his doctrine of fire. For this the strong eighteenth-century textbook tradition, maintained

† 'sGravesande, *Philosophiae Newtonianae institutiones, in usus academicos* (Leiden, 1723), p. 190. The earliest published version of Boerhaave's lectures, a spurious edition prepared without Boerhaave's consent by some of his pupils, appeared in 1724 as *Institutiones et experimenta chemiae* (2 vols., Paris). 'sGravesande's fire was similar to Boerhaave's in its properties and functions, although it was never stated to be composed of particles.

‡ This important link between Descartes's subtle fluid and the fire particles of the Lemerys, Homberg, and Boerhaave has recently been established by Dr G. R. Talbot in his Manchester University Ph.D. thesis, 'Origins and solutions of some problems in heat in the eighteenth century' (1967). See also the comment by Hélène Metzger referred to on p. 14. On p. 16.2 of his thesis Dr Talbot argues that Lavoisier's views on fire, as expressed in 1777, place him clearly in the Descartes–Boerhaave tradition.

principally by 'sGravesande and Pieter van Musschenbroek as well as by the numerous editions of Boerhaave's own writings, was chiefly responsible.† Hence, if only for this reason, Boerhaave's doctrine was still well known and widely taught in the 1770s, when Cleghorn was formulating his own views. We know, for example, that in 1775 Joseph Black referred his students at Edinburgh to Boerhaave's *Elements* for information on the theory of heat, although he studiously avoided committing himself to any particular theory.‡ Presumably the same advice was given to Cleghorn when he attended Black's lectures in 1777–8 and 1779–80, so that the appearance in the *De igne* of several references to Boerhaave's work is not at all surprising.

The strength of eighteenth-century belief in the materiality of fire is undoubtedly one of the most important elements in the background to the emergence of the caloric theory of heat in the 1770s, but it does not constitute the whole story. Certainly the belief had done much both to discredit the rival vibrational theory usually associated with the names of Bacon, Boyle, and Newton, and to make the existence of a particulate, imponderable fluid of heat thoroughly plausible. Yet there were important differences between Boerhaave's fire particles and Cleghorn's. Indeed, the late Hélène Metzger pointed out that Boerhaave's fire, both in its functions and in its structure, resembled Descartes's subtle fluid rather more closely than it did Lavoisier's caloric,§ and her observation seems well founded. As the observation implies, the Descartes–Boerhaave tradition was not alone in determining the form that the caloric theory took. There was also something strongly Newtonian about caloric and it is this Newtonian element that must now be considered.

By the time the caloric theory emerged there was already a number of highly developed theories of electricity, magnetism,

† To judge by the frequency with which their works were published and reprinted, 'sGravesande was the more influential of these two authors before 1750, but some of the most important of Musschenbroek's books appeared in the 1760s, although Musschenbroek himself died in 1761.

‡ According to a set of notes taken by an unidentified student at Black's lectures on chemistry in 1775 (Edinburgh University Library, MS. Dc. 3. 11) Black declared that the cause of the phenomena of heat was 'a subject which promises but little advantage and is very much involved in obscurity' (p. 95). Elsewhere in the lectures, however, he admitted that the theory of the 'German philosophers', according to which heat was caused by the 'tremulous motion' of a 'subtle matter', was 'most agreeable to Chymical experiments' (p. 4).

§ Metzger, *Newton, Stahl, Boerhaave et la doctrine chimique* (1930), p. 221.

THE STUDY OF GASES AND HEAT TO 1800

and light that were based on the existence of subtle, elastic, often imponderable fluids with properties remarkably similar to those of caloric. For example, in Benjamin Franklin's theory of electricity, first described in the late 1740s, the electric fluid, like caloric, was conceived as being composed of small, weightless particles which, although they were mutually repulsive, were attracted by the particles of 'common matter'.† All bodies were thought to contain a certain quantity of the fluid, even when they were electrically neutral, and they became positively or negatively charged only when an excess or deficiency of fluid was created, by friction or some other means. Although the phenomenon of repulsion between two negatively charged bodies presented serious difficulties for this theory and eventually caused modifications to be made,‡ the other important phenomena of static electricity, such as the repulsion between positively charged bodies and the attraction between those that were oppositely charged, were convincingly and elegantly explained. Certainly the theory, like the caloric theory in later years, was open to the charge that the real problem of accounting for forces acting at a distance was merely being postponed by the introduction of a hypothetical elastic fluid, but at the time it was adopted the inadequacy of existing explanations of electrical action was widely acknowledged, and advances of an experimental nature, notably the discovery of the Leiden jar in 1745, only served to make the inadequacy more obvious.§ In these circumstances a completely new theory had every chance of success and it is not surprising, therefore, that Franklin's theory attracted much attention and won many adherents. Despite a vigorous rearguard action by the Abbé Jean Antoine Nollet it quickly replaced the various doctrines based on the existence of effluvia and emanations, such as had been held not only by Nollet but also by William Gilbert,

† Franklin's views are most easily consulted in I. B. Cohen (ed.), *Benjamin Franklin's experiments* (Cambridge, Mass., 1941), where a description of the electric fluid appears on pp. 213–15. For an excellent account of the early history of Franklin's theory see Professor Cohen's *Franklin and Newton* (Philadelphia, 1956), pp. 366–573, where the slight doubts concerning Franklin's priority are also discussed.

‡ The most important modification being the introduction of the two-fluid theory; see p. 16.

§ See E. T. Whittaker, *A history of the theories of aether and electricity* (2nd edn, 1951), pp. 35–46; also Cohen, op. cit., p. 386.

Kenelm Digby, Boyle, 'sGravesande, Desaguliers, and Charles Du Fay among many others. Soon it had been adopted, albeit with some slight modifications, by the most noted 'electricians' of the period, including the German Franz Ulrich Theodor Aepinus, his pupil Johan Carl Wilcke, and the Hon. Henry Cavendish. In 1759 the position of the theory was strengthened considerably by the appearance of Aepinus's theory of magnetism, based on a fluid (of magnetism) that was analogous in nearly every respect to the electric fluid;† and even the emergence, in the same year, of a two-fluid theory of electricity‡ served only to make the imponderable, elastic fluid even more acceptable as an explanatory device.

By the 1770s the fluid theories of electricity and magnetism were well known and, in one or other of their forms, generally accepted. Full accounts of the theories were to be found in some of the most widely read scientific works of the day,§ and it is inconceivable that Cleghorn (or Lavoisier and Higgins, for that matter) would not have been thoroughly familiar with them. In fact, in one of the descriptions of his fluid theory that he gave in September 1777 Lavoisier acknowledged his debt not only to Boerhaave but also to Franklin,¶ so that the possibility that the fluids of electricity and magnetism, quite as much as Boerhaave's fire, provided the model for caloric is a strong one in this case at least.

Now all this does not prove that the early calorists looked to the existing imponderable fluids for their model—proof on this point is probably impossible anyway. But it would be plausible and consistent with the scant available evidence to suppose that the form that the caloric theory took was influenced, at least in part, by contemporary beliefs concerning elastic fluids, especially the fluids of electricity and magnetism.‖ The basic

† Aepinus, *Tentamen theoriae electricitatis et magnetismi* (St. Petersburg, [1759]).
‡ R. Symmer, 'New experiments and observations concerning electricity', *Phil. Trans. R. Soc.* **51** (1759), 371–89.
§ For example, in the numerous English, French, and German editions of Franklin's *Experiments and observations on electricity*, which had appeared since 1751, and in Joseph Priestley's *The history and present state of electricity*, which went through four English editions, as well as being translated into German, Dutch, and French, between 1767 and 1775.
¶ Lavoisier, *Mém. Acad. Sci.* (1777), p. 595.
‖ Although by the mid 1770s the extension had also been made to phlogiston,

idea for a fluid of heat, I believe, was derived from the view on fire held by Boerhaave and his followers, but certain of the characteristics of the caloric theory, most notably the mutually repulsive particles, seem to come straight from the existing fluid theories of electricity and magnetism. A tendency to extend the degree of uniformity between the various imponderables invoked to explain physical phenomena was natural enough in any circumstances but there are a number of reasons for believing that such a modification of the traditional properties of fire would have been especially welcome in the 1770s. Not the least of these is that the modification allowed the repulsive forces between gas particles to be accounted for in a far more satisfactory manner than had been possible hitherto, the explanation being, of course, the one that Cleghorn adopted in the *De igne* and that calorists recognized as an essential part of their theory until the 1850s. Naturally enough this new explanation of the inter-particle forces did much to strengthen the 'Newtonian' view of gas structure, although, as Davy pointed out some years later and as Lavoisier himself appears to have felt,† the cause of the repulsion between the particles of an imponderable fluid was really no less obscure than the cause of the repulsion between the gas particles themselves (or between similar magnetic poles or bodies having a similar electric charge).

As it happened, such objections were scarcely given the weight that was due to them, at least in the eighteenth century, and so the caloric theory and the other theories of imponderable fluids flourished as a fairly coherent body of doctrines from the 1780s until, one by one and for somewhat different reasons, they gradually fell into disrepute and were rejected in the nineteenth century. Why they should have been so generally acceptable in this period is not immediately obvious. Certainly they had some distinguished champions (in the case of the caloric theory first Lavoisier and later Laplace were probably the most important), and despite certain anomalies they did have genuine merits as theories, as I shall seek to show for the specific case of the

which Bryan Higgins, for example, described as a fluid composed of mutually repulsive particles, like light. See the *Syllabus* for his 1775 lectures, p. 16.

† See Davy's comment of 1799 in *The collected works of Sir Humphry Davy, Bart.* (9 vols., 1839–40), vol. 2, pp. 20–1, and Lavoisier, *Traité élémentaire de chimie* (1789), vol. 1, pp. 25–7.

caloric theory of gases. But I would suggest that there was one other factor that did a great deal to give all of them plausibility. This was the sanction that such theories appeared to have in the work of Newton and some other respected authorities working in the eighteenth-century Newtonian tradition. This is not to say, of course, that Newton had any conception of imponderable fluids as these were understood in the late eighteenth or early nineteenth centuries, although the static gas, his 'subtle spirit' described in the General Scholium of 1713, and, perhaps even more clearly, the ether that he described in the second and subsequent editions of the *Opticks* did have a structure, if not a function, similar to that of the later imponderables.† Moreover, physical theories involving subtle, elastic fluids, quite apart from the theories of Franklin and Aepinus, became increasingly a part of eighteenth-century Newtonianism from the 1740s. For example, in 1748 Gowin Knight, the physician and (from 1756) first principal librarian of the British Museum, described a fluid of similar structure to Franklin's in the course of his attempt to reduce all natural phenomena to the two 'active principles' attraction and repulsion, which he associated with two fundamentally different kinds of matter.‡ Knight believed that attracting matter was composed of mutually attractive particles, the particles of ordinary gross matter, while the particles of repelling matter, like Franklin's fluid, were mutually repulsive, although (again like Franklin's fluid) they were attracted by the particles of the first type and clustered round them in the form of diffuse atmospheres. The idea was used by Knight principally in his explanation of light, which he believed to be a vibration set up in the fluid of repelling matter, but it was also applied in an ingenious explanation of magnetism. It has to be emphasized that the work of Knight

† See Newton, *Principia* (2nd edn, Cambridge, 1713) p. 484 and *Opticks* (2nd edn, 1718), pp. 322–8 (Queries 17–24). In Andrew Motte's 1729 translation of the *Principia* the spirit was described as 'electric and elastic' but no further details of its supposed structure were given; in the *Opticks* (see especially Query 21) it was suggested that the postulated ether might consist of mutually repulsive particles. The function of Newton's ether, which was intended primarily as the basis for an explanation of gravitation and optical phenomena, was quite different from that of the late eighteenth-century imponderables. On Motte's interpolation of the words 'electric and elastic' see A. Koyré and I.B. Cohen, 'Newton's "electric & elastic spirit"', *Isis* 51 (1960), 337.

‡ Knight, *An attempt to demonstrate, that all the phenomena in nature may be explained by two simple active principles, attraction and repulsion* (1748).

and of the others who worked in the same branch of the eighteenth-century Newtonian tradition—notably Cadwallader Colden, Peter Dugud Leslie, James Hutton, and Adam Walker†—did not use their elastic fluids in quite the way that these were used by Franklin, Aepinus, and the calorists,‡ but they almost certainly had a contribution to make in that their work helped to establish the imponderable elastic fluids, with their mutually repulsive particles, as an acceptable explanatory device and to make them appear a perfectly natural extension of Newtonian physics.

To sum up: the caloric theory of gases seems to have emerged in the 1770s as a result of the fusion of the well-established 'Newtonian' view of gas structure with the newly devised caloric theory of heat. The caloric theory itself appears to have its roots very firmly in the doctrines of fire that enjoyed such popularity during the first three-quarters of the eighteenth century, although I believe that the precise form that the theory took was also determined in certain important respects by widely held contemporary beliefs about the imponderable fluids, particularly those of electricity and magnetism, beliefs that were first clearly expounded by Franklin in the 1740s and were seen at the time as being truly Newtonian in conception.

So the caloric theory of gases certainly had a great deal to commend it in the eyes of men working in the late eighteenth-century tradition of Newtonian science, but the acid test was whether or not it really worked. Did it yield a convincing explanation of existing experimental data? Did it prove consistent with new knowledge or, failing that, was it at least adaptable? On these counts the caloric theory in general and the caloric theory of gases in particular must be accounted a

† For the identification of this tradition and of the individuals who worked in it I am indebted to Mr J. E. McGuire and Mr P. M. Heimann, who kindly allowed me to see the typescript of their forthcoming paper, 'Newtonian forces and Lockean powers: concepts of matter in eighteenth-century thought'. In the paper it is argued that Bryan Higgins too has a place in the tradition, as does the nineteenth-century mathematician Thomas Exley, whose ideas may have influenced Faraday. Cavendish also belongs in the tradition, although he emphasized the differences between the various types of repelling matter more than the others mentioned here, who stressed that one type of repelling matter could cause a wide range of seemingly different phenomena.

‡ Notably in that they did not usually see the mere accumulation of repelling matter as the most important factor in their explanation. Higgins is a transitional figure in this respect.

20   THE STUDY OF GASES AND HEAT TO 1800

success in the period when they were most widely accepted, i.e. roughly from 1780 to 1820. Indeed, in this period they did far more than allow the phenomena to be 'saved', for they possessed that other characteristic of any really successful theory, namely the capacity to point to new and fruitful problems for investigation. As has been pointed out in the Introduction, the caloric theory of gases provided the basis for a most lively experimental and theoretical research tradition, and it is with the early history of this tradition that the rest of the chapter is concerned.

Gases and heat before 1783

The growth of interest in the thermal properties of gases after 1780 was the natural sequel to some of the most important developments in physical science during the eighteenth century. Perhaps the most fundamental of these was the discovery of methods for the preparation and identification of gases that were recognized to be of a quite different nature from ordinary atmospheric air. Although Robert Boyle had unwittingly prepared both hydrogen and nitric oxide as early as 1659,† it was not until Stephen Hales described his greatly improved methods for the collection and handling of gases in 1727‡ that the techniques necessary for the science of pneumatic chemistry became available. Hales, however, was not concerned to identify the different types of gas collected in his experiments, and another quarter of a century elapsed before Joseph Black prepared the first of what he termed the 'factitious airs'. The gas was carbon dioxide. In his celebrated doctoral dissertation of 1754 and in the enlarged version of it that appeared in English shortly afterwards,§ Black described how fixed air, as he called the new gas, had been liberated when he heated *magnesia alba* (basic magnesium carbonate). The identification of virtually all the common permanent gases followed quickly as a result of the work of Henry Cavendish, Joseph Priestley, and

† Boyle, *New experiments physico-mechanicall*, pp. 176–9.
‡ Especially important among Hales's contributions was the apparatus for the collection of gases depicted in the plate facing p. 262 of his *Vegetable staticks*.
§ Black, *Dissertatio medica inauguralis, de . . . magnesia alba* (Edinburgh, 1754), and 'Experiments upon magnesia alba', *Essays and observations, physical and literary. Read before a society in Edinburgh and published by them* (Edinburgh, 1756), vol. 2, pp. 157–225.

the Swedish chemist Carl Wilhelm Scheele,† and a period of intense activity on the part of all these men culminated in the discovery of oxygen, made independently by Scheele in 1773 and Priestley in 1774.‡ The discovery, quickly and brilliantly applied by Lavoisier, led to an entirely new theory of combustion which by the early 1780s had already reached a high stage of development and was beginning to replace the rival phlogiston theory. It is well known that the victory was not easily won. Pierre Joseph Macquer and Priestley were two who remained phlogistonists until their deaths, in 1784 and 1804 respectively, and such influential chemists as Claude Louis Berthollet, Antoine François de Fourcroy, and Louis Bernard Guyton de Morveau were converted only in the mid-1780s. In Britain Lavoisier's doctrines do not seem to have been taught publicly until the winter of 1787, when Thomas Charles Hope first expounded them at Edinburgh.§ But by this time the debate was going very much Lavoisier's way and gases, which played such a prominent part in his system, were already the subject of ever-increasing attention.

By itself, of course, the rise of pneumatic chemistry would not have stimulated any special interest in the thermal properties of gases, and we turn now to consider two other developments of the eighteenth century that served to direct attention to this particular study. The first development was the discovery of the concept of specific heat by Black about 1760.¶ Now it is hardly necessary to point out that the newly discovered gases possessed other thermal properties that deserved investigation: the extent to which they expanded when heated, for example, or the temperature changes that accompanied their rapid

† Cavendish, 'Experiments on factitious air', *Phil. Trans. R. Soc.* **56** (1766), 141–84; Priestley, 'Observations on different kinds of air', *Phil. Trans. R. Soc.* **62** (1772), 147–264, also the three volumes of his *Experiments and observations on different kinds of air* and the three volumes of *Experiments and observations relating to . . . natural philosophy* which appeared between 1774 and 1786; Scheele, *Abhandlung von der Luft und dem Feuer* (Uppsala and Leipzig, 1777), conveniently translated in *The collected papers of Carl Wilhelm Scheele*, trans. L. Dobbin (1931), pp. 86–178.

‡ Scheele, *Collected papers*, pp. 105–6; Priestley, *Phil. Trans. R. Soc.* **65** (1775), 387–9 and 392, and *Experiments and observations on air* (1775), vol. 2, pp. 29–103.

§ See T. S. Traill, *Trans. R. Soc. Edinb.* **16** (1849), 420.

¶ For a full account of this see D. McKie and N. H. de V. Heathcote, *The discovery of specific and latent heats* (1935).

22   THE STUDY OF GASES AND HEAT TO 1800

expansion or compression. But although these other properties aroused a good deal of interest and could in themselves have provided the basis for a lively research tradition, the theoretical problems associated with the determination of specific heats proved to be of a particularly fundamental nature, as we shall see.

Moreover, a whole new science of calorimetry, which owed much to Black's clear understanding of the concept of quantity of heat, had grown from his discovery of specific heat and from that of latent heat that he made soon afterwards. Indeed, in its importance as a research interest, calorimetry was a close rival to the new pneumatic chemistry in the eyes of chemists and other men of science during the last quarter of the eighteenth century, for although none of Black's own writings on the subject appeared in his lifetime, his influence in this period extended far beyond his immediate contacts, and important early experiments on specific heat were conducted by, among others, Richard Kirwan in Ireland and by Johann Carl Wilcke and Johan Gadolin in Scandinavia.† While noting this work as evidence of a widespread interest in calorimetry, we shall not examine it further since it was not concerned with matter in the gaseous state.

It was natural that the increasing familiarity of the concept of quantity of heat should tend to promote the material, as opposed to the kinetic, theory. The principles of calorimetry were so much easier to grasp when heat was considered to be an elastic fluid rather than a vibration or some other motion of the particles of ordinary matter that implied support for the material theory was all too easily read into contemporary accounts. In fact in the 1780s, when interest in calorimetry was beginning to mount rapidly, the vibrationalists' cause was still by no means lost, but its position was weakening. It was at this time that it lost two of its most influential supporters—Macquer, by his death in 1784,‡ and Fourcroy, Macquer's

† Wilcke, 'Über die specifische Menge des Feuers ...', *Der Königl. Schwedischen Akademie der Wissenschaften neue Abhandlungen*, **2** (1781; published Leipzig, 1784), pp. 48–79; Gadolin, 'Über der Körper absolute Wärme', *Der Königl. Schwedischen ...*, **5** (1784; published Leipzig, 1786), pp. 222–39, and *De theoria caloris corporum specifici* (Abo, 1784), pp. 13–16. On Kirwan's work see J. H. de Magellan, *Essai sur la nouvelle théorie du feu élémentaire* (London, 1780), pp. 176–9.

‡ For evidence of Macquer's support see his *Dictionnaire de chymie* (2nd edn, 1778), vol. 2, pp. 168–92. In earlier writings Macquer had adopted the view that

pupil, by his conversion to the material theory about 1786†—
and the cause as a whole became increasingly marked by a
shortage of new ideas. I shall have more to say on this point in
the next chapter, where I shall argue that even the most
renowned of the eighteenth-century supporters of the vibrational theory, Count Rumford, said little, if anything, that was
really original and did hardly more than keep the debate alive
when he described his famous cannon-boring experiments in
1798. The vibrationalists as a whole seemed unable to relate
the wealth of new discoveries in heat to their own theory,
while the calorists by contrast rose most impressively to the
challenge.

The growing dominance of, and interest in, the caloric theory
during the last quarter of the eighteenth century is the second
of the two developments that were cited above as being chiefly
responsible for directing some of the general preoccupation
with pneumatic chemistry towards the study of the thermal
properties of gases. In this connection it is curious that, despite
all that his discoveries did to promote both the theory and the
study of specific and latent heats, Black himself is reputed to
have had positivist leanings on the question of the nature of
heat.‡ John Robison, who later succeeded Black as lecturer in
chemistry at Glasgow and who, in 1774, as the new professor of

heat was caused by the rapid motion of the particles of a material fire; see his
Élémens de chimie-théorique (1749), pp. 12–14, and the first edition of the
Dictionnaire (1766), vol. 1, pp. 498–507.

† Cf. Fourcroy's very different opinions in his *Élémens d'histoire naturelle*
(2nd edn, 1786), vol. 1, p. 117, and in the 'Discours préliminaire' to the same
work (vol. 1, p. xxiv). According to Dr W. A. Smeaton, in his *Fourcroy, chemist
and revolutionary, 1755–1802* (Cambridge, 1962), p. 102, the former passage, in
which the vibrational theory is supported, was written in 1784, while the
'Discours préliminaire', in which the material theory is adopted, was presumably written in 1786. On Fourcroy's change of opinion see also the third
edition of the *Élémens* (1789), vol. 1, p. 127, and [P. A. Adet], *Supplément à la
2e édition d'histoire naturelle* (1789), pp. 11–14. About this time the theory lost
yet another sympathizer in the person of Richard Watson, Bishop of Llandaff
since 1782 and earlier professor of chemistry and then Regius professor of
divinity at Cambridge, who announced, in 1786, that he had destroyed all his
'chemical manuscripts'; see the Preface to vol. 4 (1786) of his *Chemical essays*.
In vol. 1 of these widely read *Essays* (1781, pp. 149–64) Watson declared himself ignorant of the cause of heat but gave considerably more prominence and
favour to the vibrational theory (of Newton) than he did to its rivals, notably
Boerhaave's doctrine of fire.

‡ On Black's 'positivism' see McKie and Heathcote, *Specific and latent
heats*, pp. 27–30.

natural philosophy, became his colleague in Edinburgh, gave the following recollection of their first meeting in 1758:

> Gently and gracefully checking my disposition to form theories, he [Black] warned me to suspect all theories whatever, pressed on me the necessity of improving in mathematical knowledge, and gave me Newton's *Optics* to read, advising me to make that book the model of all my studies, and to reject, even without examination, every hypothetical explanation, as a mere waste of time and ingenuity.†

From the posthumously published *Lectures on the elements of chemistry*, edited by Robison from Black's own lecture notes and from notes taken by a student and corrected by Black himself, we obtain the same impression of caution, but there are at least hints of a slight preference for the material theory of heat. Of the version of this theory proposed by Cleghorn, Black said: '... such an idea of the nature of heat is ... the most probable of any that I know', although he added: 'It is, however, altogether a supposition.'‡ Elsewhere in the *Lectures* he reaffirmed his caution by firmly dissociating himself from a view that had been widely adopted by the time he revised his notes, in the late 1790s. This view, which illustrates clearly how acceptance of the discovery of latent heat could lead also to acceptance of the material theory, was that changes of state occurred because the fluid of heat entered into some sort of chemical combination with ordinary matter, thereby becoming imperceptible to the thermometer, or 'latent'.

> This [Black said] will please the imagination, but does not advance our knowledge. I therefore avoid such speculations, as taking up time which may be better employed in learning more of the general laws of chemical operations. I content myself with saying, that in liquefaction and vaporisation, water absorbs a great quantity of heat, because this expression immediately raises the notion of a sudden, and somewhat copious, accumulation of heat.§

---

† Black, *Lectures on the elements of chemistry*, ed. J. Robison (Edinburgh, 1803), vol. 1, p. vii. For a similar view of Black see A. Ferguson 'Life of Black', *Trans. R. Soc. Edinb.* **5** (1805), part 3, p. 111.

‡ Black, op. cit., vol. 1, p. 34. Now, in the last years of his life, Black was somewhat less cautious on the question of the nature of heat than he had been in 1775 (see p. 14). The published notes were revised in the late 1790s (see Black, op. cit., vol. 1, p. x) and presumably represent his views at that time.

§ Black, op. cit., vol. 1, p. 193.

THE STUDY OF GASES AND HEAT TO 1800   25

There is certainly a danger of being misled by Black's public show of caution into underestimating the closeness of the relationship between his work and the development of the material theory of heat. Black, as I have argued, did a great deal to further the theory, however indirectly or unwittingly; and it is also hard to believe that he himself thought of heat as anything but a substance when he was arriving at and elaborating the concepts of specific and latent heat. The latter point is most easily seen in his discovery of latent heat, for in this case there was an obvious analogy between the fixed air that Black supposed to exist in *magnesia alba* before its release by heating and the heat that entered a solid or liquid to effect liquefaction or vaporization and thereupon lost all its normal properties. For example, did not both fixed air and the heat that had been latent regain their original properties when they were released ? Unfortunately, though quite predictably, Black's own comments do not confirm or deny that he made use of this analogy.

Understandably enough, many of Black's contemporaries were not satisfied with his cautious attitude to the processes of liquefaction and vaporization, and his position became characterized by a number of writers as one of support for the very opinion that he had explicitly rejected in the last quoted passage.† This mis-interpretation of Black's comments is important in the early history of the caloric theory, for it was largely responsible for the sharp distinction that came to be drawn between the version of the theory supposed to be adopted by Black and that of his former pupil William Irvine— despite the claims of both men that their views were essentially identical.‡ Irvine had first come into contact with Black when he was a medical student at Glasgow University about 1760 and he had assisted Black in many early experiments, notably some important ones on the latent heat of steam in 1764.§ Irvine's explanation of the emission and absorption of heat during changes of state must also date from about this time, since

† See, for example, M. Landriani, *J. Phys.* **26** (1785), 94; A. Seguin, *Annls Chim.* **3** (1789), 179; R. Heron, *Elements of chemistry* (1800), pp. 76–7; W. Irvine, in *Essays . . . by the late William Irvine* (1805), p. 50; T. Thomson, *A system of chemistry* (Edinburgh, 1802), vol. 1, pp. 323–4, and in *Encyclopaedia Britannica* (3rd edn, Edinburgh, 1801), supplement, vol. 1, p. 269.
‡ Irvine, op. cit., pp. 51 and 54, and Black, op. cit., vol. 1, p. 195.
§ Black, op. cit., vol. 1, pp. xliv–xlv and 171–3.

Black records that it originated almost immediately after his own discovery of latent heat.† Unfortunately Irvine, like Black, published nothing in his own lifetime, but it is not difficult to reconstruct his views from contemporary comments and from the collection of essays published by his son nearly twenty years after his death.‡ He supposed that heat was absorbed by a body during melting or vaporization simply because at the melting- and boiling-points sudden changes took place in the ability of the body to contain heat. Although this did not explain why the change of state occurred in the first place, as Black pointed out,§ it won wide acceptance and was to influence the supporters of both of the main theories of heat. It was fundamental to Irvine's treatment that the relative quantities of heat contained in equal weights of different substances at any given temperature, their 'absolute heats',¶ were proportional to their 'capacities' at that temperature. 'Capacity', we should note, was used by both Black and Irvine to indicate specific heat, the term *chaleur spécifique* being introduced by Jean Hyacinthe de Magellan only in 1780. As a convinced supporter of Irvine's theory, Magellan naturally saw the specific heats that he gave in that year as indicating far more than the relative quantities of heat required to raise the temperature of equal masses of the various substances by the same amount. They were, as he put it, the 'ratios (*raports*) of the specific heat, or elementary fire, contained in different substances'.||

The Irvinist ideas are, and sometimes were, conveniently illustrated by the analogy of a cylinder of uniform cross-section filled with a liquid, the liquid in the cylinder representing the heat in a body (see Plate 1). On this view the level of the liquid would then indicate the temperature of the body in degrees above the absolute zero of temperature or 'point of total privation of heat' and the cross-sectional area of the cylinder would be proportional to its specific heat. The emission and absorption of heat in such processes as changes of state and chemical reactions could then be accounted for by supposing that these processes were always accompanied by changes in

† Black, op. cit., vol. 1, p. 194.   ‡ Cited on p. 25 n. †.
§ Black, op. cit., vol. 1, pp. 194–5.
¶ A term introduced by Irvine; see Crawford, *Experiments and observations on animal heat* (1779), p. 2.
|| Magellan, *Essai*, p. 177.

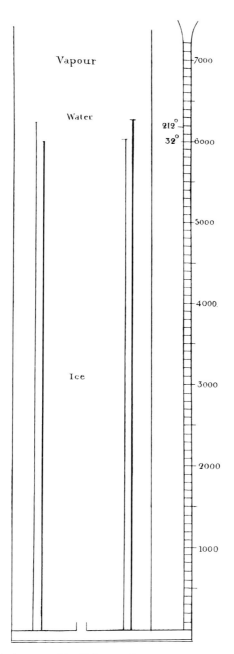

Cylinders of caloric, as used by Dalton to illustrate the Irvinist views on heat capacity. From Dalton's *A new system of chemical philosophy* (Manchester, 1808), vol. 1, part 1, facing p. 217.

specific heat, which for the Irvinist, of course, meant changes also in the ability of a body to contain heat. In the case of a change of state from ice to water, for example, an Irvinist would argue that the absorption of heat was necessitated by a sudden increase in the specific heat of the ice, and hence also in the cross-section of the cylinder, which occurred as soon as the melting-point was reached. If the temperature, or level of the liquid in the cylinder, was not to fall, it was obvious that some additional heat, or liquid, would have to be added. Although this heat was that which was commonly termed latent heat, there was no suggestion that it entered into combination with the ice, nor that it ceased to contribute to temperature, nor even that it *caused* the change of state. The absorption of heat was rather the *consequence* of the change of state.

The experimental evidence in support of the Irvinist doctrines was slight and inadequate, whether they were applied to the explanation of latent heats or to that of heats of reaction. However, both Irvine's own experiments and those of Kirwan showed that the specific heat of ice was less than that of water, so that in one very familiar case at least they appeared to be justified. Not surprisingly, the importance of such favourable evidence as could be found was easily exaggerated.

An interesting application of Irvine's theory was in the location of the absolute zero of temperature, a position of which on the temperature scale had considerable significance for Irvinists. The typical procedure is well illustrated in the following example, which is a corrected version of one given by Irvine's son in 1805.† In the calculation the latent heat of a given mass of ice was taken as the quantity of heat that would raise the temperature of the same mass of water by 140 °F and the values for the specific heats of ice and water (Kirwan's) were 0·9 and 1·0 respectively. Where the melting-point was x °F above the absolute zero, it followed from Irvine's assumptions that the total quantities of heat in the ice before melting and in the water after the change of state, both quantities being measured at 32 °F, were proportional to $0·9x$ and $1·0x$

† Irvine, *Essays*, pp. 122 and (for the result of Irvine's calculation) 127. The nature of the slight error in the 1805 calculation is best understood by comparing it with correctly performed calculations in Magellan, op. cit., p. 176, and Crawford, op. cit. (2nd edn, 1788), pp. 453–4.

respectively and hence, on the assumption that heat had been conserved, that

$$x = 0{\cdot}9x + 140$$

and $x = 1400$.

The zero in this calculation would therefore be at $-1368\,°F$. Irvine himself placed it at about $-900\,°F$, which suggests that, if he had the same figure for the latent heat, he must have taken the specific heat of ice as $0{\cdot}85$[†]—an indication, incidentally, of how very considerable variations in the absolute zero could result from small changes in specific heat.[‡] The calculation could of course be made equally well from data relating to vaporization or to a chemical reaction, and Irvine apparently confirmed his figure from observations of the heat evolved during the mixing of sulphuric acid with water. We have no details of the procedure that he used to measure either the heat evolved or any changes in capacity accompanying the mixing.

Irvine gave no indication that he was concerned to elucidate the nature and properties of the fluid, or vibration, that caused the sensation of heat. His son emphasized that Irvine's views did not imply belief in any particular theory,[§] and the Irish-born physician Adair Crawford, who developed these views so successfully in the two editions of his *Animal heat* in 1779 and 1788, expressed only tentative support for the material theory.[¶] Yet, despite this show of agnosticism, there can have been few who did not agree with Crawford that such a concept as latent heat was more readily explained in terms of a fluid than of a vibration, and many (and here we can surely include Crawford himself) must have seen this alone as strong evidence for the materiality of heat. Magellan and Tiberius Cavallo, both of whom unhesitatingly rejected the vibrational theory about 1780, were typical in this respect,[‖] although neither sought to give such detailed support to the material theory as William Cleghorn had done in 1779.

The work of Black, Irvine, and Crawford quickly became

[†] Not $0{\cdot}87$, as McKie and Heathcote suppose in their *Specific and latent heats*, p. 134, having followed the erroneous calculation given by Irvine's son.
[‡] On which see also p. 31. [§] Irvine, op. cit., p. 71.
[¶] Crawford, op. cit. (1st edn), p. 115; (2nd edn), pp. 435–6.
[‖] Magellan, op. cit., p. 167, and Cavallo, *A treatise on the nature and properties of air* (1781), pp. 20–1.

known in France. As early as 1772 an account of Black's views on latent heat, written by an unnamed Edinburgh student, had appeared there† and the first edition of Crawford's *Animal heat* aroused considerable interest some years later. For this Magellan's *Essai sur le feu*, published in London in 1780 as a commentary on *Animal heat* but incorporating details of Irvine's work that Crawford had not mentioned, was no doubt largely responsible, especially after its republication in France in 1781.‡ The intensity of French interest is brought out well in an account by Jean André De Luc of a visit to Paris that he made in 1781. Referring to his meetings with Laplace, Lavoisier, Monge, and Vandermonde during the visit, he recalled:

> Dr. Crawford's theory on the phenomena of heat observed when substances undergo certain changes in their nature, was then in great agitation among these philosophers, and the two last had, from the Academy, the special commission to examine and follow that new view.§

By this time, as we have seen, Lavoisier had already given two descriptions of a 'matter of fire' that resembled caloric in many respects, so that it is highly probable that in 1781 he was still committed to some form of the material theory. However, he had written nothing on heat since 1777 and it was not until June 1783, when he and Laplace presented their 'Mémoire sur la chaleur' before the Académie des Sciences, that he returned to the subject. For this renewal of interest Crawford's work, we may suppose, was very probably responsible.

On the question of the nature of heat Lavoisier and Laplace adopted a cautious attitude in the 1783 paper. After describing both a material theory similar to that which Lavoisier had already described in 1777 and the vibrational theory, they concluded:

> We shall not decide in favour of either of these two hypotheses. Several phenomena would appear to favour the latter: for example, the fact that heat is produced by the rubbing together of two solid

---

† The paper, dated September 1772, was reprinted in the Abbé Jean Rozier's journal *Introduction aux observations sur la physique*, **2** (1777), 428–31.

‡ In [Rozier's *Observations sur la physique*, **17** (1781), 375–86 and 411–22.

§ De Luc, *Edinb. Rev.* **6** (1805), 511.

bodies. But there are other phenomena which are better explained if we adopt the former. Perhaps both hypotheses are true simultaneously.†

Despite this non-committal statement a tradition has grown up that Lavoisier was the champion of the material theory while Laplace was a vibrationalist. It appears to have originated with E. M. Lémeray and has been accepted by most recent writers on the matter.‡ In fact, since Lavoisier was still inclined to support the material theory in 1789,§ it seems reasonable to suppose that he held similar views six years earlier. The position of Laplace, however, is far less clear. It may well be that, like Macquer and Fourcroy, he did accept the vibrational theory at this time, but there is no conclusive evidence and it must at least be born in mind that Laplace had become a convinced calorist when he next wrote on the subject about 1803.¶

Among the other members of Lavoisier's circle Gaspard Monge stands out as a particularly vigorous advocate of the material theory and he was a man of no small stature in the discussions of the early 1780s. The following account is taken from the anonymous article on caloric that appeared in the physics section of the *Encyclopédie méthodique* in 1816. After describing Monge's increasingly close contact with Parisian scientists following his election to the Académie des Sciences in 1780, the writer continued:

Accepted into the circle of the great Lavoisier, who assembled in his home every Monday the most distinguished *savants* in the capital—accepted also, along with several academicians, into the home of a friend of that famous man after the meetings of the Académie on Wednesdays and Fridays—Monge felt compelled to lay before this learned company his ideas on the caloric theory as they developed. His opinions were heard with the attention commanded by a man of genius, though sometimes they were discussed with the frankness and cordiality which are found among men whose sole concern is the pursuit of truth. Out of all this there emerged the fine and elegant

† Lavoisier and Laplace, *Mém. Acad. Sci.* (1780), pp. 355–408.
‡ Lémeray, *L'éther actuel et ses précurseurs* (1922), pp. 65–9. Among more recent writers see G. Bachelard, *Étude sur l'évolution d'un problème de physique* (1928), p. 23 n.; C. C. Gillispie, *The edge of objectivity* (1960), p. 239; J. R. Partington, *A history of chemistry* (1962), vol. 3, p. 428; Smeaton, *Fourcroy*, p. 102.
§ Lavoisier, *Traité*, vol. 1, pp. 4–5. ¶ See pp. 125–7.

ideas on caloric which have since been widely published but for which no one has claimed the credit.†

Whether Monge took the opportunity of expounding his caloric theory to a wider audience when he began teaching at the École Polytechnique in 1794 is not known, but it is interesting that it was he who in 1809 contributed the article on caloric to a short work by Hachette intended for the use of students at the École.‡

Their agnosticism concerning the nature of heat did not prevent Lavoisier and Laplace from criticizing Irvine's views on the state of heat in bodies. Incorporating many of their own results for the specific and latent heats and for heats of reaction in Irvine's argument, they made five different determinations of the absolute zero.§ Although the five figures showed no semblance of agreement and even included one that was *above* the melting-point of ice, Lavoisier and Laplace were careful to emphasize, quite rightly, that small errors in the specific heats could seriously affect the calculation. Yet they clearly had little sympathy for such basic Irvinist assumptions as the proportionality between the capacities of various bodies at a given temperature and the quantities of heat that they contained. This assumption, they said, was 'at least ill founded, and it must only be adopted after a large number of experiments'.¶

Strangely enough they made no mention of Black, but their cautious assertion that in a change of state some heat was absorbed which thereupon became imperceptible to the thermometer was one of which Black would surely have approved. There was, we note, no mention of a combination between heat and ponderable matter. But in one important respect they went beyond Black, for they suggested that the absorption of heat was necessary in order to effect not only melting and vaporization but also expansion.‖ Thus, when a body was heated, some heat would go to raise its temperature and some to increase its volume. The idea, skilfully developed by Laplace, became a most important one during the first quarter of the nineteenth century, as we shall see in Chapter 4, but it provoked little immediate reaction and so need not detain us here.

† *Encyclopédie méthodique. Physique* (1816), vol. 2, p. 170. The writer may well have been J. H. Hassenfratz, one of the editors of the *Encyclopédie*.
‡ See p. 230.   § Lavoisier and Laplace, op. cit., pp. 382–6.
¶ Lavoisier and Laplace, op. cit., p. 381.
‖ Lavoisier and Laplace, op. cit., p. 388.

The 'Mémoire sur la chaleur' provided the first intimation of the inevitable weakness of Irvine's theory in the face of reliable data for specific heats. The attack was taken up in 1790 by Armand Seguin, Lavoisier's partner in some important early work on respiration, who again pointed to the serious discrepancies between the results obtained in several different determinations of the absolute zero of temperature;† and Gadolin, arguing in a similar fashion in 1792, was another who applied the results of new experiments (his own) to the rejection of Irvinist doctrines, although only eight years earlier he had upheld them and had determined the absolute zero in a perfectly conventional manner.‡ However, the data of extreme accuracy and reliability that were required to disprove the theory beyond all doubt were not easily obtained and it was therefore experimental deficiencies as much as the true merits of his theory that allowed Irvine's influence to survive until well into the nineteenth century. Certainly this influence was declining by 1800 but in the period with which this chapter is concerned it can still be detected in nearly all aspects of the study of heat and nowhere more clearly than in the experimental work on the specific heats of gases.

Specific heats of gases

The first determination of the specific heat of a gas was made by Crawford in the summer of 1777. His main concern in this as in his later work was to derive an explanation for the heat evolved in the bodies of warm-blooded animals, and the experiments were described, appropriately enough, in the two editions of his *Animal heat* (1779 and 1788).§ Although he was educated at St. George's Hospital and spent most of his life in London, Crawford visited Scotland in 1776 and it was presumably during this visit that he heard the lectures of William Irvine.¶

---

† Seguin, 'Second mémoire sur le calorique', *Annls Chim.* **5** (1790), 231–57. Also see pp. 38–9.

‡ Gadolin, 'De theoria caloris corporum specifici', *Nova acta R. Soc. Scient. upsal.* **5** (1792), 1–49. For his earlier determination see Gadolin, *Der Königl. Schwedischen Akademie der Wissenschaften neue Abhandlungen* **5** (1784; published Leipzig, 1786) 222–39, and *De theoria caloris corporum specifici*, pp. 25–7. In 1784 Gadolin placed the zero at -800 °C.

§ Crawford, *Animal heat* (1st edn), pp. 34–43; (2nd edn), pp. 177–253.

¶ His attendance at these lectures and their influence on him are referred to in Crawford, op. cit. (1st edn), p. 17 n.

His adoption of Irvinist doctrines throughout his work is sufficient evidence that he was impressed, but he appears to have been particularly interested by Irvine's suggestion that the absolute heats of atmospheric air and fixed air (carbon dioxide) were different. According to Irvine's theory of heat all that was required to test this was a determination of the specific heats of the two gases, and Crawford duly set about the experiments in the following summer. The theory of animal heat that he expounded in 1779 was based on this work and in particular on two results. First he showed that the absolute heats of equal weights of atmospheric and fixed air were in the ratio 67 to 1, while that of phlogisticated air (nitrogen), although not precisely measurable, was extremely small. Since fixed and phlogisticated airs were known to be the major constituents of exhaled breath, it followed that a large amount of heat must be released during the process of respiration, in which inhaled atmospheric air yielded these two products. The second result was that the absolute heats of the blood which entered the lungs and that which left them were as 10 to $11\frac{1}{2}$. Hence, Crawford argued, the effect of passage through the lungs was to increase the blood's ability to contain heat, so that without any increase in temperature it could absorb the heat liberated by the conversion of atmospheric air to fixed and phlogisticated airs. Uniform distribution of this heat throughout the body was then effected through the gradual absorption of phlogiston by the blood during circulation and the consequent decrease in its absolute heat.†

Crawford's method of determining the specific heats of the gases was simple, not to say crude. The principle was essentially to heat the gases in bladders and then to compare the rises in temperature that occurred when the bladders were immersed in water. The calculation of specific heat from the data so obtained was in theory an easy matter, but the smallness of the quantities of gas that were used made his thermometer readings all but meaningless. In fact precise determinations were hardly important and it was enough to have shown, as Crawford was convinced he had done, that atmospheric air contained a greater quantity of heat than an equal mass of exhaled air. That it

† In assuming this connection between absolute heat and the quantity of phlogiston in a body, Crawford was obviously very vulnerable.

contained only about one-fifth as much as dephlogisticated air (oxygen) was nevertheless interesting confirmatory evidence for his theory, since it accorded well with Priestley's recent discovery that dephlogisticated air was five times as effective as ordinary atmospheric air in supporting life.†

Crawford's work attracted widespread attention, as we have already seen, but all the comments were by no means as favourable as Magellan's, for example. In 1781 the actuary William Morgan severely criticized the way in which the experiments had been conducted and on repeating them, with a number of modifications, he could detect no difference between the absolute heats of the various gases examined by Crawford.‡ It followed, Morgan concluded, either that equal weights of all gases contained the same quantity of heat or that the experiments conducted hitherto were too insensitive to reveal any differences that did exist. In either case Crawford's theory remained unproved, an opinion maintained by one anonymous reviewer of Morgan's work.§

De Luc, by contrast, appears to have made his criticisms personally and at Crawford's own request.¶ He particularly emphasized the smallness of the temperature changes during the mixing process and in 1786 he wrote that Crawford had acknowledged the defects in his earlier work and was now engaged on new experiments.‖ It seems that the experiments of Lavoisier and Laplace, as well as some conducted by Kirwan, about which we know nothing more than is contained in De Luc's passing reference to them,†† had been largely responsible for this admission.

The work of Lavoisier and Laplace deserves special mention, not merely for its relevance to the debate on Crawford's theory but also for its bearing on the practice of calorimetry in general. Chief among the innovations that were announced in the joint paper of 1783 was undoubtedly the ice-calorimeter, an instrument

† Crawford, op. cit., p. 53. On Priestley's observation see his *Experiments and observations on air* (1775), vol. 2, p. 48.
‡ Morgan, *An examination of Dr. Crawford's theory of heat* (1781), pp. 6–27 and 31–7.
§ 'C', *Critical Review*, **51** (1781), 212–16.
¶ De Luc, *Idées sur la météorologie* (London, 1786), vol. 1, p. 145, and *Introduction à la physique terrestre* (Paris and Milan, 1803), vol. 1, pp. 230–1.
‖ De Luc, *Météorologie*, vol. 1, p. 145.
†† De Luc, *Physique terrestre*, vol. 1, p. 233.

that measured quantities of heat by the weight of ice melted. In the paper Lavoisier and Laplace gave a full account of their experiments and results relating to the specific heats of solids and liquids and also on heats of reaction, but it seems unlikely that they had yet performed any successful experiments on gases. Lavoisier's laboratory notebooks, which cover the period 1772–88, contain only four references to determinations of the specific heats of gases, dated 9, 10, and 14 February 1784 (for atmospheric air) and 20 February 1784 (oxygen).† Yet the fact that a description of the relevant apparatus appeared in the 1783 paper‡ suggests that some preliminary work may have been done in the previous winter. According to this description the gas under examination was heated and then passed through a spiral tube inside the ice-calorimeter. Its temperature was measured both before and after its passage through the apparatus, so that, by measuring also the weight of ice melted, the heat lost by any mass of a gas could be easily determined and its specific heat calculated. How successful this method was we cannot know, but Lavoisier and Laplace seem to have retained it in their experiments during the winter of 1783–4 and also in subsequent years.§ Although Lavoisier wrote in 1789 that they still allowed no winter to pass without doing some work with the ice-calorimeter,¶ the experiments were seriously interrupted and it was not until 1793 that he described this later work.‖ By then his fate at the hands of the Revolutionary Tribunal was only too imminent and another twelve years elapsed before the paper appeared, as a joint memoir with Laplace. For gases only two specific heats were given, 0·65 for oxygen and 0·33031 for atmospheric air, where that of an equal weight of water was unity. Both results were, of course, quite different from those given by Crawford. But although Lavoisier and Laplace used these and some other data on heats of combustion in order to criticize the Irvinist doctrines and to reaffirm their

---

† Académie des Sciences, Lavoisier papers, box 24. The results for the work on gases are recorded on pp. 202–12 of the eighth volume of the notebooks.
‡ Lavoisier and Laplace, *Mém. Acad. Sci.* (1780), p. 368.
§ See Lavoisier, *Mémoires de chimie* (1805), vol. 1, p. 122.
¶ Lavoisier, *Traité*, vol. 2, p. 402.
‖ Lavoisier, *Mémoires de chimie*, vol. 1, pp. 121–47, where the description appeared in the joint paper with Laplace entitled '3e mémoire ... contenant les expériences faites sur la chaleur pendant l'hiver de 1783 à 1784'.

belief that part of the caloric in bodies existed in a combined and therefore latent state, they fully recognized that their experiments could not yet be considered definitive.

When Crawford described his own new experiments in the second edition of *Animal heat* in 1788 he knew nothing of this work, but certain other criticisms, both of his theory of animal heat and of the Irvinist principles on which it was based, would have been familiar enough. Crawford's confidence was nevertheless undiminished. Moreover, although he readily admitted to serious errors in his original experiments,† the principle of his method for the determination of the specific heats of gases remained unchanged. Two identical brass cylinders now replaced the bladders that he had used previously and numerous precautions were taken, principally to reduce heat losses. The cylinders, each containing a different gas, were heated to the temperature of boiling water and then rapidly immersed in separate vessels containing cold water. Accurate measurement of the resulting rise in temperature enabled the relative specific heats of the gases to be determined, at least in principle. In fact, the heat capacity of the enclosed gas was always so small by comparison with that of the cylinder that errors were large. In one typical experiment, for example, only 0·274 °F of a total temperature rise of 2·2 °F could be attributed to the cooling of the gas alone.‡

However, the experimental results were an improvement on those announced in 1779. The differences between the specific heats of dephlogisticated, atmospheric, and fixed airs were found to be much smaller than in the earlier experiments, although of the three that of dephlogisticated air was still the greatest and that of fixed air the smallest, so that the theory of animal heat was not disproved. Indeed, one interesting calculation—that of the specific heat of water vapour—gave it considerable additional support.§ Assuming that the absolute zero of temperature was 1650 °F below the boiling-point of water, Crawford argued that the absolute heat of water compared with that of steam must be as 1650 to (1650+914), or 1 to 1·55, since enough latent heat was absorbed in vaporiza-

† Crawford, *Animal heat* (2nd edn), 'Advertisement' pp. 3–4, and pp. 225–6.
‡ Crawford, op. cit., pp. 234–5.
§ Crawford, op. cit., pp. 269–70.

THE STUDY OF GASES AND HEAT TO 1800   37

tion to raise the temperature of an equal mass of water by 914 °F (Watt's figure). The result confirmed his theory on two counts. In the first place, since water vapour was known to be a product of respiration, it was to be expected that its capacity for heat should be less than that of atmospheric air (given now as 1·79). Secondly, the fact that the specific heat was greater than that of water agreed with Irvine's explanation of the phenomenon of latent heat. Another result that could be applied to good effect, in accounting for the large amount of heat liberated during the explosion of hydrogen with oxygen, was that hydrogen had a specific heat which, although considerably smaller than he had been led to believe by some earlier experiments, was still far greater than that of any other gas (see Table B).†

Crawford's later experiments were the first in which any attention had been paid to the distinction between the specific heats measured at constant pressure (c_p) and at constant volume (c_v). His awareness of the need for such a distinction almost certainly arose from his familiarity with the heating and cooling effects that accompany, respectively, the rapid compression and expansion of all gases. Although he nowhere offered a clear explanation of these temperature changes, Crawford did believe that any variation in the volume of a gas would cause a change in its capacity for heat.‡ Thus, if a gas was allowed to expand while it was being heated, its capacity, according to Crawford, would increase and it would have to absorb more heat than would be required under constant-volume conditions. Such considerations were irrelevant to his own experiments, in which the gases were contained in closed brass cylinders, but he undertook to demonstrate that the difference would have been negligible even if the gases had expanded during the heating process.§ He had observed, and no existing experiments on the subject suggested that he was far wrong, that a fall in temperature of about 5 °F accompanied the rapid evacuation of the receiver of an air pump, the cooling being approximately the same whether atmospheric,

† Crawford, op. cit., pp. 221 and 247–9. The earlier work on hydrogen was not mentioned in the first edition of *Animal heat* but reference to it appeared in Bergman's *Opuscula physica et chemica* in 1783 (vol. 3, p. 436). Kirwan was apparently the source of Bergman's information.

‡ Crawford, op. cit., pp. 249–50.   § Crawford, op. cit., pp. 250–3.

dephlogisticated, or phlogisticated air was extracted. Using Colonel William Roy's results on the expansion of air by heat in conjunction with some very suspect data on the relative expansivity of different gases obtained by Priestley,† Crawford argued that a rise in temperature of 110 °F would increase the volume of dephlogisticated air by half. Now he also believed, as a result of his experiments with the evacuated receiver, that such an expansion would cause a cooling of 3 °F if the gas were thermally insulated. Therefore, if dephlogisticated air was allowed to expand while being heated through 110 °F, it followed, as he thought, that 113 units of heat would be required to effect the temperature change, compared with 110 units when the volume did not vary. Crawford showed that such a difference would have been quite undetectable in his experiments, since even with his improved apparatus a fall in the temperature of dephlogisticated air of about 110 °F caused a heating of only 0·4 °F in the water surrounding the brass cylinder, so that an increase in the amount of heat communicated to the water of less than 3 per cent would affect this by roughly 1/100 °F, 'a quantity so inconsiderable that it could not have been distinguished by the thermometer'. Crawford's serious, if forgivable, underestimate of the magnitude of the difference was unfortunate and it must be seen as at least one factor in accounting for the neglect of the distinction between c_p and c_v over the next twenty years.

Not surprisingly, the second edition of *Animal heat* aroused less interest than the first. Crawford's ideas had now been widely discussed for nearly ten years and the new edition raised no essentially new issues. Above all, his experiments were still far too unreliable to allow the debate between the supporters of the rival versions of caloric theory to be taken further. Only Lavoisier's colleague Seguin was moved to offer a detailed criticism. In two papers published in the *Annales de chimie* in 1789 and 1790‡ he vigorously attacked both the experiments and the theory of heat adopted by Crawford, maintaining Lavoisier and Laplace's distinction between what he called

† On the work of Roy and Priestley see pp. 63-5. Crawford's debt to Priestley is obvious, though unacknowledged.

‡ Seguin, 'Observations sur le calorique', *Annls Chim.* **3** (1789), 229–42, and **5** (1790), 191–271.

THE STUDY OF GASES AND HEAT TO 1800

'interstitial caloric' (*calorique interposé*), which affected the thermometer, and 'combined caloric', which did not contribute to temperature and which combined with a solid or liquid in bringing about a change of state. Seguin's views were not original but his thorough re-examination of the evidence, especially that relating to the determination of the absolute zero, which has already been mentioned, was an important contribution.

But by 1790 it must have been clear that, for the moment, there was little more that could usefully be said on any theoretical issues which depended on an accurate knowledge of specific heats. The need for new and greatly improved experiments, especially on gases, was only too evident and yet the practical problems were daunting. Fortunately the new century was to bring with it a renewed interest in gases and new hope for solving the outstanding problems, so that after ten years of comparative neglect the debate, as we shall see later, was to be reopened.

Adiabatic heating

It is well known that a rise in temperature occurs whenever a gas is compressed in conditions such that it loses little or no heat to its surroundings. The effect, now known as adiabatic heating,† is readily explained by the principle of the conservation of energy. The work done in compression has been converted to heat energy in the gas and, for a given amount of work, the same quantity of heat will always appear. It is natural and proper, therefore, that adiabatic heating and the reverse effect, in which cooling results from the expansion of a gas against external pressure, should now be generally associated with this principle and with the closely related dynamical theory of heat. The association, in fact, goes back well into the nineteenth century. Carnot, Mayer, Joule, and a number of others who have a claim to be considered discoverers of energy conservation saw adiabatic phenomena as decisive qualitative evidence and some used them quantitatively to calculate the mechanical equivalent of heat, the constant of proportionality

† The term 'adiabatic', which strictly applies only when the thermal insulation is perfect, was first used in 1858 by W. J. Macquorn Rankine; see *Phil. Trans. R. Soc.* **149** (1859), 180.

relating work done and heat produced.† It might seem from this that adiabatic phenomena were in some important respect irreconcilable with caloric theory, but in fact the thermal effects associated with the compression and expansion of a gas had been known for nearly a century before they became indissolubly linked with energy conservation in the 1840s. In this period they had been explained almost exclusively in terms of material theories of heat and had even appeared as strong evidence in favour of such theories.

This earlier history of adiabatic heating and cooling may be considered most conveniently in three distinct periods. In the first, which lasted from 1755 until 1779, the effects were observed exclusively during the evacuation and refilling of the receivers of air pumps. So little were they understood at this time that serious misinterpretations were common and very few observers appear to have drawn the all important conclusion that a mere change in the volume of a gas could give rise to variations in temperature. In the second period, from 1779 to 1800, the effects were still normally observed during work with the air pump and the explanations offered were coloured accordingly. However, they were now far better known and it was generally recognized that temperature changes did occur both in the air-pump experiments and when the volume of a gas was altered rapidly, although serious departures from true adiabatic conditions meant that the magnitude of the changes was still seriously underestimated. A peculiar characteristic of the first two periods is that interest in the effect was almost entirely limited to those who had had contact with one of two schools: either that of the Scottish chemists and physicians, headed by Cullen and Black, or that in which the Swiss scientists Saussure and Pictet were the leading members and in which the main interest was meteorology. Of the best-known early authorities only the German Lambert worked outside these schools, but his views, written in German, untranslated and, if only for this reason, less accessible to the British and French, exerted such an influence on the Swiss that he can properly be grouped with them. The most surprising conclusion that emerges is that, until about 1800, adiabatic phenomena

† See T. S. Kuhn, 'The caloric theory of adiabatic compression', *Isis* **49** (1958), 132–3.

THE STUDY OF GASES AND HEAT TO 1800

were virtually unknown in the French school of chemistry, although by 1783 Lavoisier and Laplace had devised a theory of heat that could have accommodated them in a most convincing and elegant way. The history of the phenomena in France is thus almost entirely restricted to our third period, beginning in 1800, which is examined in Chapter 3.

We return now to the first period. The discovery is usually and rightly attributed to William Cullen, professor of medicine at Glasgow University at the time, who mentioned it briefly in a paper on evaporation read to the Edinburgh Philosophical Society in May 1755.† Although Cullen was almost certainly the first man to recognize that the movement of the liquid in a thermometer suspended in the receiver of an air pump did indicate that temperature changes had occurred, he was not, however, the first man to observe it. In his *History of cold*, published in 1665, Robert Boyle gave the following description of an experiment that he had performed on 20 November 1662:

> ... we took a Weather-glass fill'd to a convenient height with well rectifi'd spirit of Wine, and Hermetically seal'd, this we inclos'd in a glass Receiver of a Cylindrical form, of about two inches Diameter, and about a foot and a half high, and having cemented on the Receiver, we let it alone for some hours, that it might perfectly cool. Then drawing out the Air, and watching it narrowly, we observ'd, that the liquor in the Weather-glass descended a little, though but a very little upon the first Exuction of the Air, and a little, though it seem'd somewhat less, upon the second, but afterwards we did not find it sensibly to descend. This subsidence of the liquor in all amounting to about the length of a Barley corn, we attributed to the stretching of the glass by the spring of the included Air, when the ambient was withdrawn, and accordingly upon our allowing a Regress to the excluded Air, we saw the spirit in the Thermometer, rise about half a Barley-corn's length to the place whence it began to subside. Afterwards we suck'd out, and let in the Air of the Receiver, as before, with like success, as to the descent and remounting of the liquor.‡

As late as 1783 the Swedish chemist Torbern Olof Bergman independently attributed the apparent fall in temperature that

---
† Cullen, 'Of the cold produced by evaporating fluids . . .', *Essays and observations, physical and literary. Read before a society in Edinburgh and published by them* (Edinburgh, 1756), vol. 2, pp. 145–56.

‡ Boyle, *New experiments and observations touching cold* (1665), pp. 470–1.

accompanied the rapid evacuation of an air-pump receiver to deformation of the thermometer bulb.† Since he did not enlarge on the subject, we can only assume that he would have accounted for the rise in the thermometer reading after the sudden reintroduction of air in exactly the same way. Boyle was thus not alone in treating his observation as a purely mechanical effect and it seems likely that in the late eighteenth century there were others besides Bergman who shared his view. It is true that only one writer of this period (William Cleghorn) even referred to the passage in Boyle,‡ but both Erasmus Darwin and John Dalton conducted experiments that were intended specifically to show that no appreciable deformation of the thermometer bulb occurred during the evacuation and refilling of the receiver.§ Although neither Darwin nor Dalton mentioned any authorities for the view that they were here attempting to disprove, the very fact that they felt such experiments to be necessary suggests that this view was more widely held than the published literature would indicate. Such opinions, however, need not necessarily have originated with Boyle's single passing reference.

An obvious first task in undertaking this examination of the early studies of adiabatic phenomena was to ascertain whether any other observation similar to Boyle's had been made before Cullen's announcement, but a detailed search has revealed no other account. This fact, like the slowness with which knowledge of Cullen's work spread after 1755, deserves special comment in view of the extraordinary interest that had been taken in all branches of pneumatics since the middle of the seventeenth century. With the air pump and its accessories an established and important part of the 'philosophical apparatus' in any European laboratory, how had the discovery been missed? The answer seems to be simply that the range of experiments that could yield easily identifiable adiabatic phenomena by chance was very limited and that those few instances that were observed could all too easily be misinterpreted. Consider, for example, the heating that must long have been familiar to

† Bergman, *Opuscula*, vol. 3, p. 425. The section of the 'De attractionibus electivis' in which the reference appeared was added in the *Opuscula* and was not in the original version of the essay in *Nova acta R. Soc. Scient. upsal.* 2 (1775), 159–248.

‡ Cleghorn, *De igne*, p. 20. § See pp. 58 and 89.

anyone who tried to compress air with a condensing syringe. This instrument, widely used for attaining high pressures in the laboratory and as an essential component of that popular laboratory instrument the air-gun, provided the simplest example of air being heated by compression, yet never once was the rise in temperature associated with a decrease in the volume of the gas. If we are to believe one comment dating from 1799,† the heating was widely attributed to friction between the piston of the syringe and the cylinder walls—an interpretation that is quite understandable. With the air pump itself the smallness of the temperature changes obtained was probably the important obstacle to the recognition of the adiabatic phenomenon, for although it must have been an everyday practice to allow air to rush into an evacuated receiver, heat losses through the walls were nevertheless so great that temperature changes of the order of only 2 or 3 °F would occur. Since the rapid evacuation of the receiver would give rise to a cooling that was equally small, it can be seen that the presence of a thermometer inside the receiver would be essential if any effect was to be noticed. It is significant, therefore, that the passage quoted from Boyle should contain the only reference that has been found to an experiment in which a thermometer was used in conjunction with the air pump before 1755. The typical experiments of the late seventeenth and early eighteenth centuries‡ concerned such problems as respiration under reduced or increased pressure, the failure of a vacuum to transmit sound, and the striking mechanical effects of air pressure, as demonstrated by the Magdeburg hemispheres, for example; and, what is really important, none demanded that an air pump and thermometer should be used together. There was yet just one observation which, even in these circumstances, might have indicated that temperature changes were taking place. This was the appearance of mist during the rapid exhaustion of a receiver. The mist was particularly dense

† See p. 89.
‡ As described, for example, in F. Hauksbee, *Physico-mechanical experiments* (1709); Desaguliers, *Experimental philosophy*, vol. 2, pp. 375–411; Nollet, *Leçons de physique expérimentale* (5 vols., Amsterdam and Paris, 1745–55), vol. 3, pp. 185–291, *Mém. Acad. Sci.* (1740), pp. 385–432 and 567–85, and *Mém. Acad. Sci.* (1741), pp. 338–62; Musschenbroek, *Philosophiam naturalem*, vol. 2, pp. 823–905.

in a moist atmosphere and we know that it was familiar in eighteenth-century laboratories.† But not until 1787 was it explicitly associated with the cooling of the air undergoing expansion during its extraction from the receiver.‡ As we see, the conditions that were required before the discovery of the adiabatic effect could be made were clear and specific, and they were provided for the first time (if we except Boyle's observation) in Cullen's experiments on evaporation, an investigation that necessitated both a measurement of temperature and the existence of a vacuum. Cullen's interest began with an observation by a pupil§ who noticed that the reading on a thermometer fell immediately after having its bulb moistened in alcohol. This led Cullen to suspect that the very process of evaporation in itself produced cold. Now he might well have conducted his investigation of cooling in the normal manner, without the aid of the pump, had he not wished to examine one particular point, namely whether the cooling could occur in the absence of the surrounding air or whether it arose from the mixture of the liquid with the air (after the fashion of the well-known freezing mixtures in which the act of mixing substances lowered the temperature). It was while conducting his experiments on evaporation *in vacuo* that he observed a fall in temperature of 2 or 3 °F whenever the receiver was evacuated, even when the bulb of the thermometer being used was unmoistened, and a similar rise in temperature when air was readmitted. Cullen made only passing reference to the observation in his paper and offered no explanation. He related it in no way to evaporation, referring to it as merely one of 'a number of new and to me curious phenomena' that had occurred. Yet the fact that it was described together with the evaporation experiments and immediately before some observations on the cooling of a moistened thermometer *in vacuo* appears to have created the misconception, shared by several subsequent writers, that Cullen viewed it as a case of evapora-

† See, for example, Nollet, *Mém. Acad. Sci.* (1740), pp. 243–53.
‡ By Erasmus Darwin in *Phil. Trans. R. Soc.* 78 (1788), 45. It is an interesting reflection on French ignorance of adiabatic phenomena that even in 1790 Monge still failed to associate the mist with cooling; see Monge, *Annls Chim.* 5 (1790), 11–13.
§ Identified in Black, *Lectures*, vol. 1, p. 162, as 'the late Dr. Dobson', presumably Matthew Dobson, M.D., F.R.S., of Liverpool and later of Bath.

tion. The Swiss alpinist, meteorologist, and university professor Horace Bénédict de Saussure even went so far as to specify that Cullen's explanation appeared to be based on the evaporation of a thin layer of liquid from the surface of the thermometer bulb.† How Cullen was thought to explain the rise in temperature in this way is unclear, but no such difficulty need exist, for he neither gave nor implied any explanation. It seems unlikely, if he suspected evaporation to be the cause, that he would have omitted such striking evidence that evaporation in the void gave rise to cooling, a view that he was anxious to substantiate.

Saussure was not the first to interpret Cullen's views in this way, however. In 1759, in his inaugural dissertation as professor of physics at the University of Erlangen, Johann Christian Arnold described his own, and Cullen's, work on the effects accompanying the evacuation and refilling of air-pump receivers.‡ While he was convinced that Cullen had in some vague way associated the cooling with evaporation, he would commit himself no further than to assert that there was 'some connection' (*nexum aliquem*) between the two. Aware that evaporation could account for cooling alone, Arnold offered a novel explanation of the heating effect, attributing it to friction between the inrushing air and the sides of the entrance channel and of the receiver. It can be seen that Arnold's views shared one important characteristic with those of all the other writers of the first period in that he sought explanations in terms of familiar everyday phenomena, such as friction and evaporation, and not in the properties of air itself. This, of course, was exactly what Boyle and Bergman did when they invoked the mechanical deformation of the thermometer bulb.

The contrast with the period after 1779 is striking. From that year all writers, with the single exception of Bergman, recognized that a completely new type of effect had been discovered. Whether they associated it with the entry of air into a vacuum

---

† Saussure, *Essais sur l'hygrométrie* (Neuchâtel, 1783), p. 232. It should be noted that there are two editions of the *Hygrométrie*, both published at Neuchâtel in 1783 by Samuel Fauche but with different pagination and page size. Throughout this book my references are to the quarto edition (pp. xviii and 367).

‡ Arnold, *De thermometri sub campana antliae pneumaticae suspensi variationibus* (Erlangen, 1759).

or, more fundamentally, with a change in the volume of the air, they were now in no doubt that the effect was truly thermal and of such significance as to justify any appropriate modification in the theory of heat.

Between 1779 and 1800 a clear distinction can be drawn between the work of the Scottish and German–Swiss schools. An important representative of the former was William Cleghorn, who, in 1779, related the effect to the material theory of heat in his *De igne*. One strength of his account was that he used the same basic principle to explain both the heating and cooling. His view was as follows:

... since the quantity of fire distributed among bodies increases with the attraction for fire that the bodies exert and decreases with the repulsion between the fire particles themselves, it follows that if in any body the former quantity is diminished or the latter increased, then fire will flow from that body until equilibrium is again restored. Heat is then said to be generated. On the other hand, if the attraction of any body were to be increased or if the repulsion between the fire particles were diminished, more fire would flow into the body. . . . In this case cold is said to be generated.†

On this basis Cullen's experiments were easily explained:

While the air was being withdrawn, the reading on the thermometer fell. This was because, with the air in the receiver being now rarefied, the repulsion between the fire particles was diminished and so fire flowed out of the thermometer. But once equilibrium with the external air had been restored, the thermometer took up the outside temperature once again. Then, when the external air was admitted, the rarefied air inside the receiver, whose temperature was equal to that outside, was suddenly compressed, the repulsion between the fire particles was increased, fire flowed out of the air, entered the thermometer and so caused its temperature to rise.‡

Although Cleghorn had understandably misconstrued the obscure evacuated receiver experiment, this passage is an important one since it contains the first recognition by any writer that the compression and rarefaction of air were the sole cause of the temperature changes. Of course his failure to appreciate that the compression is not experienced solely by the small amount of air remaining but that air rushing into a vacuum

† Cleghorn, *De igne*, p. 20.   ‡ Cleghorn, op. cit., p. 21.

continually compresses that which enters before it made his argument somewhat implausible. How, for example, could the compression of such a small quantity of air as remained in a near-perfect vacuum give rise to so much heat? As Gay-Lussac later pointed out in his criticisms of a similar explanation given by John Leslie, the heating effect was not known to diminish, even in the best available vacua.†

Cleghorn's treatment, like Arnold's, was little known, the only subsequent published reference to it being made by Black in his *Lectures on the elements of chemistry*. This is unfortunate, since it had as much to recommend it as the far more influential explanation proposed in the same year in Johann Heinrich Lambert's *Pyrometrie*.‡ According to the account that he gave in this book, Lambert had known of the effect for some time by 1779, having been shown the experiments with the air pump by Arnold in 1761. At the time of their meeting Lambert had not seen Arnold's dissertation and he independently devised a quite different explanation, attributing the cooling to the fact that 'fire particles' (*Feuertheilchen*) were abstracted from the receiver together with the air. If the density of the air was reduced by half, for example, the density of the fire particles and hence also the temperature, which was proportional to this density, would be diminished by the same factor. In his references to temperature Lambert clearly had in mind absolute temperature, but he recognized that there was no question of attaining the absolute zero by the mere evacuation of air, since fire particles from the walls of the receiver and the base of the pump would quickly replace those abstracted. Indeed, cooling could only occur if the rate at which the particles were abstracted exceeded that at which they were replaced. Conversely, if a gas was compressed to half its original volume without heat loss, its (absolute) temperature would be doubled and this, added to the effect of an isothermal decrease in volume, would cause the pressure of the gas to be increased fourfold. But again heat exchange with the surroundings would ensure that these ideal conditions were never actually attained. Unfortunately Lambert offered no explanation for the rise in temperature accompanying the entry of air into a

† See pp. 108–9 and 130.
‡ Lambert, *Pyrometrie oder vom Maasse des Feuers und der Wärme* (Berlin, 1779), pp. 266–9.

vacuum, but a plausible reconstruction of his opinion on this matter can be made. For he certainly considered that a vacuum could contain fire particles, to the point of maintaining that temperature changes would occur if the volume of the Torricellian vacuum in a simple barometer was altered rapidly.† Hence he would surely have accounted for the rise in temperature by supposing that the entry of air with its accompanying fire particles caused an accumulation of fire particles and that it was this increase in fire-particle density that brought about heating.‡

It was through Lambert's work that adiabatic heating came to the notice of Saussure. Saussure's interest in hygrometry ensured his knowledge of Lambert's writings, among which he had been particularly impressed by the *Pyrometrie*.§ It is appropriate to note here how closely the effect was linked with meteorological studies and particularly with hygrometry from Cullen to John Dalton. Most of the really influential eighteenth-century authorities, among them Lambert, Saussure, and Pictet, as well as Cullen and Dalton, worked on hygrometry or were concerned with the study of vapours in some way.¶ Having become fortuitously connected with this subject through Cullen, the effect, before 1800, was almost always reported and discussed in literature relevant to it. In view of this association with a particular type of literature and in view also of the specific and unusual conditions that were necessary for an independent discovery, it is hardly surprising that knowledge of the effect spread slowly to other research traditions. One fruitful analogy, which could easily have provoked a far wider interest at this time and so have brought the effect into a closer association with the development of the caloric theory, was with the heat produced by the percussion of a solid. However, the general slowness to see that the compression of air rather than the filling of a vacuum was the fundamental phenomenon that required examination can only have obscured the fact that the heating effects produced in the gas and the

† Lambert, op. cit., pp. 268–9.
‡ This explanation was ascribed to Lambert by Marc Auguste Pictet in his *Essais de physique* (Geneva, 1790), pp. 19–20.
§ Saussure, *Hygrométrie*, Preface; also his *Voyages dans les Alpes* (4 vols., Neuchâtel, 1779–96), vol. 2, p. 350, where he entered a plea for the translation of the *Pyrometrie*.
¶ The name of De Luc may also be added to this list.

solid both originated in a similar fashion, and it was not until the early years of the nineteenth century that the analogy became generally accepted. De Luc and Erasmus Darwin were the only writers to recognize it before 1790† and they, significantly enough, showed less interest than their contemporaries in the work with the evacuated receiver.

The location and scarcity of accounts in the scientific literature are clearly relevant to any examination of the extraordinary degree of ignorance of adiabatic heating in France during the eighteenth century. Only one reference by a French author before 1798 has been found, in J. A. C. Chaptal's textbook, *Élémens de chimie*, which was first published in 1790. In the book Chaptal briefly mentioned Cullen's observation with the air pump but misleadingly described it as being due to evaporation.‡ Although the *Élémens de chimie* was an important and successful work (a third edition had appeared by 1796), such a minor remark, the slight importance of which was in any case diminished by Chaptal's interpretation, was all too likely to be overlooked. And it certainly had no detectable influence, as an examination of textbooks and papers on heat in the 1790s clearly demonstrates.

This failure on the part of the French to read, or at least to recognize the importance of, the early accounts of the adiabatic phenomena requires explanation. The authors of these accounts, though few in number, were, after all, well-known figures. Perhaps the most important factor, especially as far as neglect of the German–Swiss school was concerned, was that meteorological inquiry and hygrometry aroused rather less interest in France than it did in Switzerland and Germany, so that the relevant literature was presumably less widely read by the French. British and, in particular, Scottish science, on the other hand, was well known in Paris at the time when Lavoisier and Laplace were formulating their views on heat in the 1780s, but of the works that would have been discussed then Magellan's *Essai sur le feu* alone mentioned adiabatic heating, and then only briefly and in passing.§ Crawford made no published

---

† De Luc, *Météorologie*, vol. 1, p. 208, and Darwin, *Phil. Trans. R. Soc.* **78** (1788), 43. On the subsequent adoption of the analogy see p. 100.
‡ Chaptal, *Élémens de chimie* (Montpellier, 1790), vol. 1, pp. 70–1.
  Magellan, *Essai*, p. 185.

reference to it until 1788, although a passage in the minute book of the Chapter Coffee House Society of London,† dated 21 February 1783, shows that he had performed experiments on the subject by that date, and one comment even suggests that news of these might already have reached Paris. Speaking of his work with evacuated receivers, Crawford reported 'that Mr. Lavoisier had repeated his (Dr Crawford's) experiments and drawn from them similar conclusions . . .'; but of this contact we know nothing more, and the possibility that Lavoisier knew of adiabatic heating by 1783 but that he either considered it unimportant or misunderstood it must therefore remain unconfirmed. By the time Crawford's work became common knowledge five years later, caloric theory was no longer a major subject of debate in France and, in any case, the second edition of *Animal heat*, in which Crawford announced his results, attracted far less attention there than the first had done, as has already been noted. Most French chemists were now preoccupied with the task of establishing the anti-phlogistic chemistry rather than with Crawford's problems, which were essentially physical and, by 1788 at least, no longer even novel.

That the effect should have aroused so little interest in France was unfortunate, since most of the treatments given before 1800 lacked the experimental and theoretical rigour that the French calorists could have brought to bear. The explanations of both Saussure and Pictet, for example, were incomplete in that neither treated both heating *and* cooling. For both men Lambert was the most important authority, although Saussure, who wrote in 1783, had certainly read Cullen as well and he even quoted from Cullen's paper of 1755.‡ Like Arnold, whose work he must have known if only through Lambert, Saussure carelessly interpreted Cullen's account as an attempt not only to attribute the cooling effect to evaporation but also to prove that evaporation occurred more readily and produced greater cooling in the void than in air. In his own experiments Saussure had obtained a temperature change similar to that observed by Cullen and he had also shown that the cooling effects obtained with damp and with dried airs were identical, so ruling

† Museum of the History of Science, Oxford, MS. Gunther 4. The pages of the manuscript are unnumbered.
‡ Saussure, *Hygrométrie*, pp. 129 and 231–3.

out the possibility that evaporation contributed appreciably to the phenomenon. Vigorously rejecting Cullen's (supposed) explanation he instead followed Lambert in attributing the cooling to the abstraction of fire particles (*parties de feu*) together with the extracted air. Unfortunately Saussure attempted no explanation of the heat evolved when air entered a vacuum, but his chief concern was clearly to refute Cullen rather than to give a full description of adiabatic phenomena.

Saussure's comments provided the starting-point for his pupil and close friend Marc Auguste Pictet, who succeeded him in the chair of philosophy at Geneva. In 1790 Pictet wrote that he himself had observed a temperature rise of about 2–3 °C (2 °R) on allowing air to rush into an evacuated receiver and a similar rise by fitting a thermometer into the end of a condensing syringe and giving a single rapid stroke of the piston.† In a manner characteristic of our period Pictet expressed grave doubts concerning the true nature of heat but proceeded to argue in terms of the material theory, which he reluctantly admitted was preferable.‡ Although he vigorously defended the originality of his own theory, maintaining that he had developed it before Lavoisier's *Traité* was published,§ it showed close similarity to current French views. His fire (*feu*) was almost identical with caloric in its properties and he even expressed his approval of the French term.¶

Pictet's work shows a conservatism that strongly emphasizes the difficulty imposed by the way in which adiabatic heating was discovered in the eighteenth century, viz. in the course of work with the air pump. Unlike many earlier writers, Pictet was fully aware that the temperature of a gas rose when it was rapidly compressed and that this effect was related to that observed in the air-pump receiver, yet he still could not abandon belief in the ability of the void to contain heat as an essential part of his explanation. With the density of fire particles still assumed to determine temperature (and Pictet made this assumption without question) it was understandably difficult to see how heating resulted from the mere entry of a mass of air into a vacuum, unless, as Lambert had suggested, additional

† Pictet, *Essais*, pp. 19–20.   ‡ Pictet, op. cit., pp. 2–4.
§ Pictet, op. cit., pp. vii–viii.   ¶ Pictet, op. cit., p. 5.

fire particles were already present in the vacuum ready to be taken up by the air and so to increase its temperature. Although the confusing attempts to treat a changing mass of gas and a constant volume (as in an air pump) rather than a constant mass of gas in a changing volume (as in a condensing syringe) persisted throughout our second period, the realization that the rapid compression of a gas did cause heating was spreading by the 1780s, creating the need for new explanations and diverting attention from the evacuated-receiver experiments. How much the French could have contributed at this stage is obvious. They could have argued, as Laplace was later to do, that compression decreased the quantity of latent caloric that a gas could hold and that the excess caloric would thereupon become 'sensible' and so raise the temperature. But this type of explanation belonged to the nineteenth century. The much simpler view, adopted by Pictet for example, was that heat was expelled from a compressed gas like water from a sponge.

That changes in volume altered the ability of a gas to contain heat was fundamental to nearly all the explanations of the 1779–1800 period, and the explanation offered by De Luc in 1786† was typical in this respect. De Luc's account nevertheless marks an important stage in the history of the effect. Although he was Swiss by birth, his connections with the Swiss school were tenuous, since he had travelled widely and, as 'reader' to Queen Charlotte, had lived in England since 1773. He still believed that the void did have a definite capacity for heat, but for reasons quite unrelated to adiabatic heating.‡ Seeing the change in volume of a gas as the basic phenomenon to be explained, he was able to free his work completely from the influence of the evacuated-receiver experiments. It is significant in this respect that it was he who first publicly identified adiabatic heating with the heat evolved in the percussion of a solid, as has already been pointed out. He maintained that a simple change in volume, i.e. in the space between the particles of ponderable matter that could be occupied by fire (*feu*), would affect only what he termed the 'geometric capacity' (*capacité géométrique*), while the other contribution to the total

† De Luc, *Météorologie*, vol. 1, pp. 142–8.
‡ De Luc, op. cit., vol. 1, pp. 143–4.

capacity, the 'physical capacity' (*capacité physique*), determined solely by the nature of the substance, might or might not vary as well. Capacity, for De Luc, was synonymous with both specific heat and ability to contain fire, so that any decrease in volume, and hence also in the geometric capacity, would be sufficient to cause the liberation of fire and hence heating. De Luc's explanation may seem similar to that of an Irvinist, but in fact he showed no sympathy for Irvine's views and criticized Crawford.†

Accounts by eighteenth-century Irvinists are disappointingly rare, with the second edition of Crawford's *Animal heat* the only major source before 1800.‡ Here Crawford described experiments in which he had observed temperature changes of roughly 5 °F on evacuating and filling the receiver of an air pump. The work was carefully performed and throughout he had tried to obviate the effect of moisture. Yet it was original in only one respect, namely in his use of gases other than air. When phlogisticated and dephlogisticated air were used, the temperature changes were, respectively, slightly smaller and slightly greater than those obtained with atmospheric air, but the differences were small. Although Crawford gave no explanation, he appears to have associated the effect with the changes in capacity that he supposed to occur when the volume of a gas varied. Indeed, as we have seen, he almost certainly inferred this dependence of capacity on volume from the adiabatic phenomena.§ We can assume, moreover, that he saw the heat evolved during the admission of a gas into a vacuum as resulting in some way from compression, since he considered the volume capacity of the void to be either zero or very small by comparison with that of a gas¶ and so could not have interpreted the effect as a case of a gas of low capacity replacing a vacuum of high capacity, as Dalton, another Irvinist, was later to do.||

† De Luc, op. cit., vol. 1, pp. 142–8.
‡ Crawford, *Animal heat* (2nd edn, 1788), pp. 250–2. The results laid before the Chapter Coffee House Society in 1783 (which had been obtained in experiments at the house of the instrument-maker Edward Nairne) were similar to those given in *Animal heat*. Temperature changes of between 4 and 5 °F were obtained when air at atmospheric pressure was pumped rapidly from the receiver of an air pump, when air at 2 atmospheres was allowed to escape freely, and also when a 'second atmosphere' was introduced (presumably indicating that work had been performed with a condensing pump).
§ See pp. 37–8.   ¶ Crawford, op. cit., pp. 266–7. || See p. 88.

An interesting anomaly relating to Crawford's work occurred in December 1787, not long before the publication of the second edition of *Animal heat*, when Erasmus Darwin independently described his own full experimental investigations of adiabatic phenomena to the Royal Society in London.† Darwin made no reference to earlier sources and so appeared to be claiming an independent discovery. This seems strange when Crawford, in 1788, could refer to the observations made with the air pump as being 'well known',‡ but Crawford of course was speaking as one familiar with the Scottish school of chemists and of this community his comment was no doubt true. The extent to which the effect was known in England, on the other hand, is less easily gauged. The absence of any reference to it in some prominent scientific works published in London before the turn of the century§ certainly suggests that Darwin's might well have been an original contribution. And the fact that his paper was read to the Royal Society and published in the *Philosophical Transactions of the Royal Society* would imply that it was seen as such by members of that body at least.

Yet the situation is an interesting one. In his paper Darwin claimed that he had begun his experiments twelve or fourteen years earlier, i.e. in 1773 or 1775. Then, working with the itinerant teacher John Warltire, he had observed the cooling that accompanied the evacuation of an air-pump receiver and, with Richard Lovell Edgeworth, his friend in the Lunar Society of Birmingham, and the great Scottish geologist James Hutton, he had investigated the fall in temperature that occurred when the blast of an air-gun was directed on to a thermometer bulb. He had also noticed the heat evolved during the compression of air in the receiver of the gun. There is no means of confirming whether these experiments were in fact performed in the period mentioned by Darwin, but Hutton was certainly engaged on a tour

† Darwin, 'Frigorific experiments on the mechanical expansion of air', *Phil. Trans. R. Soc.* **78** (1788), 43–52.
‡ Crawford, op. cit., p. 250.
§ It was not mentioned in, for example, W. Nicholson, *The first principles of chemistry* (3 editions between 1790 and 1796) and *An introduction to natural philosophy* (4 editions between 1782 and 1796); C. Hutton, *A mathematical and philosophical dictionary* (1st edn, 1796). In G. Adams, *Lectures on natural and experimental philosophy* (1st edn, 1794), vol. 1, p. 309, and (2nd edn, 1799), vol. 1, p. 383, it was evidently misunderstood, being cited without explanation in a section on cooling by evaporation.

THE STUDY OF GASES AND HEAT TO 1800

of England about this time and in 1774 he was introduced to the Lunar Society circle by Darwin.† It seems reasonable to suppose, then, that Darwin's somewhat unspecific recollection was correct on this point, and we may be quite sure from other evidence that he knew of adiabatic heating by about 1776. This evidence appears in Darwin's commonplace book‡ in some notes on winds. Among these notes, on p. 45 of the book, is the following passage, which is of special interest not only as evidence of Darwin's familiarity with certain adiabatic phenomena but also in that it is the earliest known instance of changes in volume being seen as the sole cause of the thermal effects.§ It reads:

Now if air be added suddenly rarefied without the addition of heat, as in going into an exhausted receiver, or by being pumped out of it cold is produced. . . . On the contrary when air is condensed heat is produced. . . .

Now the position of the notes in the book enables us to date this comment fairly precisely. They are followed immediately, on p. 47, by a brief reference to a wind gauge which, Darwin states, was described in 'one of the last volumes' of the *Philosophical Transactions of the Royal Society*. The article to which Darwin was presumably referring, by the physician James Lind, was in fact published in 1775.¶ Moreover, the notes are preceded by a quite separate section on heat headed 'From Charles' notes at Dr. Black's lectures', 'Charles' being almost certainly Darwin's eldest son by his first wife. Charles studied medicine at Edinburgh, where Black was professor of medicine and chemistry, from 1775 until his early death in 1778,‖ so that the notes that

---

† See J. Playfair, 'Biographical account of the late Dr. James Hutton', *Trans. R. Soc. Edinb.* 5 (1805), part 3, p. 47, and R. E. Schofield, *The Lunar Society of Birmingham* (Oxford, 1963), p. 102.

‡ Now preserved among the Darwin papers at Down House, Kent.

§ However, it is likely that Bryan Higgins interpreted adiabatic phenomena in this way by 1775. In his *Syllabus of the discourses and experiments* of that year, p. 61, he referred to 'Experiments shewing, that, when a body is suddenly expanded, by reason of its own elasticity, or by any power except that of fire; then cold is produced in that body, and in the vicinity of it.' He also mentioned experiments showing that compression produced heat. No further details of the experiments were given but it must be assumed that they were experiments on gases.

¶ Lind, 'Description and use of a portable wind gauge', *Phil. Trans. R. Soc.* 65 (1775), 353–65.

‖ See Charles Darwin's biographical sketch in E. Krause, *Erasmus Darwin* (2nd edn, 1887), pp. 80–3.

he took were presumably entered in the book within this period. In this way it is possible to date, with a fair degree of accuracy, the entries in the commonplace book that precede and follow Darwin's notes on winds. In the light of this evidence, and bearing in mind also that the entries in the book follow an approximately chronological order, we must conclude that the passage quoted above was written in 1776, or at all events no more than a year or two after that date. Another factor that tends to confirm this dating is that the book is inscribed at the front '1776' and although some subsequent entries date from several years later, it is clear that the early part of it, in which the notes on winds appear, was completed quickly.

Even if we accept the date of 1776 that I have suggested and accept also that Darwin had performed experiments on the adiabatic effect before then, the question of whether he made the discovery independently still remains unsolved, since there are a number of possible sources from which he could have learnt of it. The first and strongest possibility is that he had read Cullen's account, which by 1770 had been published in two editions of the Edinburgh Philosophical Society's *Essays*.[†] The source was by no means obscure and the volume in which Cullen's paper appeared was all the more notable since it included Black's 'Experiments upon *magnesia alba*', so that it would be strange indeed if it had escaped Darwin's attention. He would certainly have known of Cullen about the time of the discovery, for it was in the autumn of 1754 that he began his medical studies in Edinburgh.[‡] At this time Cullen was still at Glasgow as professor of medicine, but the work of such a renowned physician would have been known far beyond his own university. In 1755 Darwin spent some time at Cambridge, taking his M.B. at St. John's College in that year, but he returned to Edinburgh before setting up as a physician, first in Nottingham in September 1756 and then in Lichfield in November 1756. Cullen had been elected to the chair of chemistry at Edinburgh by November 1755 and he began lecturing there in the following January, though to a small audience,[§] so that it is

[†] In 1770 it was reprinted in the second edition of the *Essays* (vol. 2, pp. 159–71).

[‡] Krause, op. cit., pp. 12–16.

[§] J. Thomson, *An account of the life, lectures, and writings of William Cullen M.D.* (2nd edn, London and Edinburgh, 1859), vol. 1, pp. 90–7.

even possible that Darwin heard of adiabatic heating from Cullen himself. A letter from Black to Cullen, dated 7 August 1755,† indicates that evaporation under an evacuated receiver was still of current interest three months after the reading of Cullen's paper, and so the matter might well have been mentioned in lectures. In fact, since Cullen's class register for 1755–6‡ makes no mention of Darwin, a personal contact between the two men remains no more than a possibility.

Of the other possible sources available to Darwin perhaps the most likely is his friend James Hutton. Before he was introduced to the Lunar circle by Watt in 1774, Hutton had been closely associated for a number of years with the Scottish chemists, notably Black,§ among whom the adiabatic effect was probably common knowledge. Even if he did not hear of it in this way, we know that he read and was impressed by Black's paper on *magnesia alba*¶ and so is unlikely to have ignored the paper by Cullen that immediately preceded it in both editions of the Philosophical Society's *Essays*.

Among other possible contacts we should also mention James Keir, Darwin's lifelong friend and contemporary in the Edinburgh medical school, who in a letter to Darwin in 1766, referred to Cullen's evaporation experiments.‖ James Watt too could well have told Darwin of the adiabatic effect. For after setting up as an instrument maker to the University of Glasgow in 1757 he came into close contact with such men as Black and Cullen†† and it is hard to believe that he would not have learned of adiabatic heating in this lively community. In 1767 Watt met Darwin in Birmingham and, although he did not move to Soho until 1774, he immediately joined the Lunar circle, corresponding vigorously with Darwin and the other members, especially Dr William Small.‡‡

† Thomson, op. cit., vol. 1, p. 579.
‡ The register, headed 'Students in the College of Chemistry' and covering the period 1755–65, is in a notebook (MS. Da. 3) in the Edinburgh University Library.
§ Playfair, op. cit., p. 46. ¶ Playfair, op. cit., p. 59.
‖ See A. and J. K. Moilliet, *Sketch of the life of James Keir* (n.d.), p. 46.
†† On this contact see S. Smiles, *Lives of Boulton and Watt* (2nd edn, 1866), pp. 112–13.
‡‡ For this correspondence see especially J. P. Muirhead, *The origin and progress of the mechanical inventions of James Watt* (1854), vol. 1, pp. 15–104, and vol. 2, pp. 1–78.

A further point seems relevant to the question of the originality of Darwin's work. In the paper of 1787 Darwin described two experiments that he had conducted specifically to prove that the cooling was not the result either of distension of the bulb of the thermometer or of evaporation from its surface.† As we have seen, both of these rather improbable explanations had been proposed by earlier writers and it seems an extraordinary coincidence that Darwin should have chosen quite independently to refute them in the first announcement of his 'discovery'. Since neither of the experiments in question was conducted before December 1784, as Darwin himself stated, this evidence throws little light on the way in which he first learned of adiabatic heating, but it strongly suggests, contrary to the general impression conveyed in his paper, that he had read at least some of the literature on the subject by 1787. Indeed, it would be surprising if such members of the Lunar circle as Watt, Josiah Wedgwood, or Priestley, among whom the effect was probably well known,‡ had not brought the earlier references to his attention.

With so much of the evidence only circumstantial, we must inevitably leave the question unresolved. But, whatever our views on the originality of Darwin's account—and this originality must at least be in serious doubt—there can be no denying its importance. Darwin's was unquestionably the fullest study of adiabatic heating that had yet been undertaken and it was full not simply in respect of the number and variety of his experiments. He was quick to see the possible relevance of his work to two other apparently unrelated phenomena that were badly in need of explanation. The first was the mysterious deposit of snow round the stopcock where damp compressed air

† Darwin, *Phil. Trans. R. Soc.* **78** (1788), 45–6.

‡ By 1787 the effect had already been mentioned in correspondence involving Wedgwood and Priestley; see Darwin to Wedgwood, 11 March 1784, in Krause, op. cit., pp. 100–1, and Priestley to Benjamin Vaughan, 26 March 1780, in H. C. Bolton, *Scientific correspondence of Joseph Priestley* ed. H. C. Bolton (New York, 1892), pp. 19–21. In his letter Priestley referred to a demonstration of the effect by the itinerant lecturer John Arden, who was also the source of Magellan's knowledge (Magellan, *Essai*, p. 185). Magellan and Vaughan, as well as James Keir, Edward Nairne, and James Hutton, attended the meeting of the Chapter Coffee House Society on 21 February 1783 at which Crawford's experiments with the receiver of an air pump were described (see p. 50); in fact, as was noted on p. 53 n. ‡, the experiments were performed at Nairne's house in London.

THE STUDY OF GASES AND HEAT TO 1800

rushed from a mine pump at Schemnitz in Hungary.† The observation had been made as early as 1758, but no satisfactory explanation had yet been offered. By close analogy with an experiment that he and his friend Fox had conducted in December 1784, in which they had examined the cooling that occurred when compressed air was suddenly released from a vessel, Darwin concluded, correctly, that the snow resulted from the freezing of water vapour by the rapid expansion of the emerging air. He evidently had a considerable interest in the cooling effect, since the notes concerning his experiments with Fox in the commonplace book are followed immediately by a diagram that seems to represent a device for attaining low temperatures.‡ Whether it was ever constructed is not clear, but in an accompanying note Darwin expressed the hope that it might be possible even to freeze mercury with it. The second phenomenon that Darwin was able to explain in his Royal Society paper was the cold experienced in the upper atmosphere, an effect that had long provoked much speculation. In his notes on winds, which were dated above to 1776 or shortly afterwards, Darwin was already applying his knowledge of the adiabatic effect to meteorological phenomena§ and in his paper to the Royal Society he ascribed the cold at high altitudes to the expansion of air as it rose and experienced a diminishing pressure. Despite its essential correctness, the explanation attracted little attention, James Hutton and later Tiberius Cavallo being the only contemporary writers who adopted it.¶

If it did nothing else, the appearance of Darwin's paper in the

† See pp. 79–80.

‡ The diagram, which appears on p. 119 of the book, is reproduced as Plate 9 in Schofield, *Lunar Society*, facing p. 268. The notes on his work with Fox are on p. 118 of the commonplace book.

§ See also Darwin, *The botanic garden* (1791), 'The economy of vegetation', additional notes, pp. 82, 88–9, and 92, where he repeated many of his earlier opinions. On p. 15 of these additional notes Darwin also applied his knowledge of the adiabatic effect in an ingenious explanation of the heat produced by friction in solid bodies. He argued that friction caused the particles of a solid to vibrate and that when they receded from one another, causing a temporary expansion, heat was absorbed from the surroundings. The surroundings, thus deprived of heat, immediately absorbed more heat 'from other bodies behind them', so that by the time the vibrating particles moved together again there was a net accumulation of heat. The argument, which had obvious weaknesses, does not appear to have aroused much interest.

¶ See Playfair, op. cit., pp. 66–7 n., and Cavallo, *The elements of natural or experimental philosophy* (1803), vol. 3, p. 94.

Philosophical Transactions ensured that henceforth there could be no excuse for ignorance of the adiabatic effect, and the paper was understandably an important source, in fact probably the most important eighteenth-century source, for later writers. What in Britain at least had been the concern of fairly restricted groups, such as the Scottish school of chemists, the Lunar Society in Birmingham, and the Chapter Coffee House Society in London, now became common knowledge. Nevertheless, interest in the effect died down after the publication of the paper, and this account of the second of our three periods therefore ends with Darwin and Crawford in Britain and with Pictet on the Continent.

The work of the third period was to be of a very different nature, with the true magnitude of the effect, which before 1800 had been seriously obscured by heat losses, being at last appreciated. The most important factor in these advances was to be the entry of the French into the field, and it was they who were chiefly responsible for transforming something that was little more than a curiosity, without apparent relevance to any really major problem, into a phenomenon at the very centre of the study of heat. Those eighteenth-century textbooks, both British and French, which either ignored it or relegated it to a minor position could never have been written after 1800.

Expansion by heat

Although the first experiments designed specifically to measure the extent to which air was expanded by heat were described only in 1699, by the gifted French physicist Guillaume Amontons,[†] the effect had long been a familiar one. It was known to such ancients as Philo of Byzantium and Hero of Alexandria,[‡] and during the seventeenth century it had already been applied in the early air thermometers or thermoscopes of Galileo and Santorio Santorii,[§] but in no case before the time of

[†] Amontons, 'Moyen de substituer commodement l'action du feu à la force des hommes ...', *Mém. Acad. Sci.* (1699), pp. 112–26.

[‡] See, for example, the devices in Hero's *Pneumatica* and the references to the effect in Philo's *De ingeniis spiritualibus*. These are consulted most easily in *Heronis Alexandrini opera quae supersunt omnia*, ed. and trans. W. Schmidt (Leipzig, 1899), vol. 1, pp. 176, 224, and 474–6.

[§] F. S. Taylor, 'The origin of the thermometer', *Ann. Sci.* 5 (1942), 129–56, and W. E. Knowles Middleton, *A history of the thermometer and its use in meteorology* (Baltimore, 1966), pp. 3–26.

THE STUDY OF GASES AND HEAT TO 1800

Amontons had the need for a closer quantitative study of the behaviour of heated air been felt.

Amontons's immediate incentive was a purely practical one, concerned with the development of his 'fire-mill' (*moulin à feu*), a novel engine that utilized the pressure of hot air to drive a mill-wheel. In his experiments Amontons in fact measured not the expansion of air but the increase in its pressure when its volume remained constant. The apparatus consisted of a U-tube, one limb of which (the longer) was open to the atmosphere while the other was closed by a bulb, the air under examination being contained in the bulb by a short mercury column in the tube. By observing the change in the levels of mercury Amontons concluded that the pressure of air increased by approximately one-third when it was heated from room temperature to the temperature of boiling water, and hence, using Boyle's law, he showed that a similar increase in volume would occur if the pressure rather than the volume was maintained constant. Further similar experiments performed in 1702 and described before the Académie des Sciences in the same year† only confirmed Amontons's conclusions, although he now saw clearly how variations in the pressure of air could be used to indicate temperature and how his U-tube apparatus might be employed for the purpose.

It was in his Académie paper of 1702 that Amontons gave the first hint of his understanding of the existence of an absolute zero far below any temperatures previously encountered,‡ and in April 1703 he formally defined what he termed the 'extreme cold' (*extrême froid*) of his thermometer as that point at which air would exert no pressure.§ Although he did not make the calculation himself, his data indicate a zero at $-239 \cdot 5$ °C, if we assume the air pressure to decrease uniformly with temperature. Despite references to it by George Martine, J. J. d'Ortous de Mairan, Lambert (who was very favourable), and Alessandro Volta,¶ Amontons's definition was never widely accepted

† Amontons, 'Discours sur quelques proprietez de l'air . . .', *Mém. Acad. Sci.* (1702), pp. 161–80; also pp. 1–8 of the *Histoire* at the front of this volume.
‡ Amontons, op. cit., p. 171.
§ Amontons, 'Le thermometre réduit à une mesure fixe & certaine . . .', *Mém. Acad. Sci.* (1703), p. 52.
¶ Martine, *Essays medical and philosophical* (1740), pp. 291–2; Mairan, *Mém. Acad. Sci.* (1765), pp. 204–9; Lambert, *Pyrometrie*, pp. 29–30. For Volta's

during the eighteenth century and, when interest in the absolute zero revived with the advent of Irvinist doctrines, it was not even mentioned in the debate of the 1780s and 1790s. Even immediate comment, such as that by Lazare Nuguet and Amontons's fellow academician Philippe de la Hire, was concerned with the experimental rather than theoretical issues.† There can be little doubt that the effect of heat on the pressure or volume of air would have aroused a far greater and more profound interest at this time had thermometers of the Amontons type come into general use. Two observers who did adopt this form of thermometry in their meteorological work were Nicolaas Samuel Cruquius in Holland and the Italian mathematician Giovanni Poleni and, as we should expect, both men checked (and substantially confirmed) Amontons's observations.‡ But air thermometry made little headway against the increasingly refined methods involving the use of alcohol and mercury in closed tubes and, although occasional measurements of the expansion of air were made despite the lack of any obvious incentive,§ it was not until the last quarter of the eighteenth century that the subject became one of major importance. The chief reasons for this later development of interest seem to be twofold. First, De Luc's demonstration that air temperature had to be taken into account in measuring altitude with the aid of a barometer created the need for a

reference, in a paper in Brugnatelli's *Annali Chim.* 4 (1793), 227–94, see *Le opere di Alessandro Volta* [Edizione Nazionale] (7 vols., Milan, 1918–29), vol. 7, p. 357. Of these writers only Mairan was severely critical. He attacked Amontons's definition of the zero, arguing that a cold greater than any envisaged by Amontons might exist. He believed that the zero, if it existed at all, could not yet be located, and his position was strongly supported by the writer of the official *Histoire* of the Académie for 1765 (p. 10), who described the zero as 'a chimera that exists nowhere in nature'. There was clearly a considerable, though diminishing, scepticism with regard to the absolute zero in the eighteenth century. Boerhaave, for example, considered that the 'limits' of cold could not be ascertained; see Boerhaave, *Elementa chemiae* (1732), vol. 1, pp. 151 and 189.

† Nuguet, *Mém. Hist. Sci.* (1705), pp. 1790–1807, and de la Hire, *Mém. Acad. Sci.* (1708), pp. 274–88.

‡ Cruquius, *Phil. Trans. R. Soc.* 33 (1724–5), 4, and Poleni, *Phil. Trans. R. Soc.* 37 (1731–2), 205.

§ See, for example, Hauksbee, *Physico-mechanical experiments* (1709), pp. 170–5, and *Phil. Trans. R. Soc.* 26 (1708–9), 93–6; Musschenbroek, *Philosophiam naturalem* (1762), vol. 2, pp. 883–4; J. C. P. Erxleben, *Anfangsgründe der Naturlehre entworsen . . . Fünfte Auflage* (Göttingen, 1791), p. 394.

THE STUDY OF GASES AND HEAT TO 1800   63

reliable value for the expansion coefficient of air. And secondly, the growth of pneumatic chemistry meant that some method was required for correcting the volumes of gases, and not only of air, to a standard temperature. These two new incentives were felt by such clearly distinguishable groups of workers that they will be considered here separately.

De Luc published his account of altitude measurement in 1772.† Using a rule first derived by Edmund Halley, according to which the difference between the heights of two points was proportional to the difference between the logarithms of the barometric readings taken at the points,‡ he made a large number of determinations of altitude and compared his results with those obtained by trigonometrical methods. By making the comparisons several times at different temperatures he estimated that the discrepancy between the results obtained by the two types of method disappeared when the mean temperature of the points at which barometric measurements were made was 21 °C ($16\frac{3}{4}$ °R). Further examinations showed that the effect of any departure from this mean temperature was, for every °C, an error of 1/269 of the height calculated by the simple logarithmic law, this figure (or 1/215 per °R, as De Luc gave it) being also the coefficient of expansion of air in the region of 21 °C.§

In view of the growing interest in mountaineering and, from 1783, in ballooning, in both of which pursuits portable barometers were commonly carried, it is not surprising that De Luc's work became well known. Within five years of its appearance it had moved both Sir George Shuckburgh and Colonel William Roy in England to make direct observations of the expansion of air and, as a result of their experiments, to modify some of De Luc's figures, though without rejecting the principles of his treatment.¶ The Swiss physicist Jean Trembley, by contrast, appears to have performed no special experiments but, by applying De Luc's indirect method to barometric observations made by Roy and Shuckburgh, he concluded, in a paper presented to the Académie des Sciences in Paris in 1781, that

† De Luc, *Recherches sur les modifications de l'atmosphère* (Geneva, 1772), vol. 2, pp. 43–242.   ‡ Halley, *Phil. Trans. R. Soc.* **16** (1686), 105–6.
§ De Luc, op. cit., vol. 2, pp. 90–2.
¶ Shuckburgh, *Phil. Trans. R. Soc.* **67** (1778), 562–8, and Roy, *Phil. Trans. R. Soc.* **67** (1778), 689–715. Their coefficients were the equivalents of, respectively, 1/228 and 1/207 per °C.

modifications were required both in the temperature at which the barometric and trigonometrical methods were supposed to be in agreement and in the expansion coefficient.† Although it is difficult to cite other work that was definitely inspired by the problem of altitude measurement,‡ later writings, notably those by Laplace, indicate that both this subject and the closely related one of atmospheric refraction remained important well into the nineteenth century and that therefore, if only for this reason, the demand for accurate data relating to the expansion of air continued.

But altitude measurement, as has been noted, was not the only issue that stimulated interest in the expansion of gases by heat, and we turn now to consider the other important incentive for work of this type, namely the rise of pneumatic chemistry in the last quarter of the eighteenth century. It was quickly recognized that if quantitative work in the chemistry of gases was to be meaningful, volumes had to be corrected for variations in both pressure and temperature. For pressure Boyle's law was unquestioningly adopted, while for temperature De Luc's figure of 1/269 per °C (or 1/215 per °R) appears at first to have been considered the most acceptable, if we judge from Lavoisier and Laplace's application of it in their treatment of combustion and respiration in 1783.§ Yet De Luc's work was not entirely to be relied upon, as subsequent comment had suggested, and in any case pneumatic chemistry was not solely concerned with air. As early as 1777 Joseph Priestley had performed some crude experiments on the expansion of all the gases that he knew,¶ but when Vandermonde, Berthollet, and

† Saussure, *Voyages dans les Alpes* (1786), vol. 2, pp. 616–41. Trembley's figure was the equivalent of 1/229 per °C.

‡ Volta, however, may be an exception on this point; see p. 66. So may Saussure—a noted alpinist—who performed a 'large number' of experiments on the expansion and variation in pressure of air with temperature; see Saussure, *Hygrométrie*, p. 108. He concluded that air at room temperature expands by 4·24383 parts in 1000 for a rise in temperature of 1 °R. This yields a coefficient of expansion of about 1/295 per °C.

§ Lavoisier and Laplace, *Mém. Acad. Sci.* (1780), p. 396. See also Guyton de Morveau's comments in his article on 'Air' in *Encyclopédie méthodique. Chimie* (Dijon, 1789), vol. 1, p. 677.

¶ Priestley, *Experiments and observations on air*, vol. 1, p. 250; vol. 3, pp. 345–8; and *Experiments and observations relating to . . . natural philosophy*, vol. 2, p. 359. The experiments showed the various gases to expand to very differing extents.

Monge came to investigate the quantities of hydrogen in various types of iron and steel in 1786, they very properly ignored Priestley's work and they redetermined the expansion coefficients for air and hydrogen.†

Although these experiments of Vandermonde, Berthollet, and Monge fulfilled their immediate purpose, they were of limited scope, and some four years later Guyton de Morveau and the army engineer and prominent revolutionary Prieur Duvernois (otherwise Prieur de la Côte-d'Or) initiated the thorough investigation of gas expansion that was so obviously needed, still with the problems of pneumatic chemistry in mind.‡ The experiments, conducted by Prieur in Guyton's laboratory, are significant, since they were concerned not only with a wide range of gases but also with variations in the rate of expansion over different parts of the temperature scale. Previous attempts to compare the expansion of air with that of a thermometric liquid had proved inconclusive. Shuckburgh, for example, saw little reason to doubt that the volume of air increased proportionally to that of mercury, but De Luc and, more especially, Roy were convinced that no such proportionality existed.§ Unfortunately the quality of Prieur's experiments was hardly worthy of the importance of the problems, and it is clear that the gases that he enclosed in his mercury manometer had not been sufficiently dried. The seven gases under examination expanded to very different extents and their volumes all increased far more rapidly at higher temperatures, a sure indication of the presence of water vapour.

Despite their shortcomings, Prieur's results had the air of authority. In Guyton's eyes they were 'free from errors'¶ and for the engineer Riche de Prony they provided the data for an elaborate attempt to derive an expression relating the volume of a gas to its temperature.‖ And they were quoted as standard in less-original work also.†† Indeed, they were so authoritative

† Vandermonde, Berthollet, and Monge, *Mém. Acad. Sci.* (1786), pp. 164–6.
‡ Guyton and Prieur, *Annls Chim.* **1** (1789), 256–99; also Guyton's description in the article on 'Air' in *Encyclopédie méthodique. Chimie*, vol. 1, pp. 675–87.
§ Shuckburgh, op. cit., pp. 565–6; De Luc, op. cit., vol. 1, p. 282; vol. 2, pp. 161–9; Roy, op. cit., p. 695.
¶ *Encyclopédie méthodique. Chimie*, vol. 1, p. 685.
‖ Prony, *Nouvelle architecture hydraulique* (2 vols., 1790–6), vol. 2, pp. 163–75, and *J. Éc. polytech.* **1** (an 4 [1795–6]), 36–49.
†† See, for example, Nicholson, *Chemistry* (1st edn, 1790), p. 522; (2nd edn,

as to go unchallenged even by such an eminent physicist as J. A. C. Charles who, with the Robert brothers, his companions in his early work on balloons, had already shown, in some unpublished experiments performed in 1786 or 1787, that all insoluble gases at least were equally expanded by heat.† Such open criticism as there was in the eighteenth century came from neither Britain nor France, but from Volta in Italy, and was consequently only too easily overlooked.

Volta's chief concern in his own work‡ was probably thermometry, although the point was never made explicitly and in examining his motives it is worth observing that he may well have had some interest in altitude measurement since he was a noted mountaineer.§ He was familiar with far more earlier work on the subject of gas expansion than any other writer of this period and he appears to have been particularly influenced by Lambert and by the less well-known German writer and cleric Johann Friedrich Luz,¶ both authorities who were never cited by the British or French. While he would not commit himself on Lambert's assertion that the expansion of air was proportional to the quantities of heat added, Volta's experiments, conducted between 1791 and 1793 with a Drebbelian thermometer incorporating oil instead of water, showed that from the freezing-point to the boiling-point of water dry air expanded by almost exactly 1/270 of its volume at 0 °C for every °C rise in temperature (or 1/216 per °R, as Volta wrote) as measured on

1792), p. 536; (3rd edn, 1796), p. 554; and Thomson, *System* (1802), vol. 1, p. 269, and *Encyclopaedia Britannica* (3rd edn, Edinburgh, 1801), supplement, vol. 1, p. 259.

† The only published account of Charles's experiments was that given by Gay-Lussac in *Annls Chim.* **43** (1802), 156–8. The participation of the Robert brothers was not mentioned by Gay-Lussac but a reference to it is to be found in Charles's notes for lesson 25 of a physics course which he gave about 1806 (Library of the Institut de France, Paris, file 2104). The dating of the experiments is based on these notes and on Gay-Lussac's account. Some of the apparatus used is now in the museum of the Conservatoire des Arts et Métiers, Paris (exhibit 1584).

‡ Volta, 'Della uniforme dilatazione dell'aria', in Brugnatelli's *Annali Chim.* **4** (1793), 227–94. This paper has been consulted by me in Volta, *Opere*, **7**, 347–75, where it is accompanied by some relevant correspondence (pp. 323–44).

§ See M. Cermenati, *Alessandro Volta, alpinista* (Turin, 1899).

¶ For Lambert's work on the expansion of air see his *Pyrometrie*, p. 47. On Luz's experiments see his *Vollständige und auf Erfahrung gegründete Beschreibung* (Frankfurt and Leipzig, 1784), p. 414.

THE STUDY OF GASES AND HEAT TO 1800   67

the mercury thermometer. Thus, with regard both to the regularity of the expansion and to his numerical result, Volta anticipated the far better known conclusions that Gay-Lussac reached independently some ten years later. It should be borne in mind, however, that Volta's claims to priority are limited by the fact that, unlike Gay-Lussac, he worked only on air and made no observations or predictions involving other gases. Volta, after all, was a physicist rather than a chemist and he was evidently not concerned with the demands of pneumatic chemistry.

So the eighteenth century came to a close with the data relating to the expansive properties of gases in quite as unsatisfactory a state as those concerning their specific heats. Moreover, the inadequacy of the situation, aggravated as it was by the fact that adiabatic phenomena were virtually unknown to those who might have applied and studied them most effectively, was all too little recognized. By most eighteenth-century writers the existing results were readily accepted, if only as fair approximations, and at least the general principles of such problems as the nature of heat and the structure of gases appeared to be settled. The need for the far higher standard of experimentation that was to give new stimulus to the study of gases so soon after the turn of the century had not yet been felt.

3

THE SPECIAL STATUS OF GASES

WORK on the thermal properties of gases before 1800 had done little to suggest that matter in the gaseous state deserved special attention. The expansion coefficients of gases had been found to be as varied and unpredictable as those of liquids and solids, and their specific heats similarly had shown no sign of regularity. Only adiabatic heating seemed in any way an unusual phenomenon and even this had hardly attracted the attention that it deserved. In France it was almost unknown, even by 1800, while elsewhere, even if it was no longer exclusively associated with the entry of a gas into a vacuum, it was all too easily dismissed as just another instance of heating by compression analogous to the more familiar and apparently more interesting effect that accompanied the percussion of a solid. Gases, it seemed, were no more than modified forms of liquids and solids, with their properties altered but not in any fundamental sense different.

Two events early in the new century served to change this opinion radically and to give gases a unique status. The first was the discovery, made independently by John Dalton in England and Joseph Louis Gay-Lussac in France, that, for any given increment in temperature, all gases expanded to the same extent. Taken in conjunction with the equally simple Boyle's law, this constituted strong evidence that a certain fundamentality was to be associated with the gaseous state and that laws, particularly those relating to caloric, might be deduced in their simplest form from observations on gases. The second event was the sudden and rapid spread of knowledge of the adiabatic effect, especially in France. Dalton again was largely responsible, but Laplace's removal of the notorious discrepancy between the experimental and theoretical values for the velocity of sound, which had existed since Newton, not only ensured lasting prominence for the effect but also gave the first reliable, and somewhat surprising, indication of its true magnitude. The intense interest in the thermal properties of gases, which marked the first three decades of the century, owed much to these two events and we shall therefore need to examine them in some detail.

Expansion by heat

In October 1801 John Dalton read four papers on various subjects relating to gases and vapours to the Literary and Philosophical Society of Manchester. It was in the last of these† that he announced his conclusion that 'all elastic fluids under the same pressure expand equally by heat'.‡ He claimed to have conducted 'a great many experiments', in which he had observed the expansion of air, hydrogen, oxygen, nitric oxide, and carbon dioxide in a manometer tube. Although he fully recognized the importance of drying the gases, he admitted that he had succeeded in this only for air and so he readily ascribed minor deviations from his 'law' to the presence of water vapour. The observations on air were therefore the definitive ones and these showed an increase in volume of 325 parts in 1000 between 55 and 212 °F.

Gay-Lussac announced a similar conclusion to the First Class of the Institute (the successor, since 1795, of the Académie Royale des Sciences) on 31 January 1802,§ when he was still only 23 years of age. Although Dalton's paper had already been read, and possibly even published,¶ there is no evidence that Gay-Lussac knew of it. Of undoubted importance, however, was his familiarity with J. A. C. Charles's experiments conducted some fifteen years earlier, and to these he paid due acknowledgement. Gay-Lussac seems to have been more successful than Dalton in drying the nine gases that he used and he concluded confidently that between 0 and 100 °C all the permanent gases expanded by 1/266·66 of their volume at 0 °C for every °C rise in temperature. Even the soluble gases were found to expand in exactly the same way as air, in contradiction to Charles's con-

† Dalton, 'On the expansion of elastic fluids by heat', *Mem. Proc. Manch. lit. phil. Soc.* **5**, part 2 (1802), 595–602.

‡ Dalton, op. cit., p. 602.

§ Gay-Lussac, 'Recherches sur la dilation des gaz et des vapeurs', *Annls Chim.* **43** (1802), 137–75.

¶ According to Thomas Thomson, in *A system of chemistry* (2nd edn, Edinburgh, 1804), vol. 1, p. 399, the paper was published 'early in 1802'.

clusion on this point. As we should expect from the remarkable consistency of his results, Gay-Lussac's experimental procedure was more sophisticated than Dalton's. The gas under examination was enclosed in a flask that was heated in a water bath and, as heating proceeded, gas was allowed to escape to the atmophere through an exit tube fitted with a tap. When the flask had stood for about twenty minutes in boiling water, the tap was closed and the whole apparatus cooled. The volume of gas expelled could now be determined by opening the flask over water and measuring the volume of water that entered. The same principle was also used in a simpler alternative method in which the tap was dispensed with. For the soluble gases no direct measurements were made and their expansion in a tube inverted over a mercury bath was merely compared with that of air similarly heated in an identical tube.

That two essentially correct measurements of gas expansion should have been made almost simultaneously and yet independently clearly requires comment. Was this the result of pure coincidence ? If not, what were the contemporary problems that required the information ? Unfortunately only Gay-Lussac gave any indication of the motives for the work, but his comments suffice to show that the results would have been seen as a very welcome addition to physical knowledge. In his paper he wrote:

. . . every day, in physics as well as in chemistry, we need to correct a given volume of gas for temperature and to measure the amount of heat released or absorbed by bodies in changes of state or as they are cooled and heated; in the arts we need to calculate the effect of a heat engine and to assess accurately the expansion of various substances; in meteorology we need to determine the quantity of water vapour held in solution by the air, a quantity that varies with temperature and density in accordance with a law of which we are still ignorant. Finally, in the drawing up of tables of astronomical refraction and in the use of the barometer for measuring altitude, it is essential to know exactly both the temperature of the air and the law governing its expansion.†

Most of this passage speaks for itself, but the last sentence is particularly interesting for its allusion to two topics with which Laplace was concerned about this time. Indeed, the expansion coefficient of air figured so prominently in Laplace's recently

† Gay-Lussac, op. cit., pp. 139–40.

published accounts of atmospheric refraction and of the use of the barometer to measure altitude† that we can be almost certain that it was he who requested Gay-Lussac to conduct the experiments.‡ Both problems, moreover, were of more than academic interest. With regard to the former, it was obviously a constant concern of astronomers that it should be possible to correct for refraction in the atmosphere and so determine the true position of stars. Laplace's treatment, which was repeated in the *Traité de mécanique céleste* in 1805 after being first published in an abbreviated form in the *Exposition du système du monde*,§ was the best known, but Johann Tobias Mayer (the son of the rather more famous astronomer of the same name) gave another detailed account in 1810 to the Royal Academy of Science in Göttingen, using his own figure of 1/266 per °C (1/213 per °R) for the coefficient.¶ The second of Laplace's problems had been a familiar one for some thirty years, but contemporary interest in mountaineering and ballooning ensured its continued importance and Gay-Lussac himself was actually to use the barometric method of altitude measurement, incorporating his own revised expansion coefficient, in two balloon flights that he made in 1804.||

In the 1802 paper Gay-Lussac warmly expressed his gratitude to Laplace and to Claude Louis Berthollet for their encouragement and material assistance in the experiments.†† His acquaintanceship with these two great French scientists had begun only recently when, in 1800, shortly after his return from Egypt,

† Laplace, *Exposition du système du monde* (1st edn, 1796), vol. 1, pp. 144–68; (2nd edn, 1799), pp. 81–6.

‡ There is further evidence for this in Laplace's *Traité de mécanique céleste* (1805), vol. 4, p. 270, and in the article 'Calorique' in *Encyclopédie méthodique. Physique* (1816), vol. 2, p. 165.

§ Laplace, op. cit., vol. 4, pp. 269–76. For references to the treatments in the *Système du monde*, see note † above.

¶ Mayer, 'Commentatio de apparentiis obiectorum terrestrium . . .', *Commentationes societatis regiae scientiarum Gottingensis recentiores*, **1**, commentationes mathematicae (1811). The figure of 1/213 per °R had been announced first in Mayer's *Physicalisch-mathematische Abhandlung über das Ausmessen der Wärme in Rucksicht und deren Anwendung auf das Höhenmessen vermittelst des Barometers* (Frankfurt and Leipzig, 1786); see especially pp. 85–6.

|| Gay-Lussac, 'Relation d'un voyage aérostatique', *Annls Chim.* **52** (1804), 75–94. For Laplace's amendment to his earlier treatment of altitude measurement (made in the light of Gay-Lussac's work) see the *Traité de mécanique céleste* (1805), vol. 4, pp. 289–93.

†† Gay-Lussac, *Annls Chim.* **43** (1802), 140.

Berthollet had asked for a pupil from the École Polytechnique to act as his assistant in the laboratory that he had established in his country house at Arcueil just outside Paris.† Gay-Lussac, newly graduated, was chosen and so entered into close contact not only with Berthollet but also with Laplace, Berthollet's close friend and his neighbour at Arcueil from 1806.‡ The collaboration was not always to prove a salutary one, as we shall see, for example, in Gay-Lussac's reluctance to accept the atomic theory and in his apparently firm belief in the physical reality of caloric;§ but during the early years of the century, when Laplacian science was pre-eminent in France, an *entrée* to Arcueil was a most valuable asset for any aspiring scientist.

Another problem in which Laplace took a particular interest was the search for a rational scale of temperature. The chief difficulty was that there was no guarantee that the mercury in a conventional thermometer expanded uniformly or, as Gay-Lussac put it in terms of current caloric theory, 'that equal divisions on its scale represent equal increments in the "tension" of caloric'.¶ The work of the eighteenth century, of course, had given no reason to suppose that gases were any more suitable in this respect. Moreover, gas thermometers were unwieldy in practice and the mercury-in-glass instrument had quite recently received influential support from both De Luc and Crawford.‖ But the discoveries of Dalton and Gay-Lussac quickly revived interest in gas thermometry. Gay-Lussac's first experiments could not help, since they did not involve the gradual variation of temperature in the water bath, but a new procedure, which he must have adopted before 1805, allowed the expansion of gases to be compared with the readings on a mercury thermometer over a range of temperatures up to the boiling-point of water. No account was given at the time, but the method was

† See Arago's biographical sketch of Gay-Lussac in *Œuvres complètes de François Arago* (17 vols., Paris and Leipzig, 1854–62), vol. 3, p. 7.
‡ It was in 1806 that Laplace bought the house next to Berthollet's at Arcueil. Before Laplace's arrival Berthollet, who had lived at Arcueil since 1801, had a door built into the wall that separated their gardens, in order to facilitate contact between the two houses. See J. B. Biot, *Mélanges scientifiques et littéraires* (1858), vol. 1, p. 9.
§ See Chapter 4, *passim*, and p. 247.
¶ Gay-Lussac, op. cit., p. 139.
‖ De Luc, *Recherches sur les modifications de l'atmosphère* (Geneva, 1772), vol. 1, pp. 285–330, and Crawford, *Animal heat* (2nd edn, 1788), p. 53.

presumably that described some years later by Biot, who recorded that the expansion coefficient determined with Gay-Lussac's new apparatus was precisely that obtained in the original experiments and was the same for all gases.† Gay-Lussac had apparently shown also that the expansion of the gas was proportional to the readings on the mercury thermometer between 0 and 100 °C, and Laplace would seem to have known something of his results when, in 1805, he recommended the air thermometer as the ultimate standard and the mercury thermometer as a reliable everyday instrument.‡ Laplace was of course far-sighted in this, but it was only after Dulong and Petit's strong advocacy of gas thermometry in 1818§ that the method became widely accepted as a standard of comparison. The delay is hardly surprising when we compare the first-class experimental work that Dulong and Petit conducted in support of their claim with the following unconvincing argument, which was the best that Laplace could offer:

... if the temperature of the air is supposed to increase while its volume is kept constant, it is quite natural to suppose that its elastic force, which is caused by heat, will increase in the same proportion as the temperature. If the pressure of the air is now altered to the value that it had initially, its volume also will increase in the same proportion as the temperature. Hence it seems to me that the air thermometer gives a precise indication of variations in quantity of heat.... ¶

Dalton had no mentor of the stature of Laplace and his writings give little indication of the problems that led him to undertake his experiments. It is possible that mere dissatisfaction with the existing results provided sufficient incentive for someone, like Dalton, who had already shown such a keen interest in gases and vapours, and this at least was the impression conveyed in the paper that was read to the Literary and Philosophical Society in October 1801.|| Yet the paper contained more than a mere account of experimental procedure and observations, for Dalton immediately applied his results in a highly original derivation of a theoretical expression relating

† Biot, *Traité de physique* (1816), vol. 1, pp. 182–90.
‡ Laplace, *Traité de mécanique céleste* (1805), vol. 4, pp. 270 and xxii.
§ See p. 238.   ¶ Laplace, op. cit., p. xxii.
|| Dalton, *Mem. Proc. Manch. lit. phil. Soc.* **5**, part 2 (1802), 595.

the volume of a gas and its absolute temperature. In examining this ingenious argument by Dalton two points must be borne in mind. First, and most important, he was a convinced Irvinist, rejecting any idea that some caloric in bodies existed in a latent state.† His commitment is nowhere more apparent than in the first part of volume one of his *New system of chemical philosophy* (1808), which he devoted largely to an account of caloric theory based on Irvinist principles and in which, significantly, one of the few illustrations (see Plate 1) shows the cylinders of uniform cross-section that Crawford had already used to explain his views on the quantities of heat in bodies.‡ The second point is that Dalton, like so many eighteenth-century authorities, accepted as a physical truth Newton's purely mathematical demonstration that Boyle's law could be accounted for if it was supposed that the adjacent particles of a gas expanding isothermally repelled one another with a force inversely proportional to the distance between them. Caloric theory, as we have seen, had given new authority to the static gas hypothesis by providing a ready explanation of the repulsive forces, but in 1801 the implications of the theory had not yet been examined in detail. Unfortunately the cryptic nature of Dalton's account prevents us from knowing exactly how he tackled the problem of relating volume and temperature theoretically, but the following is a possible reconstruction of his argument.§

Dalton naturally chose to consider the expansion of air at constant pressure between 55 and 212 °F, the range of his own experiments, and he would presumably have had to treat the process in two stages. In the first, the heating of the air at constant volume from 55 to 212 °F, the total heat content would have increased from, say, Q_{55} to Q_{212}, while the inter-particle force, originally F_{55}, would have become F_{212}. Since Dalton assumed, explicitly though without justification, that this force was proportional to Q, it followed that $F_{212}/F_{55} = Q_{212}/Q_{55}$. His

† See, for example, his comment in a letter published in *Nicholson's Journal*, 2nd ser. **5** (1803), 36.

‡ Crawford, *Animal heat* (2nd edn, 1788), p. 379. For a similar use see J. Leslie, *An experimental inquiry into the nature, and propagation, of heat* (1804), p. 529.

§ This reconstruction first appeared in my paper 'Dalton's caloric theory', in D. S. L. Cardwell (ed.), *John Dalton & the progress of science* (Manchester, 1968), pp. 197–8.

Irvinist views would now have allowed him to go one stage further, for he could argue that Q_{212}/Q_{55} was in turn equal to T_{212}/T_{55}, the ratio of the absolute temperatures before and after the heating. In order to relate this ratio to the increase in volume that would have occurred if the air had been allowed to expand at constant pressure, it was necessary to consider a second process, the isothermal expansion of the air to the point where the pressure, and hence also F, fell to the values that they had before the heating. Dalton must have supposed, and here he was using Newton's result, that the force between the gas particles varied in inverse proportion to the distance, d, between their centres, so that $F_{212}/F_{55} = d_{212}/d_{55}$, where d_{55}^3 and d_{212}^3 were, of course, proportional to the volumes that the air would have occupied at 55 and 212 °F respectively had it been heated at constant pressure in a single process. It followed simply that $T_{212}/T_{55} = d_{212}/d_{55}$, and hence, on the assumption that the particles were spherical, in contact with one another, and of volume proportional to d^3, the insertion of Dalton's own experimental values for the expansion yielded an absolute zero at -1540 °F. Dalton in fact made only an approximate calculation, obtaining a figure of -1515 °F which, he declared, was 'in more than fortuitous' agreement with Crawford's (approx. -1500 °F.).†
It should be noticed that even here he had not abandoned his belief in the correctness of the more traditional Irvinist method of determining the zero. Indeed by 1808, when he placed it at about -6000 °F, he relied solely on the latter method.‡

Although Dalton's argument provoked some discussion,§ there is no evidence that it was anywhere accepted as correct. Even such an admirer of Dalton as Thomas Thomson, who discussed it fully in the second and subsequent editions of his *System of chemistry*, expressed serious reservations.¶ But if its influence on the subsequent development of caloric theory was slight, the new relationship was of no small significance to Dalton himself. Above all, in so far as it predicted that gases

† Crawford, *Animal heat* (2nd edn, 1788), p. 265.
‡ Dalton, *A new system of chemical philosophy* (Manchester, 1808), vol. 1, part 1, pp. 82–99.
§ For example, by an anonymous writer ('Π') in *Nicholson's Journal*, 2nd ser. **4** (1803), 220–4, and by L. W. Gilbert in *Gilb. Ann.* **12** (1803), 316–17; and **14** (1803), 287–92. Dalton replied to 'Π' in *Nicholson's Journal*, 2nd ser. **5** (1803), 34–6.   ¶ See p. 106.

expanded more rapidly at higher temperatures, it strengthened his conviction that the traditional mercury thermometer was inadequate, since his own experiments had shown an expansion of 167 parts in 1000 between 55 and $133\frac{1}{2}$ °F but of only another 158 parts between the latter temperature and 212 °F. His suspicions with regard to the mercury thermometer had apparently been aroused first during his experiments on vapour pressure,† the subject of the second of his four papers of October 1801.‡ He had noticed that for equal increments in temperature the pressures at saturation of all the vapours examined increased in an approximately geometrical progression, the same progression in all cases but with a ratio which, instead of being constant, as in a true geometrical progression, diminished slightly with rising temperature as measured on the mercury thermometer. In 1808 in the *New system* Dalton recalled his immediate surprise at this approach to regularity and his concern to discover why the regularity was not perfect.§ His first thought had been to reject the traditional thermometer as inaccurate but, 'overawed by the authority of Crawford', he had hesitated. However, his subsequent work on gas expansion only confirmed his suspicions, for he now recognized that if mercury could be shown to expand not uniformly but more rapidly at higher temperatures, then the behaviour of saturated vapours would approximate more nearly to a perfect geometrical progression and the expansion of gases would obey his cube-root law more closely than on the traditional scale. The advantages of abandoning the old mercury thermometer scale were thus, in Dalton's eyes, considerable.

All this was only hinted at in 1801,¶ and by 1808,‖ when he eventually tackled the problem, the situation had been changed radically by his rejection of the theoretical treatment of gas expansion and the $T \propto V^{\frac{1}{3}}$ relationship.†† This rejection had almost certainly come about not as a consequence of any doubts

† According to Dalton in his *New system*, vol. 1, part 1, pp. 10–11.
‡ Dalton, 'On the force of steam . . .', *Mem. Proc. Manch. lit. phil. Soc.* **5**, part 2 (1802), 550–74.
§ Dalton, *New system*, vol. 1, part 1, pp. 10–11.
¶ Dalton, *Mem. Proc. Manch. lit. phil. Soc.* **5**, part 2 (1802), 601–2.
‖ Dalton, *New system*, vol. 1, part 1, pp. 9–12.
†† Dalton appears to have rejected the relationship by the time of his Edinburgh lectures of 1807; see my comment (in note 26) in Cardwell, op. cit., p. 200.

concerning the Irvinist or Newtonian principles on which he had based the treatment but because he had now perceived how a close and highly satisfying analogy between the behaviour of heated gases and the variations in the pressure of a saturated vapour with temperature might be established. To do this it was necessary simply to define a new temperature scale in which the expansion of mercury and water, and probably of all pure liquids, was taken as proportional to the square of the 'true' temperature measured from their respective freezing points. On this scale, Dalton maintained, both the pressure of vapours and the volume of gases would increase in a geometrical progression (i.e. exponentially) with temperature, and this was evidently sufficient justification for the adoption of the new scale. In fact it probably remained so even after 1818, when Dulong and Petit's experiments on expansion† had made his position untenable, at least for virtually anyone except Dalton. Not until 1827 do we learn that he had abandoned the campaign for the scale.‡

Although Gay-Lussac differed essentially from Dalton in that he found volume to increase in an arithmetical progression with increasing temperature, there were certain consequences of their work on which both men were agreed. The new law of expansion, whether Dalton's or Gay-Lussac's, gave a unique status to gaseous matter, and one comment made by Dalton in 1801 is typical of what was very quickly to become a commonplace in treatments of caloric.§ He wrote:

> This remarkable fact that all elastic fluids expand the same quantity in the same circumstances, plainly shews that the expansion depends *solely* upon heat: whereas the expansion in solid and liquid bodies seems to depend upon an adjustment of the two opposite forces of heat and chemical affinity, the one a *constant* force in the same temperature, the other a *variable* one, according to the nature of the body; hence the unequal expansion of such bodies.¶

† See p. 238.
‡ Dalton, *New system* (Manchester, 1827), vol. 2, pp. 288–9.
§ Dalton's view that the natural attractive forces between particles had been overcome in the gaseous state can also be found in, for example, C. L. Berthollet, *Essai de statique chimique* (1803), vol. 1, p. 161; R. J. Haüy, *Traité élémentaire de physique* (2nd edn, 1806), vol. 1, p. 144; J. Playfair, *Outlines of natural philosophy* (Edinburgh, 1812), vol. 1, p. 233. For an even clearer, but later, exposition see Laplace's paper in the *Connaissance des tems . . . pour l'an 1825* (1822), pp. 386–7.
¶ Dalton, *Mem. Proc. Manch. phil. lit. Soc.* **5**, part 2 (1802), 600.

Here, as Dalton immediately saw, was the first evidence since Boyle's law of the essential simplicity of gases, and the belief that he held about this time that the particles of all gases under the same conditions of temperature and pressure were equally spaced† was almost certainly a result of his discovery. In 1801, however, it was in the further investigation of caloric theory that the most exciting possibilities lay, although Dalton's comment was cautious: 'It seems . . . [he declared] that general laws respecting the absolute quantity and the nature of heat, are more likely to be derived from elastic fluids than from other substances.'‡

The French were even more impressed by the fundamentality of behaviour which it appeared was to be associated with the gaseous state. Laplace, as we have seen, was quick to advocate the use of the air thermometer as a standard and when, in 1806, Biot and Arago undertook a study of the refraction of light, it was on gases that they chose to work, citing as a principal reason for their choice the absence of intermolecular forces in this state.§ Nearly three years later the special status of gases was put beyond all doubt by Gay-Lussac's discovery that, when two gases combine chemically, they do so by volumes that bear a simple numerical relationship to each other and to that of the product.¶ This simple behaviour Gay-Lussac too associated with the absence of intermolecular forces of attraction. He wrote:

The attraction between the molecules in solids and liquids is therefore responsible for the modification of their special properties; and it seems that it is only when this attraction is completely destroyed, as in the case of gases, that substances . . . behave in accordance with laws which are simple and constant.‖

The work of Dalton and Gay-Lussac on gas expansion must therefore be seen as more than the timely discovery of a now familiar law. For a calorist it had a far greater significance and one that we can hardly appreciate today. Yet it was only partially responsible for the period of intense activity in the

† See Dalton, *New system* (1808), vol. 1, part 1, p. 188.
‡ Dalton, *Mem. Proc. Manch. phil. lit. Soc.* **5**, part 2 (1802), 600.
§ Biot and Arago, *Mém. Sci. math. phys. Inst. Fr.* **7** (1806), 301–2. Also see p. 199.
¶ Gay-Lussac, 'Mémoire sur la combinaison des substances gazeuses', *Mém. Phys. Chim. Soc. Arcueil* **2** (1809), 207–34.
‖ Gay-Lussac, op. cit., p. 208.

THE SPECIAL STATUS OF GASES

study of the thermal properties of gases which followed, and we turn now to the second of the two crucial events that served to direct attention to the special relevance of gases for the theory of heat.

Adiabatic heating

If we except the brief reference by Chaptal in 1790, the first mention of the adiabatic effect in the French literature dates from 1798, when J. C. Delamétherie, editor of the *Journal de physique*, described a report, communicated to him by Marc Auguste Pictet, of the cold observed on releasing air from a compression pump.† Delamétherie seems to have misinterpreted Pictet's suggestion that the effect could be made even more striking by placing a little water in the receiver and observing the formation of ice round the exit valve, for the explanation he offered was that the water was carried off by the escaping air in the form of a vapour and so had to absorb latent heat that it took from the rest of the water and in particular from the drops round the valve. We may be sure that the analogy with cooling by the evaporation of ether, mentioned by Delamétherie, was not suggested by Pictet. It may possibly have its origin in Chaptal's earlier mistaken interpretation.

Delamétherie's somewhat confused account could all too easily have been overlooked, but Arsène Nicolas Baillet, an inspector of mines who also taught at the École des Mines in Paris, was quick to appreciate the true significance of Pictet's communication. In a paper to the Société Philomathique‡ and a letter to Delamétherie, which was published in the *Journal de physique* early in 1799,§ he reported an analogous effect first observed in the mines of Schemnitz in Hungary by Gabriel Jars of Lyons in 1758.¶ Jars had noted that when air compressed by a water column over 40 metres high in Hoell's column-of-water engine was suddenly allowed to escape, compact ice was deposited on any object placed in the way of the issuing air. The phenomenon had already attracted considerable attention. It

† Delamétherie, 'Note sur un froid considérable...', *J. Phys.* **47** (1798), 186.
‡ A. F. Silvestre, *Rapport général sur les travaux de la Société Philomathique... depuis le 23 frimaire an VI jusqu'au 30 nivôse an VII* (1799), pp. 21–2.
§ Baillet, *J. Phys.* **48** (1799), 166–7.
¶ Jars, 'Description d'une nouvelle machine...', *Mém. prés. div. Sav. Acad. Sci.* **5** (1768), 67–71.

had been described, though not explained, in the *Philosophical Transactions of the Royal Society* in 1762,† apparently quite independently of Jars, and, as we have seen, Erasmus Darwin had used it in 1787 as evidence that the rapid expansion of air gave rise to cooling.‡ Baillet, without reference to Darwin, similarly attributed the formation of ice to the cooling of the escaping air, which through its contact with the water column was, of course, saturated with vapour. He maintained that the fall in temperature of the air accorded well with the recent work of Crawford and Lavoisier, but since Lavoisier appears to have had no views on the matter,§ he was presumably referring specifically to Crawford's conjecture that the capacity of a gas increased with rarefaction.

Even Baillet's comments did little to arouse interest and nearly three years passed before the subject was again referred to in the French journals. This time it was discussed by Jacques Michel Coupé, a former priest and Jacobin and a prolific writer on both educational and agricultural matters, who, in 1801, attempted to account for some recent instances of several days of unusual cold or heat immediately preceding periods of, respectively, mild and cool weather by reference to large-scale adiabatic effects.¶ He suggested, for example, that when a cold, northerly air-stream moved southwards it compressed and so expelled heat from the air in its path and that the mild weather caused by this heat gave way to cold when the northerly stream eventually forced its way past the impeding air. Coupé's was not in fact the first attempt to apply the adiabatic effect in meteorology‖ but it nevertheless passed virtually unnoticed.††

Of course, strange effects observed on mine pumps and large-scale meteorological phenomena had little obvious connection with traditional physical problems. The numerous existing accounts in the British, Swiss, and German literature, including

† N. M. Wolfe, 'Descriptio fontis Hieronis . . .', *Phil. Trans. R. Soc.* **52** (1762), 547–54.
‡ See pp. 58–9. § Although see p. 50.
¶ Coupé, 'De la chaleur qui précède l'arrivée d'un vent froid . . .', *J. Phys.* **53** (1801), 262–4.
‖ Erasmus Darwin had anticipated him in this respect; see p. 59.
†† Joseph Mollet of Lyons, whose work is discussed in some detail later in this section, was probably one of the few people who knew of Coupé's explanation, for he was almost certainly referring to it in his *Mémoire sur deux faits nouveaux* (Lyons, 1811), pp. 23–4.

now an important review article by the young German poet and novelist Ludwig Achim von Arnim,† continued to be ignored in France, and it was only after Laplace's success in applying adiabatic phenomena to a really major problem of physics that the French accepted the heating produced by the compression of a gas as a fact to be reckoned with in the theory of heat. The material theory, as we shall see, responded impressively to the challenge of the growing body of new evidence, and what might have proved a serious threat only served to strengthen the position of caloric.

Laplace's success was to account for the error in Newton's expression for the velocity of sound. In the first edition of the *Principia* Newton, assuming understandably, though erroneously, that isothermal conditions were maintained during the passage of a sound wave, had demonstrated that the velocity in air was given by

$$v = \sqrt{\frac{P}{\rho}}, \qquad (1)$$

where P represented the 'elastic force' (*vis elastica*), or pressure, of the air and ρ its density.‡ Calculation from this expression yielded 968 ft/s for v, a figure lying well within the wide range of existing experimental values, which varied from roughly 600 ft/s, as measured in a very crude fashion by Roberval,§ to 1474 ft/s, Mersenne's figure.¶ Moreover, it agreed with Newton's own simple observations made in the court of Trinity College, Cambridge, the results of which varied from 920 to 1085 ft/s. In the second edition, in 1713, Newton obtained 979 ft/s from his original formula,‖ the modification in the figure resulting from his use of a different value for the ratio between the densities of water and air. But new and more reliable experimental determinations of the velocity of sound had now been made and Newton was convinced that even his revised figure was too low. Corrections for the bulk of the air particles, which

---

† Von Arnim, *Gilb. Ann.* **2** (1799), 238–45.
‡ Newton, *Principia* (1687), pp. 369–72 (Book II, proposition 1).
§ M. Mersenne, *Ballistica et acontismologia* (Paris, 1636), pp. 83 and 140.
¶ Mersenne, *De l'utilité de l'harmonie* (1636), pp. 44–6. This work forms part of Mersenne's *Harmonie universelle* in two of the three Cambridge University Library copies, although it is missing from the other copies I have consulted.
‖ Newton, *Principia* (2nd edn, Cambridge, 1713), pp. 342–4.

he had previously not taken into account, and for water vapour in the atmosphere led to a final value of 1142 ft/s, in exact, and no doubt intentional, agreement with the figure reported by the Revd. William Derham of Upminster in 1708.† However, Newton's attempts to account for the discrepancy were seen to be far from satisfactory and the problem continued to attract leading mathematical physicists throughout the eighteenth century. Leonhard Euler, in 1727, calculated that the true figure lay between 1101 and 1259 ft/s, depending on the temperature,‡ but by 1759 he too recognized that the experimental value considerably exceeded the theoretical figure that he adopted at this time.§ The discrepancy, he added, was unexplained. J. H. Lambert similarly rejected existing explanations and looked for the answer in the presence of heavy foreign particles in the air which raised its density without contributing to its elasticity.¶ Joseph Louis Lagrange, who obtained a theoretical figure of 975 ft/s, would seem to have shared Lambert's view.‖ Despite the attentions of such men as these, as well as of the Bernoullis and Jean d'Alembert,†† the explanations offered carried little conviction and the problem was still wide open when Laplace turned to it, probably in 1802.

The first intimation of Laplace's solution was given in August or September of that year (*fructidor an 10*) by Jean-Baptiste Biot,‡‡ then a young professor of physics at the Collège de France. Laplace, it seems, had asked Biot to examine how the velocity was affected by the heat and cold that he supposed to occur in the successive regions of compression and rarefaction constituting the sound-wave. The effect of the temperature variations was essentially to invalidate the false assumption, which had hitherto been fundamental to all calculations, that during the passage of the sound the pressure of the air was everywhere proportional to its density. Biot made constant

† Derham, *Phil. Trans. R. Soc.* **26** (1708), 2–35.
‡ Euler, *Dissertatio physica de sono* (Basle, 1727), consulted in *L. Euleri opera omnia* (Leipzig and Berlin, 1926), 3rd ser. vol. 1, pp. 182–96.
§ Euler, *Opera omnia*, 3rd ser. vol. 1, pp. 442–3 (in 'De la propagation du son', originally published in *Hist. Acad. Sci. Berlin*, **15** (1759), 185–209).
¶ Lambert, *Hist. Acad. Sci. Berlin* **24** (1768), 70–9.
‖ Lagrange, *Méchanique analitique* (1788), pp. 510–11.
†† For a summary of this other work see C. A. Truesdell's introduction to Euler, *Opera omnia* (Zürich, 1956), 2nd ser. vol. 13, pp. xix–lxxii.
‡‡ Biot, 'Sur la théorie du son', *J. Phys.* **55** (1802), 173–82.

THE SPECIAL STATUS OF GASES

reference to Lagrange's earlier treatment, but we may summarize the argument as follows, eliminating the mathematical complexities that these references introduce.

Biot argued that the density of the air at any point in a sound wave was given by ρ', where

$$\rho' = \rho(1+s),$$

ρ being the density of the undisturbed air and s the (small) fractional change in density, taken as positive for compression. Provided that truly isothermal conditions were assumed to hold, it followed simply from this that the pressure of the air in the sound-wave, P', could be expressed by the analogous equation
$$P' = P(1+s),$$

where P was the pressure of the undisturbed air. Biot, however, rejected the assumption that the temperature remained constant and he supposed instead that the effect of the fractional compression s was to heat the air and so to increase the pressure by an additional factor $(1+ks)$, where k was a constant of proportionality. In fact a solution of this sort was not entirely novel, for Lagrange had already examined the implications of putting the pressure of air in a sound-wave proportional not to the density but to some power m of the density, i.e. $P \alpha \rho^m$,[†] a correct expression, of course, for adiabatic volume changes. Lagrange even calculated that m should be $1\frac{1}{3}$, if the experimental and theoretical velocities were to agree, and his value for m (the equivalent of γ in our modern notation) was thus very close to the present accepted value of 1·40 for air. How such an appreciable deviation from Boyle's law could have escaped detection was a difficulty which Lagrange, ignorant of adiabatic phenomena, naturally could not answer. Unfortunately Biot did not adopt Lagrange's expression, but his own expression for the pressure of a gas undergoing adiabatic volume changes,

$$P' = gHn\rho(1+s)(1+ks),$$

was a good approximation provided s was small. Here, we should note, he was supposing that air pressure at any point was given by
$$P = gHn\rho,$$

[†] Lagrange, *Mélang. Phil. Math. Soc. R. Turin* 2 (1760–1), 152–4.

where H was the reading on a mercury barometer, g the acceleration due to gravity, and n the ratio between the density of mercury and that of air. From his expression for P', neglecting terms in s^2, Biot obtained

$$v = \sqrt{\{gHn(1+k)\}},$$

which reduces to the more familiar $v = \sqrt{(\gamma P/\rho)}$ if we substitute our term γ for Biot's constant $(1+k)$. The effect of Newton's error in assuming the compressions and rarefactions to take place isothermally had thus been to put $k = 0$. Under these conditions the velocity became simply $\sqrt{(gHn)}$, the quantity that Lagrange had calculated as 975 ft/s.

An independent determination of k, which confirmation of Biot's treatment required, was unavailable in 1802, and Biot tried instead to use the data relating to the expansion of gases by heat, in a strikingly naïve fashion. First, he assumed (quite erroneously, of course) that the heat evolved in the adiabatic compression of a gas and absorbed by the gas was equal to that which was released by the gas in a similar contraction while cooling at constant pressure. Thus, to use modern notation, he supposed that a unit mass of gas, on being compressed adiabatically by, say, 1/266 of its initial volume, would yield c_p units of heat, which would go to raising its temperature. Now Gay-Lussac's recent experiments had shown that an increase in temperature of 100 °C would cause a gas to expand by 0·35 of its volume at 0 °C, so that Biot, disregarding any distinction between c_p and c_v, could argue that a compression of 0·35/100 would raise the temperature of the gas by 1 °C. Showing a somewhat misplaced confidence in the results of Amontons (see Table C), Biot now declared that such a temperature change would effect a fractional increase of 1/300 in the pressure of the gas and he equated this increase to $k(0·35/100)$, i.e. to ks, to obtain 0·95 for k and a value for v (1362 ft/s) which was considerably greater than the currently accepted experimental value. That the inequality of the coefficients of Gay-Lussac and Amontons implied the falsity of Boyle's law seems, surprisingly, to have escaped him.

An alternative, and far more accurate, value for k was obtained in the same paper simply by equating $\sqrt{\{gHn(1+k)\}}$ to 1107 ft/s, the velocity determined experimentally for the Académie des Sciences in 1738 by a group of workers under

Cassini de Thury.† The appropriate value for k in this case was 0·2869. If the velocity of 1151 ft/s determined by Cassini, Picard, and Roemer in 1677‡ was used, k, as Biot pointed out, became 0·3922, still closer, albeit fortuitously, to what we now accept as the correct value, 0·4031. In connection with this calculation it is interesting to note that since Biot followed Lagrange in using a figure for the ratio of the density of mercury to that of air which was too big (11 900 compared with the correct value of 11 240), the use of the more accurate velocity of 1107 ft/s yielded an appreciably worse value for k. The discrepancy between the figure for k deduced from the laws of gas expansion and those deduced from comparison with the experimentally determined velocities apparently aroused no serious doubts in Biot's mind concerning the assumptions that he had made in the former method. However, he admitted that there might be errors in the basic hypotheses and he went so far as to suggest that some of the heat evolved in the regions of compression might well be emitted as radiation without raising the temperature of the air and without, therefore, affecting the velocity of the sound-wave.

Despite the difficulties which remained, there could now be few who doubted that small temperature changes did occur in sound-waves.§ And five years later Biot provided further striking evidence by his demonstration that sound was transmitted through various saturated vapours.¶ The experiment was a decisive one since it was recognized that unless some heat was liberated in the regions of compression, as Laplace and Biot maintained, liquefaction would take place and hence the passage of sound would be impossible.

Biot's paper of 1802 was more than a spectacular beginning to the study of adiabatic phenomena in France. It provided also a convincing intimation that earlier authorities had seriously underestimated the magnitude of the effect, for, on the false but plausible assumption that the rise in temperature was

† Cassini de Thury, *Mém. Acad. Sci.* (1738), p. 135.
‡ J. B. du Hamel, *Regiae scientarum academiae historia* (Paris, 1698), p. 158.
§ Though John Leslie's important *Experimental inquiry into the nature, and propagation, of heat* of 1804 contained no mention of Laplace's solution and referred to the velocity of sound problem as an open one (see p. 544).
¶ Biot, 'Expériences sur la production du son dans les vapeurs', *Mém. Phys. Chim. Soc. Arcueil* **2** (1809), 94–103.

proportional to the degree of compression, Biot concluded that a reduction in volume by one-half would raise the temperature of any mass of air by either 86 °C (assuming k to be 0·2869) or 118 °C (when k was taken as 0·3922). For many years Biot's remained the most acceptable method of determining the true temperature changes, although it was Denis Poisson's figure, given in 1807,† that became standard in the literature, being used by Sadi Carnot,‡ for example, and by both pairs of competitors in the French Institute's important prize competition of 1812.§ Poisson's treatment was essentially that of Biot, though with the refinement that account was taken of the effect of air temperature on velocity. Poisson also differed in using a lower value for the ratio between the densities of mercury and air (10 476) and he concluded that, if the theoretical and experimental velocities were to agree, k must be 0·4254. By assuming, like Biot, that the temperature change was proportional to the degree of compression, though limiting his assumption to cases where the compression was small, Poisson deduced that a sudden decrease or increase in volume of 1/116 would raise or lower the temperature by 1 °C.

When Biot wrote, he would have been unaware that an attempt to obtain the much-needed direct experimental evidence for the magnitude of adiabatic heating had already been made by John Dalton. Although Dalton's paper on the subject was read to the Manchester Literary and Philosophical Society in June 1800, it was not published in England until 1802¶ and notice of it appeared in France only in January 1803.‖ Dalton

† Poisson, 'Mémoire sur la théorie du son', *J. Éc. polytech.* **7**, cahier 14 (1808), 319–92.

‡ Carnot, *Réflexions sur la puissance motrice du feu* (1824), pp. 53–4 and 79–81.

§ See pp. 140 and 144.

¶ Dalton, 'Experiments and observations on the heat and cold produced by the mechanical condensation and rarefaction of air', *Mem. Proc. Manch. phil. lit. Soc.* **5**, part 2 (1802), 515–26.

‖ In *Annls Chim.* **45** (1803), 103–7, and *J. Mines* **13** (1803), 257–67. About this time, or shortly afterwards, experiments were apparently also performed by John Southern. In a footnote to a letter of 26 March 1814 from Southern to James Watt, it was stated (presumably by Southern himself) that 'some years back' the writer of the note had shown experimentally that the expansion of air by half its initial volume caused a cooling of 19 or 20 °F. Unfortunately knowledge of the experiments was not available until the letter was published in J. Robison, *A system of mechanical philosophy* (Edinburgh, 1822), vol. 2, pp. 160–73; see especially p. 166 n.

correctly interpreted the rapidity with which the temperature variations occurred on evacuating and filling an air-pump receiver as indicating that the true temperature variations were far greater than those observed directly on a thermometer and he argued that the true magnitude of the effect was inevitably obscured by the rapid heat exchange between the air in the receiver and its surroundings (including the thermometer). Perhaps the most convincing evidence that he obtained to demonstrate this was that a thermometer with a large heat capacity indicated considerably less variation in temperature than a smaller instrument in the same circumstances.

To measure the true temperature changes Dalton used two quite separate methods. In the first he noted that the maximum rate at which a small thermometer rose when air was suddenly admitted to an evacuated receiver in which it was suspended was 1 °F in $3\frac{1}{2}$ seconds. He then showed that a similar rate of *fall* of temperature was obtained whenever the temperature of the thermometer was 50 °F above that of the surrounding air and hence he concluded, presumably on the basis of Newton's law of cooling, that the temperature of the air surrounding it in the first case, at the moment when the mercury was rising most rapidly and when the maximum temperature was reached, was 50 °F above that of the thermometer itself. His second method was far more satisfactory and it is of particular interest as a possible precursor of the familiar method of Clément and Desormes for determining γ.† The respect that Clément and Desormes were later to accord Dalton's work supports this view of their debt to him, although Dalton was using an observation which, if we are to believe Baillet and René Just Haüy,‡ was already a familiar one in laboratories even though it had not previously been satisfactorily explained. Dalton placed a capillary tube, sealed at one end and containing a short mercury thread, inside a receiver. The air in the receiver was then compressed to a pressure of 2 atmospheres, so that the length of the short air column trapped by the thread was halved; and finally it was allowed to escape rapidly until the thread regained its original position, whereupon the tap was quickly closed. It

† Clément and Desormes, *J. Phys.* **89** (1819), 323 and 324.
‡ Baillet, *J. Mines* **13** (1803), 267, and Haüy, *Traité élémentaire de physique* (2nd edn, 1806), vol. 1, pp. 139–40.

was observed that, on standing, the length of the column of air trapped in the capillary tube decreased by one-tenth, being restored to its original length only when the tap was reopened. An analogous effect occurred when air was admitted to a receiver evacuated to one-quarter of atmospheric pressure, although in this instance the length of the air column *increased* after the tap was first closed and did so by slightly more than one-tenth. Dalton's explanation of these two observations was a somewhat confused version of the truth, which was that the tap had been closed while the pressure of the air had been temporarily decreased by cold in one case or increased by heat in the other. Hence, although in the latter case, for example, the pressure in the receiver fell below that of the atmosphere once temperature equilibrium had been restored, it had exactly restored the mercury thread to its original position when the air was heated. Since his own experiments on gas expansion indicated that the temperature increment necessary to raise the pressure by one-tenth was approximately 50 °F, Dalton naturally saw the estimate of the temperature changes obtained by his first method as being confirmed.

Dalton's experimental examination of the subject, although thorough so far as it went, was restricted to work on an evacuated receiver, and the explanation that he adopted was coloured accordingly. It was an interesting union of Crawford's views on capacity with the concept of a vacuum capable of containing heat. Dalton, who accepted Crawford's explanations of heat of reaction and latent heat, was in little doubt that changes in the volume specific heat of air with compression and rarefaction accounted for the thermal effects. The heat produced by the rapid entry of air into a vacuum, he could argue, arose merely from the replacement of the void, which had the greatest† volume capacity of all, by air, with a lesser capacity. Although Dalton's conviction that his work might lead to a method of determining the capacities both of the void and of various gases was later to provide the starting-point for the work of Clément and Desormes, he unfortunately never pursued the matter himself.

† On p. 526 of the version of Dalton's paper that appeared in *Mem. Proc. Manch. phil. lit. Soc.* the heat capacity of the vacuum is erroneously stated to be *less* than that of an equal volume of air. The mistake is corrected in other versions of the paper.

THE SPECIAL STATUS OF GASES

Dalton's explanation was soon effectively attacked,† but the competence of his experimental investigation was generally recognized and, largely because it was so widely reprinted and translated,‡ his paper did much to stimulate interest at about the same time as Biot's announcement was made. Moreover, in experiments designed for the purpose Dalton sought to dispel the persistent myths that distortion of the thermometer bulb or evaporation might account for the observed effects. There is a good deal of evidence that such confirmation of the true nature of the phenomena was still required. For example, in his comments on the summary of Dalton's paper that appeared in the *Journal des mines* Baillet felt it necessary to go to the trouble of rejecting explicitly the view of 'some teachers' that the distension and contraction of the receiver was the cause, although he himself accepted an explanation in terms of the increasing capacity of air with rarefaction.§ Misunderstanding, as we shall see later in this section, was also apparent in the criticisms that certain members of the Académie des Sciences, Belles-Lettres et Arts of Lyons were soon to level at Joseph Mollet's important work on the adiabatic phenomena. And there is reason to believe that even in England false explanations were still adopted by some. In 1799, for instance, Alexander Tilloch, the editor of the *Philosophical magazine*,¶ referred to a current opinion (not his own) that the heat observed on compressing air in a cylinder arose from the friction of the piston against the cylinder wall.

However, further evidence both for the genuineness and for the true magnitude of the effect quickly became available in a series of striking discoveries that were made in Lyons. On 14 and 29 December 1802 first the Lyons Academy and then the French Institute in Paris heard a communication from Joseph Mollet, at that time professor of physics at the École Centrale in Lyons, describing a light that had been found to accompany the discharge of an air-gun in the dark and also the ignition, by the

---

† Notably by Gay-Lussac; see pp. 129–30.

‡ To the numerous references given in the Royal Society's *Catalogue of scientific papers, 1800–1863*, the following should be added: *Nicholson's Journal* 2nd ser. **3** (1802), pp. 160–6; *Bull. Soc. philomath. Paris* **3** (1801–5), 163–4; and *Repertory of arts and manufactures*, 2nd ser. **2** (1803), 118–25.

§ Baillet, *J. Mines* **13** (1803), 268.

¶ Tilloch, *Phil. Mag.* **8** (1800), 213.

rapid compression of air, of a small piece of linen fixed in the exit tube of a condensing pump.† Both observations had originally been made, by chance, by a workman at the armoury in Saint-Étienne earlier in that year.‡ Only Marc Auguste Pictet, who may even have been at the Institute when Mollet's letter was read,§ seems to have thought the matter worthy of public comment, and his brief report, dated 1 January 1803 and published almost immediately in the *Philosophical magazine*,¶ provoked nothing more than a statement by William Nicholson, in April 1803, that the effect with the air-gun had already been observed, about eighteen months previously, by a Mr Fletcher, at a weekly meeting 'for philosophical experiments and conversations' held at Nicholson's house.‖ In Paris also Mollet's communication appears to have had a somewhat cool reception. No published account appeared and the referees appointed by the Institute†† were quite unable to reproduce the observations.

Undeterred by this response, Mollet continued his experiments in collaboration with Ennemond Eynard, a doctor and member of the science section of the Lyons Academy, and with two other residents of Lyons by the name of Haex and Gensoul.‡‡ In fact it was these three friends of Mollet who first heard of the workman's observations at Saint-Étienne. Although he never managed to reproduce the flash associated with the discharge of the air-gun satisfactorily, Mollet now discovered that a similar

† *Procès-verbaux*, vol. 2, p. 606 (*8 nivôse an 11*). The manuscript of Mollet's communication to the Lyons Academy is preserved in the Academy's library (MS. 230, ff. 69–74). For accounts of the events described in this and the following paragraph see J. B. Dumas, *Histoire de l'Académie Royale des Sciences, Belles-Lettres et Arts de Lyon* (Lyons, 1839), vol. 2, pp. 232–42, and Mollet, *Mémoire sur deux faits nouveaux, l'inflammation des matières combustibles et l'apparition d'une vive lumière, obtenues par la seule compression de l'air* (Lyons, 1811), pp. 3–8.

‡ The workman is identified as a Citizen Chauvain in *Procès-verbaux*, vol. 3, p. 96 (*17 floréal an 11*).

§ Pictet is known to have been in Paris at the time (see 'Journal d'un Genevois à Paris', *Mém. Docums Soc. Hist. Archéol. Genève*, 2nd ser. **5** (1893–1901), 109) but, as I have pointed out in *Technology Cult.* **10** (1969), 360 n., his diary unfortunately contains no entry for 29 December 1802.

¶ Pictet, *Phil. Mag.* **14** (1803), 363–4.

‖ Nicholson, *Nicholson's Journal*, 2nd ser. **4** (1803), 280.

†† Presumably J. A. C. Charles and L. Lefèvre-Gineau, who were appointed on 29 December 1802. See *Procès-verbaux*, vol. 2, p. 606 (*8 nivôse an 11*).

‡‡ On the identity of Eynard, Haex, and Gensoul see Fox, *Technology Cult.* **10** (1969), 361 n.

flash occurred when air was compressed in a cylinder by means of a closely fitting piston. This new observation and the now well-authenticated ignition of an inflammable material such as tinder by the same method he described first to the Institute in November 1803† and then to the Lyons Academy in January and March 1804.‡ This time the Institute's referees, Charles and Fourcroy, reported favourably and the reading of their report in May 1804 was even accompanied by a successful performance of Mollet's experiments.§ In Lyons some scepticism persisted, despite a demonstration of the effect by Haex on the occasion of the January paper, but Mollet appears to have successfully answered suggestions that the cause of the ignition was not the rise in temperature of the air and to have refuted systematically the alternative explanations that were proposed. Having established, to his own satisfaction at least, that air was in fact heated by compression, Mollet gave his explanation on the basis of the caloric theory, to which he was unreservedly committed, even though he avoided the actual word 'caloric'. He distinguished clearly between latent and free heat, both of which he supposed to exist in a gas, but he attributed the rise in temperature on compression not to the change of some heat from the latent to the free state, as several later calorists were to do, but to the expulsion from the gas of a certain amount of free heat. The much-quoted analogy with the squeezing of water from a sponge would have been a close one here, but it was not mentioned.

Jean-Baptiste Dumas, the permanent secretary and leading historian of the Lyons Academy,¶ writing in 1839, was to hail Mollet's discoveries as some of the greatest contributions to science that the Academy ever made,‖ and while we should not be deceived by Dumas's excessive enthusiasm and by his resentment against those Parisian scientists who, he thought, had tried to steal credit for the work,†† there can be no doubt that the

† *Procès-verbaux*, vol. 3, p. 28 (*6 frimaire an 12*).
‡ The paper of 27 March 1804, which was read at a public meeting of the Academy, was published in Lyons in 1811 as the *Mémoire sur deux faits nouveaux* (cited in full on p 90. n. †).
§ *Procès-verbaux*, vol. 3, pp. 95–6 (*17 floréal an 12*).
¶ Not to be confused with his more famous namesake, the chemist.
‖ Dumas, *Histoire de l'Académie de Lyon*, vol. 2, pp. 232–42.
†† Although Dumas followed Mollet in acknowledging that no less a figure than Georges Cuvier had given public recognition to the work of the men of

academicians of Lyons were by far the most important figures in the early experimental study of adiabatic phenomena in France. Moreover, there was one very tangible debt that the nineteenth-century world owed to the work conducted in Lyons. This was for the domestic fire-making device known as the fire piston, or *briquet pneumatique*, in which the ignition of tinder in a closed cylinder was put to practical use.†

Three of the very few fire pistons of European origin that have survived are shown in Plate 2.‡ It will be seen that the instrument consists simply of a cylinder, closed at one end and usually no more than 6 inches in length and $\frac{1}{2}$ inch in diameter, and a removable, well-fitting piston. By means of the piston air is compressed in the cylinder and its temperature is raised in this way until it is hot enough to ignite tinder and so provide a useful light. Curiously enough, an instrument almost identical to the European fire piston, but usually in wood or ebony rather than metal, is known to have been in use in many parts of south-east Asia by the 1860s. Henry Balfour, in a classic article on the subject in 1907,§ inclined to the view that this use of the instrument in Asia might indicate the occurrence of a completely independent discovery in that region; and others have even suggested that the instrument was not only invented there but also introduced from there into Europe during the early years of the nineteenth century.¶ However, as I have argued in some detail elsewhere,|| the latter view at least seems

Lyons by his comments in his *Rapport historique sur les progrès des sciences naturelles depuis 1789, et sur leur état actuel* (1810), p. 35. See also Cuvier's comments in his report on the work in the physical sciences performed by the First Class during 1812 in *Mém. Sci. math. phys. Inst. Fr.* **13** (1812), p. lxxxviii. According to Dumas (op. cit., vol. 2, p. 242) credit for explaining the ignition of tinder by the compression of air had generally been given to Biot, while that for the invention of the fire piston had gone to a Monsieur Bienvenu.

† For a more detailed discussion of the history of this device and of the work which led to its invention in Europe see R. Fox, 'The fire piston and its origins in Europe', *Technology Cult.* **10** (1969), 355–70.

‡ I am grateful to Mr G. L'E. Turner for his help in obtaining this illustration. Four other fire pistons are shown in Fox, op. cit., Plate II, facing p. 357.

§ Balfour, 'The fire-piston', in N. W. Thomas (ed.), *Anthropological essays presented to Edward Burnett Tylor in honour of his 75th birthday* (Oxford, 1907), pp. 17–49, but especially pp. 42–6.

¶ The view adopted in J. Needham, *Science and civilisation in China* (Cambridge, 1965), vol. 4, part 2, p. 140 n., and, with rather less conviction, in W. Hough, *Fire as an agent in human culture*, United States National Museum Bulletin 139 (Washington, 1926), p. 110. || Fox, op. cit., pp. 366–9.

PLATE 2

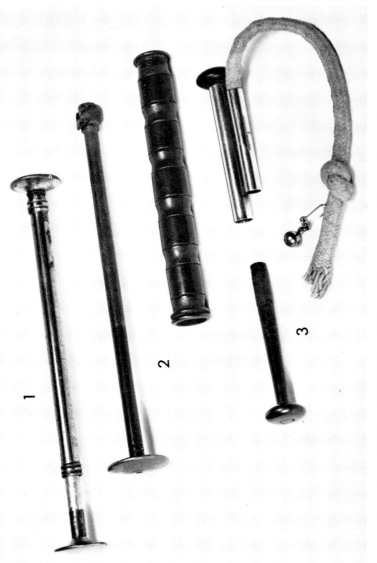

European fire pistons. (1) Probably British, c. 1820; cylinder in steel and brass, piston in brass; cylinder length 6¾ in. (2) Probably British, c. 1820; cylinder and piston in brass; cylinder length 5½ in. (3) French, c. 1900; cylinder in white metal, piston and knob in ebony; cylinder length 3 in. Photograph by courtesy of the Curators of the Museum of the History of Science and the Pitt Rivers Museum, Oxford.

THE SPECIAL STATUS OF GASES

quite untenable; and Mollet and his friends must be given undisputed priority for the discovery as far as Europe is concerned, even though the possibility of an independent discovery in Asia can still not be ruled out.

Mollet first suggested that the piston and cylinder might possibly be used for fire-making in the paper that he read to the Lyons Academy in March 1804,† but it is not until February 1806 that we find any evidence of the commercial manufacture of the instrument. The *Journal de physique* for that month contained a brief announcement that the instrument-maker Dumotiez of Paris was producing the fire piston in various sizes and designs.‡ That Dumotiez was the manufacturer is hardly surprising when we note that it was in his workshop exactly two years earlier that tinder was first successfully ignited in Paris by the compression of air.§ Another early manufacturer was a Lyons founder by the name of Dubois, who patented a design based on a prototype that Eynard constructed and demonstrated publicly in August 1806.¶ According to Mollet, writing in 1811,‖ Dubois's instruments had the advantage over Dumotiez's that the tinder was contained in a small cavity at the end of the piston. This arrangement meant that, in order to apply the glowing tinder, the piston had simply to be withdrawn after the compression stroke (or strokes).†† In Dumotiez's instruments, on the other hand, as also in Mollet's early designs,‡‡ the tinder was contained in a cavity in the end of the cylinder, which therefore had to be rapidly unscrewed before a light was obtained. No instrument by Dumotiez appears to have survived but his somewhat cumbersome type of fire piston was probably similar to one patented in London by Richard Lorentz in February 1807,§§ and the 'foreigners residing abroad' who communicated the invention to him¶¶ may well have been

---

† Mollet, *Mémoire sur deux faits nouveaux*, p. 13.
‡ 'Briquet pneumatique par Dumotiez', *J. Phys.* **62** (1806), 189.
§ Mollet, op. cit., p. 3.
¶ Dumas, *Histoire de l'Académie de Lyon*, vol. 2, p. 236.
‖ Mollet, op. cit., pp. 30–1, in a note appended to the *Mémoire*.
†† Despite certain claims that a single stroke was sufficient to ignite the tinder it seems clear that several strokes were usually required, especially when the tinder was at all damp. Indeed, the tinder often did not light at all.
‡‡ Dumas, op. cit., vol. 2, p. 233.
§§ Lorentz, 'Producing light and fire instantaneously', Patent specification no. 3007 (1807), dated 3 August 1807. ¶¶ Lorentz, op. cit., p. 1.

connected with Dumotiez. Lorentz's version was unusual in that the compressed air passed through a narrow aperture before meeting the tinder. Only two surviving examples of a fire piston constructed on this principle are known† and all the others that still exist have the tinder cavity at the end of the piston. The fact that Dumotiez's instruments were especially well known, even by 1811,‡ suggests, despite Mollet's account, that Dumotiez may well have adopted Dubois's superior design by that date. The fire piston was mentioned frequently in the first thirty years of the century, especially in scientific literature,§ but it seems to have been too inconvenient and unreliable ever to have seriously threatened the conventional tinder-box. After 1830 it could not survive the coming of the friction match and it fell rapidly into disuse, although a brief resumption of manufacture did take place in France about 1900 (see Plate 2, item 3).

The knowledge that tinder could be ignited by compression was presumably the stimulus for a perilous, though successful, attempt to form water from a suitable mixture of oxygen and hydrogen by this method. Who actually performed this experiment is in some doubt. The anonymous author of the article 'Compression' that appeared in 1816 in the physics section of the *Encyclopédie méthodique* states that it was done in the École Polytechnique by Jean Henri Hassenfratz, who was then professor of physics, at the instigation of Laplace, but that Biot, who had been present, described it to the Institute in March 1805 as if it were his own work.¶ Other existing accounts, including the paper to the Institute,‖ certainly convey the

† These are described and illustrated in Fox, op. cit., pp. 363–4 and 369–70, and Plates I and II.

‡ J. P. Dessaignes, *J. Phys.* **78** (1811), 50.

§ See, for example, H. Davy, *Elements of chemical philosophy* (1812), p. 90; Thenard, *Traité de chimie élémentaire* (1813), vol. 1, p. 82; N. L. S. Carnot, *Réflexions sur la puissance motrice du feu* (1824), p. 30 n.; and J. J. Berzelius, *Lehrbuch der Chemie*, trans. F. Wöhler (4 vols., Dresden, 1825–31), vol. 1, p. 60.

¶ *Encyclopédie méthodique. Physique* (1816), vol. 2, p. 507. Since Hassenfratz was still named as one of the editors for this section of the *Encyclopédie* (although he had now lost his chair at the École Polytechnique), it seems probable that he wrote the article himself. By 1816 there was at least one other reason why Hassenfratz might have viewed Biot with some bitterness, for in 1809 Biot had been highly critical of the fifth edition of a work by Hassenfratz entitled *De la formation et de la décomposition des corps*. Biot's vitriolic review appeared in the *Mercure de France* and is reproduced in his *Mélanges scientifiques et littéraires* (1858), vol. 2, pp. 117–39.

‖ Biot, 'Note sur la formation de l'eau par la seule compression . . .', *Annls*

impression that Biot did in fact perform the experiment and that Hassenfratz was little more than an assistant, so that, if the author of the article in the *Encyclopédie méthodique* is to be believed, Hassenfratz was the victim of a successful piece of deception.† According to Biot, the gases were contained in the chamber of an air-gun and a single stroke of the piston sufficed to effect a combination accompanied by an explosion. One particularly violent explosion, which shattered the chamber, provided sufficient warning of the danger for the experiments to be abandoned. The results of further experiments by Hassenfratz, which were not mentioned by Biot and in which Hassenfratz believed he had formed nitrogen dioxide and ammonia gas,‡ were presumably too unreliable to be published, but the possibility that other combinations between gases might be effected was not entirely overlooked by the readers of Biot's account. For example, in a letter to *Nicholson's journal* dated 17 December 1805 the writer, geologist, and antiquarian Thomas Northmore described how he had produced a number of different compound gases by merely compressing their constituent elements.§ The experiments had been conducted with the aid of the 'late chemical operator' of the Royal Institution in London,¶ but nothing more is known of them.

Chim. **53** (1805), 321–7. A generous acknowledgement to Hassenfratz for his assistance appears on p. 322 n.

† For another case in which Biot appears to have tried to take the full credit for joint work see *Œuvres complètes de François Arago*, vol. 11, pp. 702–3. This time Arago was the victim when, in 1806, Biot submitted a joint paper for publication under his own name, though with an acknowledgement to Arago. Eventually, after protests from Arago, the two names were appended to the paper (an important one on refraction in gases, discussed on pp. 198–202).

‡ These were described only in the *Encyclopédie méthodique* article.

§ Northmore, *Nicholson's Journal*, 2nd ser. **12** (1805), 368–74.

¶ I am grateful to Mr G. Smith, assistant librarian at the Royal Institution, for his assistance in identifying Northmore's collaborator. Mr Smith points out that it could well have been either Friedrich Accum, who was 'assistant chemical operator' at the Royal Institution from 1801 to 1803, or James Sadler, who was assistant in Davy's laboratory from 1800 to 1804. It might also have been the Scottish geologist Thomas Webster, who was employed as architect and clerk of works at the Royal Institution until 1802, since he was sometimes referred to as assistant to Thomas Garnett, professor of natural philosophy from 1799 to 1801 (see, for example, J. Imison, *Elements of science and art*, a new edition by Thomas Webster (1808), vol. 1, p. vi). It is curious that Davy appears to have given no indication that he knew anything of these or the French experiments when, some years later, he was consulted by Charles Babbage on the probable effect of compression on mixtures of gases. Wollaston,

It remains to say something of Mollet's other observation, that light was emitted in the compression of air, for the effect was to attract considerable attention before Thenard finally showed it to be spurious in 1830.† Biot was one writer who was quick to see the possible relevance of this appearance of light to a current problem. In 1803 Berthollet had suggested that the function of the electric spark in effecting the combination of gases in a eudiometer was essentially to compress, and consequently to heat, the gas in the vicinity of the spark to the point where a reaction occurred.‡ Writing late in 1804, when he knew of Mollet's work, Biot was able to enlarge on the idea and to suggest that the *light* of the spark also arose from the rapid compression of the adjacent gas.§ Since Biot's own experiments had failed to show that light was evolved by compression, his extreme reserve in making the suggestion was more than justified.

Mollet's observation soon became involved in another, even more fundamental, problem, namely the possible identity of caloric and light. The idea that the phenomena of heat and light might arise from one and the same cause was by no means a new one, of course. In the eighteenth century it had been frequently discussed and expounded by writers on the subject of 'fire', and in the *Traité élémentaire de chimie* Lavoisier himself had been very hesitant about adopting the separate terms *calorique* and *lumière* for what he believed might be one and the same fluid.¶ In Mollet's day, however, and at least in France, the most influential exponent of the view that the two fluids were essentially identical was Monge, who in 1790 had expressed his opinions on the matter very briefly in an article on heat in the revolutionary *Journal gratuit*,∥ which was apparently based

who was also consulted, seems to have been no less ignorant. See Babbage, *Reflections on the decline of science in England* (1830), pp. 206–9.

† Thenard, *Annls Chim. Phys.* **44** (1830), 181–8.

‡ Berthollet, *Essai de statique chimique* (1803), vol. 1, p. 307.

§ Biot, op. cit., pp. 324–5. On the dating of Biot's work see the brief account of it that appeared in *Bull. Soc. philomath. Paris* **3** (1801–5), 259.

¶ Lavoisier, *Traité élémentaire de chimie* (1789), vol. 1, p. 6. Cf. the similar uncertainty expressed by Guyton de Morveau in Guyton et al., *Système de nomenclature chimique* (1787), pp. 30–1.

∥ *Journal gratuit*, X^e classe (physique), pp. 26–32, 41–4, 49–53, 65–7, and 81–3. The reference to Monge's view that heat and light were identical appeared on p. 83.

THE SPECIAL STATUS OF GASES

on an unpublished contribution to the *Encyclopédie méthodique* first written in 1782. Most of this encyclopaedia article was soon lost and even the *Journal gratuit* paper only became easily available in 1816 when it was reproduced in the article on caloric that eventually did appear in the *Encyclopédie méthodique*.† Yet the view held by Monge was familiar to many of his contemporaries, and Riche de Prony, Fourcroy, Berthollet, Étienne Barruel, and Biot, for example, were among those who took it seriously, though not all of them mentioned Monge.‡ The discovery that light as well as caloric was emitted in the compression of gases might have been expected to strengthen Monge's case, as Mollet pointed out,§ but owing largely to Jean Antoine Saissy, a doctor from Lyons, it was the contrary opinion that gained ground. In 1811 J. P. Dessaignes of Vendôme announced that he had been able to produce light simply by compressing either oxygen, nitrogen, or hydrogen in the course of his work on phosphorescence,¶ the subject for which, two years earlier, he had won the Institute's prize for physics.‖ He was satisfied that the effect was not one of phosphorescence and that it arose from the expulsion of caloric by the compression of gas molecules. By 1813, moreover, he was convinced that light was produced when air entered a vacuum.†† In papers presented to the Lyons Academy and the Institute Saissy disputed

---

† *Encyclopédie méthodique. Physique*, vol. 2, pp. 170–4, where the circumstances surrounding the loss of the contribution of 1782 and the publication of fragments of it in the *Journal gratuit* are described (on p. 170). In fact the versions published in the *Journal gratuit* and the *Encyclopédie méthodique* were by no means identical, certain additions having been made between 1790 and 1816.

‡ Prony, *Nouvelle architecture hydraulique* (1790), vol. 1, p. 550; Fourcroy, *Système de connaissances chimiques* (1802), vol. 1, pp. 131–4; Berthollet, op. cit., vol. 1, pp. 191–205; Barruel, *La Physique réduite en tableaux raisonnés* (1806), Table XXVII; Biot, *Traité de physique* (1816), vol. 4, pp. 600–17. In Britain John Leslie was probably the best-known exponent of the opinion held by Monge, although there is little doubt that he arrived at his conclusion independently of French writings on the matter. The justification for his view that 'heat is only light in the state of combination' appeared in his *Experimental inquiry into the nature, and propagation, of heat*, pp. 151–62. Cf. the view that caloric 'in a rarefied and projectile state . . . probably constitutes light', put by Dr George Gregory, the London preacher and later Rector of West Ham, in his *Economy of nature* (2nd edn, 1798), vol. 1, p. 99.

§ Mollet, *Mémoire sur deux faits nouveaux*, pp. 27–8.

¶ Dessaignes, *J. Phys.* 73 (1811), 50–1.

‖ Dessaignes's prize-winning paper appeared in *J. Phys.* 68 (1809), 44–67, and 69 (1809), 5–35.

†† *Procès-verbaux*, vol. 5, p. 250 (4 October 1813).

Dessaignes's evidence, claiming that the light was only observed when gases that contained oxygen were used.† According to Saissy it occurred most readily in pure oxygen, rather less so in air and hydrogen chloride (or muriatic acid, which, like all acids on Lavoisier's theory of acidity, was thought to be a compound of oxygen), and not at all in nitrogen or hydrogen, for example. Saissy's results were confirmed by experiments conducted in Berthollet's laboratory‡ and by Mollet, Eynard, and two other referees appointed by the Lyons Academy. Although, in reporting, Mollet was clearly reluctant to abandon the view that the emission of caloric was the cause of the light,§ he accepted as highly likely Saissy's suggestion that the 'base' of oxygen was a combination of light with some unknown ponderable substance but that light was a constituent of no other gas. Since there was no doubt that caloric, on the other hand, existed in all gases, this was strong evidence for the dissimilarity of the two imponderables, as Georges Cuvier pointed out.¶ It is worth noting that Saissy's view was not unlike an early opinion of Humphry Davy, who had supposed oxygen gas to be the base oxygen combined with light,‖ but it seems improbable that Saissy would have read the obscure work (Thomas Beddoes's *Contributions to physical and medical knowledge* of 1799) in which Davy wrote.

One of the most striking characteristics of these early discussions of adiabatic phenomena is the absence of any attempt to interpret them in terms of a kinetic theory of heat. In fact, far from supporting the latter theory, the phenomena merely strengthened the position of caloric. For this, of course, the extraordinary ease with which they were accommodated on the assumption of the materiality of heat was chiefly responsible, and any study of this process of accommodation, such as has

† Dumas, *Histoire de l'Académie de Lyon*, vol. 2, pp. 237–9, and *Procès-verbaux*, vol. 4, p. 552 (11 November 1811). See also J. L. N. F. Cuvier, 'Histoire de la Classe . . . Partie physique', *Mém. Sci. math. phys. Inst. Fr.* **13**, part 2 (1812), p. lxxxviii. Saissy's paper for the Institute is preserved, as MS. 230, ff. 221–32, in the library of the Lyons Academy.

‡ *Procès-verbaux*, vol. 5, pp. 87–8 (24 August 1812). Here in their official report on Saissy's paper, Thenard and Gay-Lussac described experiments performed at Arcueil in the presence of Laplace, Berthollet, Chaptal, Bosc, and others. The report was highly complimentary. See also Thenard, *Annls Chim. Phys.* **44** (1830), 182.

§ For evidence of this see his comments in *J. Phys.* **90** (1820), 114–15.

¶ Cuvier, loc. cit. The same point was made by Thenard and Gay-Lussac in their report; see note ‡ above.   ‖ Davy, *Works*, vol. 2, pp. 36–49.

THE SPECIAL STATUS OF GASES

been undertaken in this and the previous chapter, can therefore serve only to illustrate the great strength of the caloric theory in the last years of the eighteenth century. At the turn of the century, in fact, the theory was almost unassailable, with such criticisms as were made proving uniformly ineffectual.

Among the critics of caloric about this time the best known was undoubtedly the versatile Benjamin Thompson, Count Rumford, but even he failed in his attack, and most of his present renown dates from a later period, the middle of the nineteenth century, when the vibrational theory that he championed was revived in the light of the newly discovered principle of the conservation of energy.† Indeed, his failure in his own lifetime was such that the fortunes of the caloric theory were virtually unaffected by his writings, and for this reason it is conceivable that a history of the theory could be written with scarcely any reference to Rumford. But some examination of the causes of his failure does throw light on the state of the caloric theory about 1798, when Rumford described his famous cannon-boring experiments to the Royal Society.‡ And it is instructive also because of the still-persistent myth that by these experiments he succeeded in disproving the existence of a fluid of heat and that his contemporaries, adhering blindly to the old theory, inexplicably overlooked this fact.

Rumford

Rumford directed his attack at what was probably the most vulnerable point in the whole caloric theory. No calorist had ever explained the production of heat by friction satisfactorily

† Rumford's work was not forgotten during the first half of the nineteenth century (see, for example, Dulong's reference in his letter of 15 January 1820 to Berzelius quoted on p. 244), but it seems to have exerted a negligible influence on the emergence of the new theory of heat in the 1850s. Many of the pioneers of the new theory certainly referred to Rumford's work with respect. John Tyndall, for example, believed that it had served to 'annihilate' the caloric theory (see his *Heat considered as a mode of motion* (1863), p. 25 n.). However, the man who seems to have been chiefly responsible for the exaggerated praise that has been accorded Rumford in the last hundred years was P. G. Tait. See especially his *Sketch of thermodynamics* (2nd edn, Edinburgh, 1877), pp. 5–10. Like his friend William Thomson (*Encyclopaedia Britannica* (9th edn, Edinburgh, 1880), vol. 11, p. 557, art. 'Heat'), Tait believed that Davy's ice-rubbing experiments of 1799 were 'completely decisive', whereas Rumford's work was open to some objection.

‡ Rumford, 'An inquiry concerning the source of heat which is excited by friction', *Phil. Trans. R. Soc.* **88** (1798), 80–102.

and, in fact, no calorist was ever to do so. It is not easy to ascertain with absolute certainty how the heat of friction was usually accounted for in the 1790s, since references to the problem are few. But it seems probable that it was the view of most calorists then, as also in the first decade of the nineteenth century (when, thanks partly to Rumford, the matter was more frequently discussed), that this heat originated in exactly the same way as the heat produced in the percussion of a solid. For the latter phenomenon a convincing explanation was available, since it could be supposed in this case that a slight decrease in volume (a plausible enough consequence of hammering) reduced the solid's capacity (taken in its loose sense of ability to hold caloric) and so caused heat to be expelled, rather like water from a sponge.† That compression must diminish capacity in this sense seemed only too obvious to calorists, and their conviction on this point was greatly strengthened by their growing familiarity with the adiabatic phenomena in gases, which provided striking evidence for the close and seemingly even necessary correlation between decrease in volume and rise in temperature. It is notable that the analogy between percussion and the rapid compression of a gas, which had been first pointed out by De Luc and Darwin in the 1780s, was referred to again by Thomas Wedgwood in an important paper to the Royal Society in 1792,‡ and soon after 1800, at the very latest, it had become a commonplace.§

But, however convincing such an explanation may have been

† Influential experimental support for the already widespread belief in the association between heat of percussion and decrease in volume came in 1809, when Berthollet published his paper 'Sur la chaleur produite par le choc et la compression' in *Mém. Phys. Chim. Soc. Arcueil* **2** (1809), 441–8. In some experiments that he had conducted in collaboration with Pictet and Biot, presumably some years earlier when Pictet was in Paris (see Pictet's diary for 1802–4 referred to on p. 90 n. §), Berthollet had shown that heating only occurred when the volume of a metal undergoing hammering decreased. The most important result was that when, after a few strokes, little further decrease in volume was observed, the heating effect became very small.

‡ Wedgwood, *Phil. Trans. R. Soc.* **82** (1792), 280.

§ See, for example, T. Cavallo, *The elements of natural or experimental philosophy* (1803), vol. 3, p. 97; Berthollet, *Essai de statique chimique* (1803), vol. 1, p. 165; Leslie, *Experimental inquiry* (1804), pp. 167–8; Haüy, *Traité de physique* (2nd edn, 1806), vol. 1, pp. 141–2; Dalton, *New system* (1808), vol. 1, part 1, p. 97; Cuvier, *Rapport historique sur les progrès des sciences naturelles depuis 1789* (1810), pp. 34–5; Mollet, *Mémoire sur deux faits nouveaux* (1811), p. 15; Thenard, *Traité de chimie, élémentaire, théorique et pratique* (1813), vol. 1, pp. 81–2.

THE SPECIAL STATUS OF GASES

when applied to the appearance of heat in percussion, it was not obvious that it could be applied in the case of friction, since there was no reason to believe that mere rubbing, however vigorous, caused a decrease in volume. Rumford was presumably aware of this serious weakness in the caloric theory but his attack lost much of its effectiveness through his preoccupation with a particular version of the calorists' explanation of the heat of friction that was unmistakably Irvinist in its basic assumptions. For an Irvinist the expulsion of heat from a body had to be accompanied not only by a decrease in the body's ability to contain heat (something that was not, of course, directly measurable) but also by an observable change in its specific heat. That this latter change was in fact brought about during percussion had been claimed by Thomas Wedgwood in his Royal Society paper of 1792, where he described how the specific heat (by weight) of the clay used in his father's celebrated pyrometer had been observed to fall by one-third when its volume was halved by strong heating.† Not surprisingly, perhaps, this evidence had little effect and only Thomas Thomson of the major calorists seems to have been impressed by it.‡ Moreover—and this is the important point as far as Rumford is concerned—it was quite irrelevant to any case, such as friction, in which a decrease in volume did not seem to occur.

Rumford's demonstration that the specific heat of the small metal chips produced in boring was the same as that of the bulk metal from which they were bored should therefore have appeared as strong evidence in the eyes of Irvinists, if not to other calorists. Some Irvinists, the Scottish chemist John Murray, for example, were obviously impressed,§ although none of them seems to have abandoned the caloric theory completely; and even other calorists who were not Irvinists, such as Thomas Thomson, confessed, presumably in the light of Rumford's findings, that at present they could not account for the heat produced by friction.¶ But the great majority of

† Wedgwood, loc. cit.
‡ He referred to it in the works cited in note ¶ below.
§ Murray, *Elements of chemistry* (Edinburgh, 1801), vol. 1, pp. 164–5, and his *System of chemistry* (Edinburgh, 1806), vol. 1, pp. 434–9.
¶ See Thomson's article on chemistry in the supplement to the third edition of the *Encyclopaedia Britannica* (Edinburgh, 1801), vol. 1, pp. 277–9, and his *System of chemistry* (Edinburgh, 1802), vol. 1, pp. 361–8. Alexandre Haldat,

calorists remained unmoved and most proceeded either to deny the important distinction between percussion and friction† or simply to ignore Rumford's evidence on the point. Others maintained that the absence of a change in specific heat did not show that the total heat content of the chips was undiminished and so they essentially criticized Rumford's Irvinist assumptions.‡ Some, it must be pointed out, did go so far as to devise an alternative explanation,§ but the alternative was always consistent with the materiality of heat, so that even those who recognized most clearly that Rumford had exposed a serious anomaly in the caloric theory were not converted to his views. The weight of the counter evidence in favour of caloric, especially that concerning the origins of chemical heat, was just far too great.¶

So Rumford failed. By his concentration on the Irvinist version of the caloric theory he had made his attack considerably less effective than it might otherwise have been, and most of his experimental evidence, apart from that concerning the specific heat of the bored chips, was manifestly unoriginal. This was true even of his demonstration that the heat produced in boring was seemingly inexhaustible, about which few could

the physician and secretary of the Académie des Sciences of Nancy, was more obviously influenced by Rumford and, although he did not go so far as to adopt the vibrational theory, he was stimulated to perform some admirable experiments on friction as an extension of Rumford's work; see Haldat, *J. Phys.* **65** (1807), 213–22.

† The most notable calorist who adopted this position was Dalton; see Dalton, *New system*, vol. 1, part 1, p. 97.

‡ Of these the most notable were Berthollet (*Recherches sur les lois de l'affinité* (1801), pp. 73–4 n.) and Dalton's friend William Henry (*Mem. Proc. Manch. lit. phil. Soc.* **5**, part 2 (1802), 603–21). By 1803 Berthollet's criticism of Rumford had been extended and slightly modified. He now argued that the heat of friction came from the whole of the cannon and not just from the chips of metal, and also that any change in the specific heat that resulted from compression could not be observed in an experiment of the 'method of mixtures' type, since the compressed metal would then expand and regain its initial specific heat; see Berthollet, *Essai de statique chimique*, vol. 1, pp. 247–50.

§ See, for example, pp. 280–1 of the *Encyclopaedia Britannica* article cited on p. 101 n. ¶, where Thomson speculated on the possible analogy between the accumulation of electricity by rubbing and that of caloric by the same method.

¶ This, for example, was the view of Haldat, who admired and extended Rumford's experiments on friction but who, because of the weight of other evidence, could not bring himself to abandon the caloric theory; see Haldat, op. cit., p. 221.

THE SPECIAL STATUS OF GASES

have been in serious doubt in 1798 and which was certainly not in need of elaborate experimental verification.† For all this, Rumford had accurately identified, or rather re-identified, a genuine weakness in the theory he was criticizing; moreover, his paper was widely read and discussed; and it is therefore all the more significant, even if it can scarcely be held surprising, that calorists, virtually to a man, rejected his case.

In this chapter we have seen how, about 1800, the caloric theory withstood shocks on two fronts which, as it might seem to modern observers familiar with the interconvertibility of heat and work, could perhaps have proved fatal. In fact, the theory was never in danger, either from Rumford's attacks or from the wealth of new evidence concerning the adiabatic phenomena in gases. For at this time it was not only a theory that had already proved successful and hence won general acceptance, but also one that was still in the process of development, still growing and therefore flexible enough to make the processes of refutation (in the case of the Rumford episode) and assimilation (in the case of the adiabatic phenomena) comparatively easy.

But the situation in which the heats of friction, percussion, or compression could be satisfactorily explained by some vague mechanical analogy, for example with the expulsion of water from a sponge, did not last long.‡ The sudden advances in the study of gases and heat that have been examined in the first two sections of this chapter soon stimulated new experimental and theoretical work, and it was this new work, far more careful and rigorous than anything that had gone before, which allowed the caloric theory to develop still further and eventually to reach its most sophisticated form during the mid-1820s in Laplace's caloric theory of gases.

† For example, the ignition of the wheel of a cart through the friction between it and its axle, was well known and was frequently referred to in seventeenth- and eighteenth-century works.

‡ Although the analogy was still used in Mary Somerville's *On the connexion of the physical sciences* (1834), p. 249.

4

THE RIVAL CALORIC THEORIES

IN the decade 1800–10 the caloric theory was probably more widely accepted than at any other time in its history. The questioning of the materiality of heat that had been conducted about the turn of the century, not only by Rumford but also, as we shall see, by Humphry Davy and Thomas Young, had had remarkably little effect, and the view that '[heat] is almost universally believed to be the effect of a fluid' was one that most men of science would have found no less acceptable in 1800, in 1810, or even in 1815 than it had been in 1797, when it appeared in the third edition of the *Encyclopaedia Britannica*,† shortly before Rumford, Davy, and Young put their criticisms. Certainly it would be an exaggeration to state that the existence of caloric was taken to be axiomatic in the early years of the nineteenth century, for lengthy and generally well-balanced discussions of the relative merits of the fluid and vibrational theories appeared in most of the textbooks of chemistry and physics of the period, and in these the shortcomings of the caloric theory were not infrequently acknowledged. But the preference for caloric was nevertheless very marked, especially after 1801 when William Herschel described his important work on the radiant heat spectrum,‡ so that this and the following two chapters are concerned not so much with the debate on the nature of heat as with the discussions concerning the behaviour

---

† *Encyclopaedia Britannica* (3rd edn, Edinburgh, 1797), vol. 8, p. 350. For equally dogmatic assertions in standard reference works see *Encyclopédie méthodique. Chimie* (1792), vol. 2, p. 698, and Charles Hutton's *Mathematical and philosophical dictionary* (1796), vol. 1, p. 473. It is interesting to note that the writer of the *Encyclopaedia Britannica* article appears to have favoured the view that heat was caused by a fluid *in motion*, so indicating the lingering influence of the theories so popular earlier in the eighteenth century.

‡ Herschel, 'Experiments on the solar and on the terrestrial rays which occasion heat . . .', *Phil. Trans. R. Soc.* **90** (1800), 293–326 and 437–538. The effect of this paper was to demonstrate the similarity between radiant heat and light and hence to associate the fortunes of the theories of heat and light more closely than ever before.

and properties of caloric itself. The issues in these discussions were not ones that set the supporters of Rumford against the supporters of Lavoisier and Black, but issues debated, for the most part, by and for calorists.

Theories of heat in Britain

In both Britain and France much of the most interesting controversy among calorists after 1800 was stimulated by the Irvinist doctrines as expounded and developed first by Crawford and then by Dalton. As might be expected, these doctrines were now generally interpreted and discussed in terms of the caloric theory, and the most important point at issue was whether or not some of the caloric in bodies existed in a combined or latent state.

Most of those who opposed the Irvinist teachings on this point in Britain claimed that they were following Black. The well-known Scottish writer Robert Heron, who argued strongly for the existence of caloric in two states, the 'free' and the 'combined', in his *Elements of chemistry* of 1800,† was an exception in this respect, since he made no specific mention of Black. But a far more notable and typical critic of Irvine was Thomas Thomson, a convinced calorist in the early years of the century.‡ Thomson strongly supported Black's claim that Irvine had not adequately explained changes of state, maintaining that it was the quasi-chemical combination of caloric with a solid or liquid that brought about fluidity or vaporization accompanied by a *consequent* change in specific heat.§ Moreover, the fact that some caloric thus existed in a latent state in bodies naturally invalidated Irvine's method for determining the absolute zero of temperature.

† Heron, *Elements of chemistry* (1800), pp. 71–8.
‡ See Thomson, *A system of chemistry* (Edinburgh, 1802), vol. 1, p. 259, where he wrote that Herschel's work on the radiant heat spectrum had 'at last put an end to the dispute' (in favour of caloric). Cf. the similar view expressed in Charles Hutton's *Philosophical and mathematical dictionary* (2nd edn, 1815), vol. 1, p. 263. In 1801, when he wrote in the first volume of the supplement to the third edition of the *Encyclopaedia Britannica* (p. 258), Thomson apparently knew nothing of Herschel's work and was slightly less confident. By 1830 he believed that the nature of heat was unknown; see the comment by him quoted on p. 276.
§ *Encyclopaedia Britannica* (3rd edn), supplement, vol. 1, p. 269, and Thomson, *System*, vol. 1, pp. 326–8.

Dr. Irvine's theorem [Thomson declared in 1801] . . . is insufficient for ascertaining the real zero; and hitherto no method has been discovered which can solve this problem. We are therefore entirely ignorant of the quantity of caloric which exists in bodies.†

This was not to say, of course, that he had no interest in the problem, and in 1804 he commented that the most promising attempt so far had been Dalton's deduction from his law of gas expansion, although he added that such a method yielded a measure only of the caloric that contributed to the interparticle repulsion in the gas state and so threw no light on the quantity of caloric that remained even if the gas was liquefied.‡

Like Thomson, William Henry of Manchester had attended Black's lectures on chemistry at Edinburgh in the 1790s. He and Dalton were good friends and they had at least one common interest, in the solubility of gases.§ Yet they were opposed on the subject of heat. The most detailed account of Henry's views appeared in a paper that he read to the Manchester Literary and Philosophical Society in June 1801.¶ His real purpose in this paper was to answer the criticisms of caloric recently made by Rumford and Davy, but he made it quite clear that it was not caloric theory in general but Black's version only that he was defending.‖ In fact Henry went still further than Black since he supposed that the caloric that existed in a latent form in bodies and brought about changes of state was in a truly chemical combination and had a characteristic affinity, exactly like ordinary ponderable matter. Irvine, he maintained, had merely indicated the point of total privation of *uncombined* caloric.

There can be little doubt that the supporters of Irvine and Crawford were losing ground rapidly in the early years of the nineteenth century, and by 1806 even Dalton felt it necessary to write in defence of Crawford's theory of animal heat.†† But

† *Encyclopaedia Britannica* (3rd edn), supplement, vol. 1, p. 272, and Thomson, *System*, vol. 1, p. 337.
‡ Thomson, *System* (2nd edn, Edinburgh, 1804), vol. 1, pp. 400–4.
§ See Henry, *Phil. Trans. R. Soc.* 93 (1803), 33, and *Nicholson's Journal*, 2nd ser. 8 (1804), 297–301.
¶ Henry, 'A review of some experiments that have been supposed to disprove the materiality of heat', *Mem. Proc. Manch. lit. phil. Soc.* 5, part 2 (1802), 603–21.
‖ See also his *Epitome of chemistry* (1801), pp. 7–13.
†† Dalton, 'On respiration and animal heat', *Mem. Proc. Manch. lit. phil. Soc.*, 2nd ser. 2 (1813), 15–44 (including additions to the paper made in 1810 and 1811).

it was not until the French Institute's prize competition of 1812 that the really decisive blow was delivered, and before that date perfectly conventional statements of the Irvinist position were by no means uncommon. Two such statements were made in important textbooks of the period by John Murray and John Leslie, both of whom taught for considerable periods at Edinburgh. In his work, published in 1804, Leslie was particularly forceful in rejecting the view that some latent caloric existed in a state of combination in bodies,† and both men expressed approval for at least the principles of Irvine's method for the determination of the absolute zero, Leslie placing it at $-1320\,°F$, Murray at about $-1500\,°F$ or rather higher.‡ They were quite satisfied that a sudden change in capacity adequately explained the absorption of heat on melting and vaporization and they argued against Black's opinion on this point with a thoroughness that emphasizes the importance of the dispute over the state of caloric in bodies during the first decade of the nineteenth century.

The easy assimilation of the adiabatic phenomena by the caloric theory continued after 1800, thanks not only to the great calorists such as Dalton and Laplace but also to lesser authorities whose influence was chiefly felt through their textbooks. Murray's treatment was typically Irvinist, with changes in capacity proposed as the cause of the effects.§ Indeed, he even adopted Dalton's explanation of the evacuated-receiver experiment, although he wrongly attributed it to Crawford. Thomson, as we should expect, rejected such explanations, even though he accepted that variations in specific heat did accompany volume changes.¶ Instead he presented a simple mechanical picture, which could also be applied to account for the heat

† Leslie, *An experimental inquiry into the nature, and propagation, of heat* (1804), p. 532; see also his preliminary dissertation to the seventh edition of the *Encyclopaedia Britannica* (Edinburgh, 1842), vol. 1, pp. 644–6. For the similar views put by Murray see his *Elements of chemistry* (Edinburgh, 1801), vol. 1, pp. 143–52, and his *System of chemistry* (Edinburgh, 1806), vol. 1, pp. 406–12, and 'Notes', pp. 66–72.

‡ Leslie, *Experimental inquiry*, pp. 170 and 532–3. In the posthumously published *Encyclopaedia Britannica* dissertation of 1842 he modified the figure to $-1500\,°F$. For Murray's figures see his *Elements*, vol. 1, pp. 155–6, and his *System*, vol. 1, pp. 386–423, and 'Notes', pp. 66–106.

§ Murray, *System*, vol. 1, pp. 366–9.

¶ *Encyclopaedia Britannica* (3rd edn), supplement, vol. 1, pp. 277–8, and Thomson, *System* (1st edn), vol. 1, pp. 361–2.

produced in the percussion of a solid. 'When the particles of a body are forced nearer each other [he wrote], the repulsive power of the caloric combined with them is increased, and consequently a part of it will be apt to fly off.'†

Leslie's explanation was similar to Murray's, for although he criticized the term 'capacity' as misleading,‡ it is difficult to see how his 'specific attraction', which was the property that determined the quantity of heat in a body, differed essentially from the more familiar expression. Since it was a change in this 'attraction' that occasioned the emission and absorption of heat in both adiabatic phenomena and changes of state, Leslie's treatment showed little originality. In his interpretation of the evacuated-receiver experiments, however, Leslie put forward an unusual view, although even here the extent of his originality is open to doubt, since in attributing the heating solely to the compression by the incoming gas of the small quantity of gas that remained in even the best obtainable vacuum he was merely repeating the explanation proposed by Cleghorn in 1779.§ Whether or not he had read Cleghorn is unclear, but the possibility that he had done so must remain, especially in view of a marked reluctance on his part to acknowledge other authorities.¶ His attempt to apply the explanation to the determination of specific heats, on the other hand, certainly was original.‖ He argued that if various gases at the same pressure were admitted to a receiver in which the air pressure had been reduced to, say, 1/10 or 1/100 of atmospheric, the extent to which the remaining air was compressed, and hence the quantity of heat evolved, would be identical, whatever gas was admitted. The temperature changes occurring with different gases would thus be determined solely by their volume specific heats, once correction for the very considerable heat losses through the walls of the receiver had been made. Unfortunately Leslie put his method to the test for hydrogen and air only, observing the same temperature change in each case. The conclusion that the

---

† *Encyclopaedia Britannica* (3rd edn), supplement, vol. 1, pp. 279, and Thomson, *System*, vol. 1, p. 362.

‡ Leslie, *Experimental inquiry*, pp. 529–30.   § Leslie, op. cit., pp. 533–4.

¶ A reluctance that is apparent throughout the *Experimental inquiry*.

‖ Dalton's earlier suggestion that the evacuated receiver experiments might be used for this purpose had been based on a quite different analysis; see Dalton, *Mem. Proc. Manch. lit. phil. Soc.* **5**, part 2 (1802), 526.

volume specific heats of the two gases were equal in fact accorded well with Crawford's results.

One of the most extravagant of all the calorists of the period about 1800 was Alexander Tilloch, a Scot living in London, who in December 1799 delivered a far from convincing attack on the view that caloric could exist in two states to the Askesian Society of London, 'a select number of gentlemen' who since March 1796 had 'associated for their mutual improvement in the different branches of natural philosophy'.† A paper read before a small private society such as this would normally have been lost, but Tilloch, as editor of the *Philosophical magazine*, had a ready vehicle for his opinions and of this fact he took full advantage.‡ In November 1800 he repeated his views and enlarged on a suggestion, which had first appeared in his paper of 1799, that the volume of the 'matter of heat' expelled in a compression or, indeed, in a chemical reaction might be equal to the change that occurred in the volume of the ponderable matter. He argued, furthermore, that heat not only possessed volume but was also subject to gravitational attraction and so could be weighed.§ In this, of course, he was reviving a persistent eighteenth-century idea and neglecting Rumford's recent experimental refutation.¶

For all their interest and novelty, Tilloch's arguments generally betray an inadequate understanding of the issues involved and they are therefore of only minor importance in the history of caloric theory, a theory that had exponents of far greater weight than anyone whose work has been examined so far in this chapter. On the Irvinist side the greatest name in the nineteenth century was undoubtedly that of Dalton. We have already seen how his commitment to Irvine's views influenced his work on adiabatic phenomena and also on the absolute zero of temperature, which to him, as to any Irvinist, was 'an object of primary importance in the doctrine of heat'.|| In all

---

† *Phil. Mag.* **7** (1800), 355 n.

‡ He published his paper, as 'A brief examination of the received doctrines respecting heat or caloric', in *Phil. Mag.* **8** (1800), 70–8, 119–26, and 211–21.

§ Tilloch, 'An attempt to prove that the matter of heat . . . possesses not only volume but gravity', *Phil. Mag.* **9** (1801), 158–67.

¶ Rumford, 'An enquiry concerning the weight ascribed to heat', *Phil. Trans. R. Soc.* **89** (1799), 179–94.

|| Dalton, *New system*, vol. 1, part 1 (1808), p. 83.

this work there was much that was original, but Dalton's originality is probably nowhere more apparent than in his views on the size of gas particles. His novel attempt of 1801 to relate particle diameter and the absolute temperature of a gas expanding at constant pressure was only one aspect of this interest, for by the time he abandoned the attempt, between early 1805 and March 1807, the subject of particle size and in particular the dependence of this quantity on the nature, rather than the temperature, of a gas had gained a new and far greater significance with the emergence of his atomic theory.

From the start size was as much a characteristic of the Daltonian atom as weight. By the time the first table of particle weights first appeared in his notebook, on 6 September 1803 or very shortly afterwards,† Dalton had already rejected his earlier 'confused idea' that under similar conditions of temperature and pressure the particles of all gases were equally spaced and hence also were of the same size;‡ and by 19 September the first list of fourteen gas-particle diameters had been drawn up.§ In view of his later descriptions of gas particles composed of huge but weightless atmospheres of caloric surrounding a small, heavy central atom,¶ it may seem surprising that it was apparently not until the early months of 1806 that he began to examine the implications of his new concept in the light of his Irvinist theory of heat. Was it not immediately obvious that the diameters were somehow related to the quantities of heat in different gases and hence also to their specific heats? The problem of deriving the relationship was certainly a difficult one, but an additional reason for the delay may lie in Dalton's 'first theory of mixed gases',‖ which he held at this time. According to this theory, which was devised by Dalton in the

† H. E. Roscoe and A. Harden, *A new view of the origin of Dalton's atomic theory* (1896), p. 28.

‡ Roscoe and Harden, op. cit., p. 27, where the rejection is clearly implied in the passage from p. 246 of the first volume of Dalton's notebook.

§ Roscoe and Harden, op. cit., p. 41; also pp. 24–5 (on the method used for the calculation of the diameters).

¶ See, for example, his comments dated 23 May and 17 June 1806 in Roscoe and Harden, op. cit., pp. 71 and 73–4; also *New system*, vol. 1, part 1, pp. 143–4, and vol. 1, part 2 (1810), p. 548, and Plates 4 and 5 in this book.

‖ The term used by L. K. Nash in *Isis* **47** (1956), 101–16, where the theory is described. The theory was first announced by Dalton in *Mem. Proc. Manch. lit. phil. Soc.* **5**, part 2 (1802), 535–50.

PLATE 3

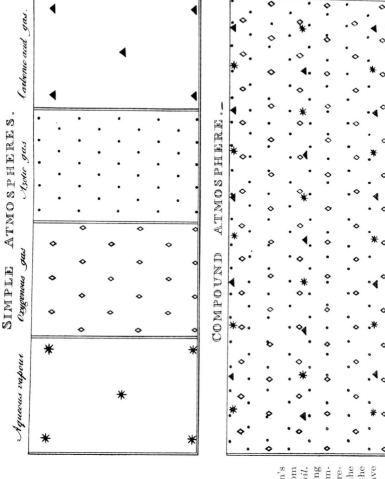

Diagram illustrating Dalton's first theory of mixed gases. From *Mem. Proc. Manch. lit. phil. Soc.* **5**, part 2 (1802), facing p. 602. In the mixture or 'compound atmosphere' mutual repulsion still exists between the particles of a like kind but the particles of different gases have no effect on one another.

autumn of 1801 to explain the uniform diffusion of gases through one another, the particles of one gas repelled only particles of a like kind and had no effect on those of any other type of gas (see Plate 3). The problems raised by such a view were considerable, and it was particularly difficult to reconcile the selective repulsion postulated by Dalton with the generally held opinion that the repulsive force between particles was due to the action of caloric. Dalton himself was conscious of this difficulty and he even examined the possibility that the inter-particle force might resemble magnetism, a suggestion that Newton himself had made, though for a quite different reason.† It may well be, therefore, that Dalton's belief that caloric was the cause of the repulsion between gas particles was weakened just at the time when his views on particle size were being formulated, for he was still defending his 'first' theory against its numerous detractors as late as November 1804, and possibly even in October 1805.‡ Towards the end of 1805, however, he abandoned the theory and in the amended or 'second' theory of mixed gases, which he proposed in its place, he was able to restore to caloric its customary role in the static theory of gas structure.§ According to the 'second' theory, diffusion was a necessary consequence of the characteristic sizes of the particles or 'atoms' of the various gases, since, as Dalton showed in 1808, if a large number of spherical particles of uniform diameter were placed in contact with others of a different diameter, equilibrium would be impossible and the particles would 'diffuse' uniformly through one another (see Plates 4 and 5).¶ It is not difficult to see how this new theory would have done rather more than simply

† Dalton, *Mem. Proc. Manch. lit. phil. Soc.* 2nd ser. **1** (1805), 436, and *New system*, vol. 1, part 1, p. 189. See also the notes for his Royal Institution lectures of 1810 reproduced in Roscoe and Harden, op. cit., p. 16. Newton's highly tentative suggestion had appeared in *Principia* (1687), pp. 302–3 (Book II, proposition xxiii).

‡ See his letter of 15 November 1804 in *Nicholson's Journal*, 2nd ser. **9** (1804), 269–75, and the comment in his paper of 4 October 1805 in *Mem. Proc. Manch. lit. phil. Soc.* 2nd ser. **1** (1805), 436.

§ On the date of Dalton's adoption of the 'second' theory see A. W. Thackray, *Isis* **57** (1966), 44–5. Here Dr Thackray argues strongly that Dalton adopted the theory in September 1805, although the passage in the paper of 4 October 1805 cited in the previous note suggests that Dalton may have retained his 'first' theory until a little later.

¶ Dalton, *New system*, vol. 1, part 1, pp. 188–93, where the earliest published description of the 'second' theory appeared.

renew Dalton's faith in caloric; it also gave the variation in particle size between different gases an importance that they scarcely had before.

So it was with this new and eminently satisfactory application of particle size very fresh in his mind, and with the difficulties that his first theory raised for caloric now removed, that Dalton returned to the subject of heat between March and June 1806. According to Roscoe and Harden, some fifty pages of notes (now unfortunately destroyed) attest to his interest at this time and show that he paid considerable attention to the relationship between the sizes of the particles of various gases and the quantities of caloric that they contained.† The earliest comment appeared in a short manuscript article 'On heat' dated 23 May 1806. Here Dalton made his first attempt to relate particle size to the characteristic attraction for heat, an attraction that he was careful to distinguish from chemical affinity. He summarized his views as follows:

> The virtual diameters of atoms of matter will therefore vary in like circumstances according to their attraction for heat; those with a strong attraction will collect a large and denser atmosphere around them, whilst those possessing a weaker attraction will have a less dense atmosphere, and consequently the virtual diameter, or that of the atom and its atmosphere together, will be less, though the atmospheres of both have precisely the same disposition to receive or to part with heat upon any change of temperature.‡

Such an assertion that the density of caloric was greatest in those atoms with the largest diameters§ clearly required further examination, and another comment in the notebook, quite separate from the above article but evidently written about the same time, shows Dalton in search of a more specific relationship. It reads: 'Query, does not the quantity of heat in a given volume of gas vary as the diameter of the atoms under like pressure?'‡ Unfortunately Dalton offered no explanation for this rejection, so that we can only guess at the considerations

---

† Roscoe and Harden, op. cit., p. 71.
‡ Roscoe and Harden, op. cit., p. 72.
§ A view similar to that adopted by Bryan Higgins in his *Experiments and observations relating to acetous acid* (1786), p. 315.

PLATE 4

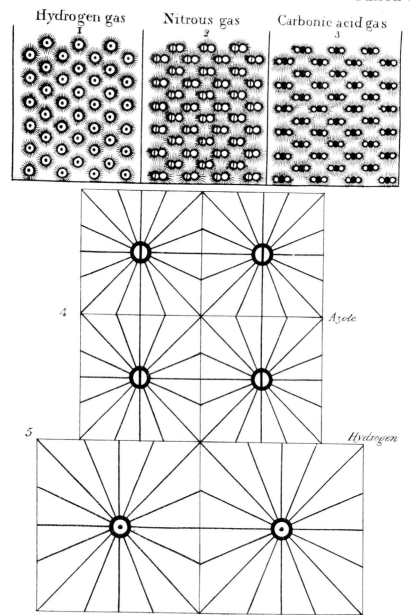

Atmospheres of caloric according to Dalton. From *A new system of chemical philosophy* (Manchester, 1810), vol. 1, part 2, facing p. 548. Diagrams 4 and 5 illustrate Dalton's second theory of mixed gases, in this case for a mixture of nitrogen and hydrogen. Dalton explained the diffusion, or 'intestine motion', of the gases by noting that the rays that represented the atmospheres of caloric surrounding the nitrogen atoms did not meet the corresponding rays for the hydrogen atoms; hence, he believed, equilibrium was impossible.

PLATE 5

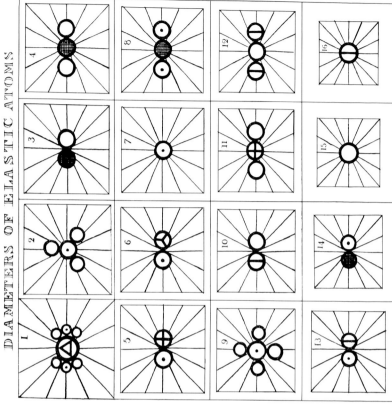

Diameters of the atoms of gases calculated by Dalton and drawn to scale. Gas 4 is carbon dioxide (carbonic acid); 7 is hydrogen; 9 is chlorine (oxymuriatic acid); 13 is ammonia; 15 is oxygen; 16 is nitrogen. A full key is on p. 548 of the *New system*, vol. 1, part 2.

which, by 1 June and on the very next page after this last note, had led him to try a quite different approach. He now wrote:

I. It seems clear that the attraction of any atoms for heat must be as their diameters in an elastic state, whatever be the law of attraction.

II. It seems also true that the law of attraction, whatever it may be, will not affect the relative quantities of heat around different kinds, in like circumstances.

III. Further, the absolute quantities of heat around different atoms must be as the cubes of the diameters of these atoms. Consequently all elastic fluids of given pressure and temperature contain the same heat in the same bulk.†

His revised opinion, given in paragraph III, that under similar conditions the density of caloric was the same in all gases, irrespective of the size of their particles, seems to have been preferable to that which he had held a few days earlier in just one respect, namely in that it predicted specific heats in rather better, if not good, agreement with Crawford's experimental values. It seems, therefore, that it was probably a concern to reconcile the consequences of his caloric theory of gases with the results of Crawford's experiments, or at least with Crawford's explanation of heats of reaction, which determined the course of Dalton's thought. It is impossible, of course, to be certain on this point, but his views on the state of caloric in gases appear to have led to no other consequences that were capable of independent verification. Admittedly the volume specific heats obtained by Crawford were by no means equal, as would have been expected if Dalton were correct, but still less did they increase with increasing particle size, the corollary of Dalton's earlier opinion.‡

Since Dalton was still working on the problem during the printing of the *New system*,§ it is not surprising that the definitive version of his views, which appeared in the first part

† Roscoe and Harden, op. cit., p. 73.
‡ The particle diameters which Dalton adopted about this time were probably very similar to those listed in his notebook on 14 September 1805 (Roscoe and Harden, op. cit., p. 65; see also Thackray, *Isis* 57 (1966), 44–5, on the dating). Slight modifications in the particle weights between November 1805 and August 1806 (see Thackray, op. cit., p. 55) would have had little effect on the calculated diameters.
§ Dalton, op. cit., vol. 1, part 1, p. 66.

of the book in 1808, showed yet another fundamental change. He now proposed three main possibilities:

either   equal weights of elastic fluids may have the same quantity of heat under like circumstances of temperature and pressure ...

or   equal bulks of elastic fluids may have the same quantity of heat with the same pressure and temperature ...

or   the quantity of heat belonging to the ultimate particles of all elastic fluids, must be the same under the same pressure and temperature.†

In his conviction that a relationship of this type existed at all he was evidently guided by his belief in the essential simplicity of gases,‡ but the justification for the particular choice that he made was slight. The first possibility was convincingly dismissed, however, by reference to any exothermic reaction between gases. For, with the conservation of caloric accepted as axiomatic by all calorists, it was obvious that the total quantity of heat in the compound resulting from such a reaction must be less than that present initially in the reactants, even though there had been no over-all change in weight. The second possibility was less easily disposed of, but Dalton, arguing of course as an Irvinist, showed that it predicted a specific heat for carbon that was quite incompatible with experiment. It is interesting to note that in his comments of 1 June 1806, discussed above, this evidence, although available, had apparently not constituted an objection. The third possibility, he now thought, was the acceptable one. In justification Dalton began by dismissing any idea that the numbers of particles in either equal weights or equal volumes of different gases were equal. At a time when the polyatomic elementary molecule was unknown Dalton's argument on this point was unanswerable, for how could it be supposed that the size of a (compound) atom of nitric oxide (or nitrous gas, as he called it) was the same as those of its constituent gases when it was observed that virtually no decrease in volume occurred in the reaction between equal volumes of nitrogen and oxygen? Was the total number of 'atoms' not halved in the process? The argument thereafter was less convincing. Dalton's arbitrary claim that the effect of an imaginary increase in the affinity of an atom for heat would be to reduce the size of its existing atmosphere, and hence also the

† Dalton, op. cit., vol. 1, part 1, pp. 68–70.   ‡ On which see pp. 77–8.

THE RIVAL CALORIC THEORIES

volume of the gas, rather than to attract more caloric, was quite unfounded and in clear disagreement, as we see, with the views that he had held in 1806. Nevertheless Dalton proceeded to the obvious conclusion, namely that equal quantities of caloric were contained not in equal volumes or equal weights of different gases but in equal numbers of particles and that therefore the specific heat of any gas must be inversely proportional to its 'atomic' weight.† A comparison between experiment and the specific heats predicted on this basis was possible for six gases, with the specific heat of water being taken as the basis for the calculations, but only in the case of air did Dalton find real confirmation. Not surprisingly, he now took good care to exclude air from the general criticism of the accuracy of Crawford's results that had appeared in a chapter of the *New system* printed off earlier!‡ His confident, if negative, claim that there was no 'established fact' that could disprove his treatment was only too true in 1808 but it reflected the standard of earlier experiments on the specific heats of gases rather than any intrinsic merit in his own work.

Within a few years this unsatisfactory state of affairs had changed. In particular, the French Institute's prize competition for 1812 established as correct a number of results which could not conceivably be reconciled with Dalton's and which, moreover, refuted the basic principles of his theory of heat. Irvinist doctrines and Crawford's explanations of animal heat and heats of reaction thereafter became untenable, at least for those less committed and more open-minded than Dalton, so that the possibility that anyone might take his theoretical treatment of specific heats seriously became even more remote after 1812 than it had been before.

The considerable attention that has been devoted in this chapter to the caloric theory fairly reflects its dominant position in the first decade of the nineteenth century. But, at least in England, there was dissent. Rumford's views, of course, were the best known and the most widely discussed, but Davy's celebrated experiment designed to demonstrate that melting could be effected simply by rubbing together two pieces of ice

---
† I naturally use 'atomic' here in the loose sense in which it would have been understood by Dalton, i.e. to signify molecular as well as truly atomic weights.
‡ Dalton, op. cit., vol. 1, part 1, p. 67. The earlier comment appeared on p. 63.

in a vacuum was also familiar to some of his contemporaries, despite the obscurity of Thomas Beddoes's journal in which the description was published early in 1799.† As Professor E. N. da C. Andrade has pointed out, the fact that melting occurred where the pieces of ice were in contact with each other in no way disproved the caloric theory nor did it establish that heat was motion, as Davy claimed.‡ But the serious deficiencies of the experiment were not apparent at the time and even P. G. Tait later described Davy's evidence as 'completely decisive',§ although in his opinion Rumford's experiments left something to be desired in terms of rigour. In 1801 Davy acquired a more prominent forum for the exposition of his kinetic theory, when he began lecturing at the Royal Institution in London, and he appears to have taken full advantage of his position.¶

Thomas Young was another critic of caloric and a supporter of the vibrational theory who lectured at the Royal Institution about this time, in his capacity as professor of natural philosophy between 1801 and 1803. In the syllabus for his course in 1802 and again in the revised version of the lectures, which he published five years later, Young emphasized that, although the material theory was especially successful in accounting for heats of reaction, yet all the principal thermal phenomena could be explained equally well on the vibrational theory and, in the important case of friction, very considerably better.‖ Young's views on the undulatory nature of light lent additional support,

† Davy, 'Essay on heat, light, and the combinations of light', in *Works*, vol. 2, pp. 5–86, where the ice-rubbing experiment appears on pp. 11–12. The paper appeared originally, in 1799, in *Contributions to physical and medical knowledge, principally from the West of England, collected by Thomas Beddoes, M.D.* For evidence that Davy's work was known see Henry, *Mem. Proc. Manch. lit. phil. Soc.* **5**, part 2 (1802), 603, and T. Young, *A course of lectures on natural philosophy* (1807), vol. 1, p. 653. Joseph Priestley probably knew of the work also; see his letter to Davy, 31 October 1801, in R. E. Schofield (ed.), *A scientific autobiography of Joseph Priestley (1733–1804)* (Cambridge, Mass., 1966), p. 313. I am grateful to Dr D. M. Knight for drawing my attention to the Priestley reference.   ‡ Andrade, *Nature, Lond.* **135** (1935), 359–60.
§ Tait, *Sketch of thermodynamics* (2nd edn, Edinburgh, 1877), pp. 5–9.
¶ See, for example, his *Syllabus of a course of lectures on chemistry* (1802), pp. 53–4, where he outlined and commented favourably on the view that radiant heat was an undulatory motion in an 'elastic ethereal medium' and that the motion was caused by the 'vibratory motions' of the particles of ordinary matter.
‖ Young, *A syllabus of a course of lectures on natural and experimental philosophy* (1802), pp. 53–4, and *Lectures on natural philosophy*, vol. 1, pp. 653–7.

since they suggested a ready explanation of radiant heat in terms of vibrations in the all-pervading subtle elastic medium that were similar to but 'larger and stronger than those of light'.† One of the most interesting features of Young's theory of heat is the lingering influence of Irvine and Crawford. Young described their work in some detail in his lectures and gave a particularly sympathetic account of their method for determining the absolute zero.‡ The disagreement between the various results obtained by this method inevitably gave rise to serious reservations and, while Young felt that there could be no objection to accounting for the heat absorbed or emitted in a change of state or of volume by reference to changes in capacity, he maintained that the heats evolved in friction and combustion defied explanation in this way. Nevertheless he accepted a value of about $-1400\ °F$ for the absolute zero as 'a tolerable approximation', and proceeded to apply it in a novel but uncritical treatment of the adiabatic effect.§ In this he adopted the explanation of the heat produced by air entering a (partial) vacuum which both Cleghorn and Leslie had already given. He supposed that, if the density of the small amount of air remaining in the receiver before the tap was opened was $1/x$ of that of atmospheric air, then its capacity would be diminished by a factor of x^n when it was recompressed to its original density, and a quantity of heat proportional to $1450\{(1-1/x^n)\}$ would therefore be expelled, where $50\ °F$ (i.e. $1450\ °F$ above the absolute zero) was taken as the initial temperature. If all this heat was now absorbed by the compressed gas alone, its temperature would increase by $1450\{1-(1/x^n)\}x^n\ °F$, but since in fact the heat would be shared by x times as much air, the rise in temperature would be only

$$1450[x^{n-1}-x^{-1}]\ °F. \qquad (1)$$

He went on to show that if n was assumed to be small, this expression had a maximum value for $x = 2\cdot7$, so that the greatest rise in temperature would be obtained by opening a receiver evacuated to about two-fifths of atmospheric pressure. Although he recognized the unreliability of Dalton's data on the subject, Young showed that a temperature change of $50\ °F$, which was

† Young, *Syllabus*, p. 149, and *Lectures on natural philosophy*, vol. 1, p. 654.
‡ Young, *Lectures on natural philosophy*, vol. 1, pp. 650–2.
§ Young, op. cit., vol. 1, pp. 632 and 652, and vol. 2, pp. 408–9.

the maximum figure observed by Dalton in his evacuated-receiver experiments, yielded a value of (1/11·2) for the constant n—compared with the figure of 1/8 which was required to reconcile the experimental and theoretical velocities of sound. With $n = 1/8$, expression (1) gave 131·2 °F for the temperature change produced in doubling or halving the density of air, without heat loss, and 1 °F for a compression of 1/180 (or 1 °C for a compression of 1/100). The general agreement between this and the similar figures derived by Biot and Poisson† is noteworthy but, as has already been mentioned, it was Poisson's figure that became standard, and Young's treatment passed unnoticed.

Davy's objections to fluid theories were to be far more comprehensive than those of Young, who adopted them in his accounts of both electricity and magnetism.‡ Yet Davy's first attack, which accompanied his description of the ice-rubbing experiment in 1799, was directed specifically at caloric.§ He claimed that to attribute the mutual repulsion between particles of ponderable matter to caloric was merely to postpone the difficulty and to leave the repulsion between the particles of caloric themselves unexplained. Some years later his attitude had hardened still further, and in some notes for a lecture that he delivered in 1809 we find the following stricture on all fluid theories:

Vulgar idea—like that of the peasant, every thing done by a spring; so every thing must be done by a fluid. The ether was the ancient fluid; then there was a phlogistic fluid: we have had the magnetic fluid, the vitreous fluid, the resinous fluid; and within the last few years there has been a fluid of sounds; and, in a book which I lately received from France, published by M. Azaïs,¶ all the phenomena of nature are explained by a gravic fluid.||

† See p. 86.
‡ Young, op. cit., vol. 1, pp. 658–95. Young favoured the one-fluid theories of both electricity and magnetism.
§ Davy, *Works*, vol. 2, pp. 20–3.
¶ This was the French philosopher Pierre Hyacinthe Azaïs. Davy was probably referring to Azaïs's *Essai sur le monde* (1806). On pp. 67–141 of this book there is a lengthy exposition of Azaïs's theory of gravitation based on the existence of an all-pervading *substance compressive*. Azaïs does not appear to have been taken seriously at the Institute. On 15 September 1806 he was asked to terminate the reading of a paper outlining his theories after only a quarter of an hour. He later published the paper as *Mémoire sur le mouvement moléculaire et sur la chaleur* (1806) and in it complained bitterly of his treatment (pp. 39–40).
|| Davy, op. cit., vol. 8, p. 348.

By this time Davy's confidence in his attitude to fluids would already have been strengthened by his realization that chemical heat might be electrical in origin,† and in a lecture in 1812 he again referred specifically to caloric:

> I do not think we have at present any means of deciding upon this question of the nature of heat; and its effects may be studied—and it may be employed as an instrument of experiment, without the necessity of adopting any hypothetical views respecting its cause. It is my intention merely to give a caution with respect to the adoption of the chemical solution—which is by far the most generally received. Indeed the matter of heat or caloric is sometimes talked of with the same confidence as water, or any common ponderable bodies. . . . The truly philosophical inquirer into nature will not consider it as a disgrace, that he is unable to explain every thing; he will wait, and labour with hope, tempered by humility, for the progress of discovery—and he will feel that truth is more promoted by the minute and accurate examination of a few objects, than by any premature attempts at grand and universal theories.‡

In fact, at this date Davy was probably not so uncommitted on the nature of heat as he seems to be in the above passage. In his *Elements of chemical philosophy*, which also appeared in 1812, he cited the phenomena of friction and percussion, as well as certain unspecified exothermic reactions in which no decrease in capacity occurred, as evidence against the caloric theory, concluding: 'The immediate cause of the phenomena of heat then is motion, and the laws of its communication are precisely the same as the laws of the communication of motion.'§

In this same discussion Davy incidentally provided revealing evidence for the role of adiabatic heating in the dispute over the nature of heat in his reference to the effect as one that was particularly well explained on the material theory.¶

---

† See p. 242. In 1799 the explanation of heats of reaction had appeared to Davy 'by far the most difficult part of the philosophy of heat' (*Works*, vol. 2, p. 20). This comment by one who was seeking to break with the caloric theory serves to emphasize that the great strength of the theory lay in its ability to account for chemical heat; it had no serious rival in this respect until the advent of the electrochemical theory.

‡ Davy, op. cit., vol. 8, p. 350.

§ Davy, *Elements of chemical philosophy* (1812), pp. 94–5.

¶ Davy, op. cit., p. 93. Likewise, in 1790, in his *Essais de physique*, pp. 20–1, Pictet had interpreted his experiments with an evacuated receiver as strong evidence in favour of the materiality of fire. Cf. also Erasmus Darwin's *The botanic garden* (1791), 'The economy of vegetation', additional notes, pp. 14–15.

Nevertheless, he could offer a physical picture for this and, indeed, for all other thermal phenomena by supposing, quite gratuitously, that the particles of gases and liquids possessed, in addition to the familiar vibrational motion supposed to occur in solids, a rotational motion about their own axes, this motion being more rapid in gases.†

Temperature [he explained] may be conceived to depend upon the velocities of the vibrations; increase of capacity on the motion being performed in greater space; and the diminution of temperature during the conversion of solids into fluids or gasses, may be explained on the idea of the loss of vibratory motion, in consequence of the revolution of particles around their axes, at the moment when the body becomes fluid or aeriform, or from the loss of rapidity of vibration, in consequence of the motion of the particles through greater space.‡

Unfounded speculation of this type was clearly vulnerable but, possibly as a result of the security of the position of the caloric theory, contemporary comments, whether favourable or unfavourable, are hard to find. The most distinguished critic was Berzelius, but his claim, made in a private communication, that Davy had not explained how the motion of compression was converted to rotational motion in the gas particles§ seems to miss the point, since it was an increase in the rapidity of vibrational, and not of rotational, motion that was manifested as a heating effect, even in gases. Davy's views nevertheless had little to recommend them to chemists of the time and it was only when they were revived by James Joule over thirty years later¶ that they exerted any appreciable influence.

Davy's attacks on caloric both in 1799 and 1812 contained much that could be convincing only to someone who agreed with the basic Irvinist principles. Indeed, Davy even accepted that the specific heat of a gas decreased with compression, a view with which his theory of heat was in good agreement.‖ The

† Davy, op. cit., pp. 95–8. ‡ Davy, op. cit., p. 96.
§ See Berzelius's undated letter to Davy, in *Jac. Berzelius Bref*, ed. H. G. Söderbaum (6 vols., Uppsala, 1912–32), vol. 1, part 2, p. 41. The hypothesis of rotational motion also raised difficulties for J. B. Emmett in *Ann. Phil.* 9 (1817), 423. ¶ Joule, *Phil. Mag.*, 3rd ser. 26 (1845), 381–2.
‖ Davy, op. cit., p. 90. However, in 1821 Davy denied any 'necessary connexion' between specific heat and heat content; see his letter to John Herapath, 6 March 1821, quoted by S. G. Brush in *Notes Rec. R. Soc. Lond.* 18 (1963), 171.

Irvinist influence was still more marked on another non-calorist, the Westphalian chemist Friedrich Christian Accum, who had lived in London since 1793 and who, in 1803, had lectured at the Surrey Institution. Writing in 1805 in a revised edition of William Nicholson's *Introduction to natural philosophy*, Accum showed no preference for either the material or vibrational theory but he was none the less satisfied that the phenomena of latent heat, chemical and animal heats, friction, and percussion were adequately explained by changes in capacity.†

So, throughout the first decade of the nineteenth century, the attitudes of British scientists to the theory of heat continued to be dominated by the legacy of Black, Irvine, and Crawford. Neither calorists nor vibrationalists strayed far from the basic principles laid down by these men and this at least gives a certain unity to the work of the period in Britain.

Theories of heat in France

In France theories of heat lacked the unifying influence that they had in Britain and the systems adopted there tended to be more varied and also more complex. With caloric being called upon to explain such a wide diversity of phenomena as radiation, temperature, specific heat, expansion, change of state, and the repulsive forces between gas particles, some degree of complexity was to be expected, but the multiplicity of forms that it was thought caloric could adopt was truly formidable about 1800. The easiest of solutions was, after all, simply to postulate as many sorts of caloric, each of them with its own characteristic properties, as there were phenomena to explain, and although no one went quite so far as this, the current terminology reflects how attractive this type of solution was. Thus we find such terms as combined, latent, radiant, sensible, and free heat (or caloric) all in common use, to say nothing of the quantities absolute and specific heat. However, it is important to note that in all versions of caloric theory only one fluid was concerned and its properties were merely modified by its state of combination with ordinary matter. The situation was in this respect quite different from that prevailing in the two-fluid theories of electricity and magnetism.

† Nicholson, *Introduction to natural philosophy* (5th edn, 1805), vol. 2, pp. 114 and 122, in the section on chemistry, which Accum rewrote.

The arbitrary way in which characteristic properties had to be attributed to each of the forms of caloric was an obvious weakness, and it was very properly the concern of early nineteenth-century writers to try to justify this procedure. One approach is seen in the attempts that were made to draw a close analogy between the behaviour of caloric and that of ponderable matter, and it was in this way that the chemist and doctor Joseph Marie Socquet tackled the problem. Although French by birth, Socquet never made much impact on Parisian science and, after serving as a doctor in the Sardinian and French armies and from 1809 to 1818 as the first professor of industrial chemistry in the Faculté des Sciences at Lyons, he eventually returned to Turin, where he had been educated at the university.† His lengthy *Essai sur le calorique*, published in Paris in 1801, does not therefore belong to a really major tradition in French science and it is significant that he acknowledged as his mentor on the subject, not a leading Parisian scientist, but Joseph Montgolfier.‡ However, the principles that he adopted in treating caloric were not untypical. He wrote:

> The action of caloric on bodies produces different effects which must be carefully distinguished and which we shall try to account for by using only the indisputable principles of physics and chemistry, treating caloric as a real substance and not as a mere modification of matter and applying to it the laws of affinity, pressure, etc. that we apply to other substances.§

In accordance with the resolve expressed in this passage, Socquet distinguished four states in which caloric could exist: as caloric of composition, of capacity, of temperature, or as radiant caloric.¶ Caloric of composition was chemically combined with the ultimate particles of matter, forming one essential constituent of them. A molecule of water, for example, would contain a certain quantity of caloric of composition that would remain unchanged so long as the molecule continued to exist. Caloric of capacity, on the other hand, was retained in a

---

† For biographical details see *Biographie universelle ancienne et moderne* (*Michaud*) (2nd edn, 45 vols., Paris and Leipzig, 1842[–65]), vol. 29, pp. 514–15. It is interesting to note that Socquet was Berthollet's nomination for the chair at Lyons.

‡ Socquet, *Essai sur le calorique* (1801), p. viii n.

§ Socquet, op. cit., p. 6.   ¶ Socquet, op. cit., pp. 9–16 and 99.

body by a purely physical force, 'affinity of adhesion', the mechanism being exactly that which bound dissolved gases to their solvent. Particles of caloric, like those of a dissolved gas in water, were supposed to penetrate any substance and to adhere to the surface of the molecules composing it with a force inversely proportional† to the square of the distance from the surface, so that capacity was determined not simply by the innate affinity of the molecules for caloric but also by the extent of the surface area of the molecules to which caloric could adhere.‡ Since he adopted an inverse square law, it is not improbable that Socquet considered the force of adhesion to be gravitational and he certainly believed that caloric had weight.§ Finally, the pores that necessarily existed between the molecules were filled by caloric of temperature. Although this was not subject to the force of adhesion, it was still not in a state of perfect freedom, being retained in a body by external pressure, which counteracted the tendency of all caloric to expand under the influence of its natural elasticity and hence to escape as radiant caloric. Caloric of temperature, unlike all other types of caloric, could be detected by a thermometer and in this important respect it differed from radiant caloric, which was supposed to move so rapidly that it was detectable only if it was slowed down by the material of the thermometer and so essentially ceased to be radiant caloric.

Socquet clearly fulfilled his aim to the extent that the attributes he bestowed on caloric were those that he supposed to be established in physics and chemistry, but the weakness of his system is only too apparent. The properties that he supposed the different types of caloric to possess were assigned in a patently arbitrary fashion and, furthermore, some of the principles that he classed as 'evident', notably the mechanism by which gases were retained in a liquid, were by no means beyond dispute.

Yet it was some time before the analogy between caloric and ponderable matter was seen to be a fruitless one, and the most

† Socquet appears to have made a slip throughout the *Essai*, for he constantly refers to this force as being *directly* proportional to the square of the distance, when it is clear from the context that this is not his meaning.

‡ Socquet, op. cit., pp. 29 and 43–4. Cf. the similar view put by Torbern Bergman in his *Opuscula physica et chemica*, (1783) vol. 3, p. 423.

§ Socquet, op. cit., p. 108.

successful attempt to establish it was still to be made, by Claude Louis Berthollet in his *Essai de statique chimique* of 1803. Although the underlying principle was similar to Socquet's, his treatment was far more convincing. Indeed, the success that he considered he had had in justifying the analogy might even be seen as additional evidence for the materiality of heat, as he claimed in this passage:

> If one hesitates to regard this similarity between the properties of caloric and those of a substance entering into chemical combination as a conclusive proof of its [i.e. caloric's] materiality, one cannot but agree that the hypothesis that it exists presents no difficulties and has the advantage of involving only general and consistent principles in the explanation of phenomena.†

The strength of Berthollet's approach was that the phenomena with which he compared the behaviour of caloric formed part of the single coherent system of chemical affinities that he had so recently devised and that he was now expounding fully in the *Essai*. Of course the truth of the system itself was by no means proved, but its obvious thoroughness and consistency, to say nothing of Berthollet's personal authority, ensured that it could not be lightly dismissed. The behaviour of acids provided the model for caloric, so that the combination of caloric with ponderable matter became no more (and no less) mysterious than familiar chemical reactions. Just as the quantities of a given acid that were required to produce neutralization varied according to the base with which the acid combined, so different quantities of caloric were required to produce the same effect, for example a certain rise in temperature, in different bodies, the quantities of caloric (or acid) in the two cases being proportional to the mass of the body (or base) and to its innate affinity for caloric (or for the acid).‡ This approach imposed one salutary restriction, since Berthollet could now abandon any attempt to explain the precise mechanism that held caloric in a body, and he was therefore able to reduce the range of his problems. In this he was breaking with a minor research tradition that went back well into the eighteenth century and of which Socquet's unfruitful physical picture of the various ways in

† Berthollet, *Essai de statique chimique* (1803), vol. 1, p. 180.
‡ Berthollet, op. cit., vol. 1, pp. 176–7.

which caloric was held in bodies was only one highly developed example. It was a tradition in which Bergman, De Luc, and even Lavoisier had participated by their attempts to devise physical explanations for the difference between the specific heats of various substances in terms of the supposed structure and properties of their ultimate particles.†

Berthollet's analogy between the behaviour of caloric and that of acids allowed certain other simplifications to be made also. Above all, it was no longer necessary to ascribe special properties to each of the forms of caloric, and a single affinity now sufficed, as it did, of course, for acids. Some of the consequences of this simplification are seen particularly clearly in his explanations of latent heat and of temperature changes in general, which in his view were simply the result of opposition to the naturally expansive force of caloric in ponderable matter.‡ In the case of solids and liquids it was the forces of cohesion between the molecules which resisted this expansive force and so caused heating when caloric was added. In the case of gases, in which the intermolecular forces of attraction had been overcome, the resistance was provided by external pressure. It followed that, as soon as the resistance was removed, caloric would enter a solid or liquid without affecting its temperature, and this, Berthollet supposed, was what occurred during a change of state. When the resistance was increased, on the other hand, as in the compression of a gas or in the percussion of a solid, a rise in temperature resulted, even without the addition of more caloric.

It would be interesting to know to what extent Laplace was associated with the development of Berthollet's views on caloric, for he and Berthollet were already good friends. At all events, Laplace contributed two important notes on caloric to the *Essai*. In these he showed a particular interest in adiabatic heating, as was only to be expected so soon after his success with the velocity of sound problem, but the carelessness of his

† See Bergman, op. cit., vol. 3, p. 423; De Luc, *Idees sur la météorologie* (London, 1786), vol. 1, p. 141; Lavoisier, *Traité élémentaire de chimie* (1789), vol. 1, pp. 18–21. The tradition did not die out completely with Berthollet's treatment, for evidence of it can still be found in the work of Joseph Mollet (*J. Phys.* 99 (1820), 124–8), and in William Prout's Bridgewater Treatise (the eighth), *Chemistry meteorology and the function of digestion* (1834), pp. 62–3.

‡ Berthollet, op. cit., vol. 1, pp. 155–6.

argument in the first, hastily written note† suggests that he had not yet given serious thought to his theory of heat. He argued (quite mistakenly) that if gas in a vessel fitted with a manometer was compressed to half its initial volume, then the number of gas molecules pressing on the surface of the mercury in the manometer would be doubled. Since Laplace supposed moreover that the compression in question took place isothermally, it was known that the pressure and hence the total force on the mercury surface were also doubled in the process, so that the repulsive force exerted on the mercury by *each molecule* must remain the same. Such a conclusion, implying as it did that the force between two gas particles at any given temperature was independent of the distance between them, was of course quite irreconcilable with Newton's view of gas structure. Disagreement of this sort, which was based purely on a careless slip, is surprising in such a good Newtonian as Laplace, for it is scarcely conceivable that he had not read the *Principia* on this point.

Laplace's conclusion had at least one important consequence for the caloric theory, since he assumed, apparently quite arbitrarily, that the magnitude of the inter-particle force depended solely on the quantity of what he termed free heat (*chaleur libre*) in a given volume of gas, so that in any isothermal compression (i.e. one in which, as he now thought, the interparticle force did not change) the quantity of free heat expelled would be proportional to the change in volume. It followed from Laplace's theory that temperature, as the only factor that could alter the force between gas particles, was wholly determined by the density of free heat.

Laplace recognized the fundamental error of his argument so quickly that there was still time to insert a revised note before the printing of volume one of the *Essai* had been completed.‡ He now maintained that the compression of a gas to, say, one-eighth of its original volume would increase the number of molecules pressing on a given surface by a factor not of 8, as he had previously supposed, but of 2. The argument he used on this point was essentially Newton's and it led to the familiar law of inverse proportionality relating the force and the distance between gas molecules. Laplace's law of force was thus in no

† Berthollet, op. cit., vol. 1, pp. 245–7.
‡ Berthollet, op. cit., vol. 1, pp. 522–3.

THE RIVAL CALORIC THEORIES

way original but, as Dalton also had realized, it now had to be integrated with caloric theory. The intimate relationship between temperature, the repulsive force between molecules, and the density of free heat, which Laplace had established in his earlier note, could no longer be maintained. There was now no mention of free heat and he supposed instead that the force was determined simply by the density of 'heat' (*chaleur*). If the Newtonian law of force was to be accounted for on this basis, it followed that the density of 'heat' must increase in any isothermal compression. When a gas was compressed to half its original volume, for example, *less* than half of its heat must be expelled. Although Laplace did not mention it, his argument had shown incidentally that the temperature of a body was not determined by the density of heat in it.

Laplace was apparently content with this rather vague conclusion, and it was left, unexpectedly perhaps, for Joseph Mollet of Lyons to take the argument a stage further in the paper that he read to the Lyons Academy in March 1804.† To do this he had to make one additional supposition, namely that the repulsive force between two particles not only depended on, but was also proportional to, the density of heat in the gas. He could then argue that since this force was doubled when a gas was compressed to one-eighth of its original volume, then the density of heat likewise must be doubled in the process. Three-quarters of the free heat present in the gas originally must therefore have been expelled.

Mollet's paper was not published until 1811, and even then it attracted little attention. Yet his argument concerning the expulsion of heat from a gas undergoing isothermal compression was used in 1806, though without acknowledgement and without any indication of its source, by the Abbé René Just Haüy, then professor of mineralogy at the Muséum d'Histoire Naturelle in Paris.‡ Whether or not Haüy had seen Mollet's work by this date is unfortunately not known, but his views on

† Mollet, *Mémoire sur deux faits nouveaux*, pp. 20–1. The delay in the publication of this paper until 1811 naturally leaves some doubt concerning the precise contents of the paper read in March 1804 (which has not survived), but it is assumed that no changes were made before publication. In the manuscript of his paper of 14 December 1802 (see p. 90 n. †) Mollet simply attributed the rise in temperature accompanying compression to an increase in the elasticity of the *feu libre* or *matière de la chaleur* of the gas. He gave no further details.

‡ Haüy, *Traité élémentaire de physique* (2nd edn, 1806), vol. 1, p. 138.

caloric, as expounded in three successive editions of his *Traité élémentaire de physique* (1803, 1806, and 1821), show such originality that a quite independent repetition of Mollet's argument by Haüy cannot seem improbable. Haüy was one of the most gifted writers of his day on physics, as well as on mineralogy and crystallography, and the section on caloric in his *Traité*, especially in the second edition, certainly deserves far more attention than would normally be accorded to a work intended primarily for the use of teachers in the *lycées*.†

Berthollet's chemical analogies and the search for mechanisms had no place in Haüy's system and throughout he maintained a strict agnosticism in his attitude to the nature of heat,‡ as also to the fluids of electricity and magnetism, emphasizing that he was using the language of materiality merely for its convenience. He tried to preserve a similar impartiality in his treatment of the rival versions of caloric theory which, for a Frenchman, were naturally seen as those of Irvine on the one hand and of Lavoisier and Laplace on the other. In the first edition, in 1803, he succeeded in presenting the views of both sides on such topics as change of state and adiabatic heating without committing himself,§ but three years later, when the second edition appeared, there had been a slight though significant change. He now went so far as to admit that the version adopted by Lavoisier and Laplace in 1783 presented a clearer picture of liquefaction and vaporization. Moreover, he showed in detail how their clear distinction between the caloric that brought about expansion and that which caused temperature changes could be used to account for adiabatic phenomena.¶ It seems clear that the latter application was not entirely original. In 1802, for example, Biot had mentioned how in the compression of air 'a part of its latent heat . . . changes its state to become sensible heat',‖ and

† On the aims of the work, which was commissioned by Napoleon himself, see Haüy, op. cit., vol. 1, pp. xxviii–xxx.
‡ Haüy, *Traité élémentaire de physique* (1st edn, 1803), vol. 1, p. 90; (2nd edn, 1806), vol. 1, p. 80; (3rd edn, 1821), vol. 1, p. 80.
§ Haüy, op. cit. (1st edn), vol. 1, pp. 109–10 and 116–18.
¶ Haüy, op. cit. (2nd edn), vol. 1, pp. 128 and 135–41. The view adopted by Haüy, which soon became common, was that a gas in expanding needed to absorb latent caloric. It gained this caloric either by the conversion of its own free caloric to the latent state (in which case cooling occurred) or by the absorption of caloric from the surroundings.
‖ Biot, *J. Phys.* **55** (1802), 176.

in the first edition of the *Traité* Haüy had referred in passing to certain unnamed physicists who maintained that some heat entered into combination with any gas undergoing expansion.† Nevertheless, his account of 1806 must stand as the first clear exposition of a principle that was to become quite crucial in Laplace's subsequent development of the analytical caloric theory. Also, as will be pointed out in Chapter 5, it led him to the soundest understanding that had yet been obtained of the distinction between c_p and c_v, the specific heats at constant pressure and constant volume.

In their conviction that, for a given temperature, the density of caloric was greatest in a gas at high pressure Laplace, Mollet, and Haüy were contradicting a fundamental point in Dalton's Irvinist theory,‡ but it was left for Gay-Lussac to deliver the first explicit criticism of Dalton's opinion in a paper on the adiabatic effect that he read to the Institute in September 1806.§ The motive for Gay-Lussac's interest was the vexing problem of specific heats. About two years earlier, in the course of their work on the proportions in which hydrogen and oxygen combined to form water, Gay-Lussac and his friend Alexander von Humboldt had observed that a complete union of the two gases did not occur if a large excess of oxygen was present.¶ When the oxygen in excess of that which actually entered into combination was replaced by nitrogen, the reaction ceased at very nearly the same point. Assuming that the rapid absorption of the heat of reaction by the excess gas had brought the temperature below that required for ignition, they concluded that oxygen and nitrogen, since they were equally effective in this, had identical volume specific heats. It was a natural extension, although one hardly confirmed by some further experiments with excess hydrogen and carbon dioxide, to suppose that all gases might be similar in this respect and, after returning from a year's tour of Italy, Switzerland, and Germany with Humboldt, Gay-Lussac set about examining the possibility at Arcueil, with the approval of Berthollet and Laplace, who appear to have shown considerable interest.||

† Haüy, op. cit. (1st edn), vol. 1, pp. 116–17. ‡ See p. 88.
§ Gay-Lussac, 'Premier essai pour déterminer les variations de température qu'éprouvent les gaz en changeant de densité et considérations sur leur capacité pour le calorique', *Mém. Phys. Chim. Soc. Arcueil* **1** (1807), 180–203.
¶ Humboldt and Gay-Lussac, *J. Phys.* **60** (1805), 136–45.
|| Gay-Lussac, op. cit., p. 182.

While paying due acknowledgement to Dalton's experimental work,† Gay-Lussac proceeded to demolish the English chemist's theoretical analysis of the adiabatic effect. He did not deny that the cooling that accompanied the escape of compressed air from a vessel indicated that heat would have to be absorbed in order to restore the remaining air to its original temperature. But the conclusion that there must have been an overall increase in the density of caloric in the vessel once the initial temperature had been regained was quite unfounded, unless it could be shown that the amount of caloric absorbed exceeded that carried off by the escaping air. Although this question could not be resolved simply by direct measurement, Gay-Lussac set out to demonstrate that a vacuum, far from containing more caloric than an equal volume of gas, in fact contained none at all. He did this by enclosing one bulb of a sensitive air thermometer in the Torricellian vacuum above the column of a mercury barometer and then showing that no temperature changes occurred when the volume of the vacuum was changed by tipping the tube.‡ Leslie's analysis of the evacuated receiver experiments was likewise dismissed, for Gay-Lussac observed that the temperature changes that followed the opening of the tap increased, rather than decreased, if the density of the small amount of air present initially in the receiver was reduced.

Because of the significance that it holds for readers familiar with the principle of energy conservation, one of the experiments described in Gay-Lussac's paper has attracted particular attention. In the experiment dry air at atmospheric pressure was contained in one of two 12-litre spherical vessels which were connected by a narrow tube fitted with a tap. The air was then suddenly allowed to escape into the other vessel, which had been evacuated, and any temperature changes were observed on alcohol thermometers in each of the vessels. This and other experiments conducted with air at initial pressures of one-half and one-quarter of an atmosphere and with different gases left Gay-Lussac in no doubt that the rise in temperature in the vessel that was initially evacuated was exactly equal to the fall observed in the other and that the temperature changes occurring with any particular gas were proportional to its initial pressure.

† Gay-Lussac, op. cit., pp. 182–3.   ‡ Gay-Lussac, op. cit., p. 191.

But it was from his observation that the changes in temperature were greatest for gases of low density that Gay-Lussac drew his most important and ill-founded conclusions. His reasoning was as follows:

... if we observe that all gases expand equally when heated and also that in our experiments the gases which absorbed the greatest quantities of heat when expanding by a given amount were those that had the smallest density (*pesanteur spécifique*), we shall draw the important conclusion that the capacity for caloric of equal volumes of different gases is greatest for those with the smallest density.†

The assumption made here that the large temperature variations observed with gases of low density indicated that they required the greatest quantities of caloric in order to expand was, of course, quite unjustified. However, once the assumption was made, the rest followed easily. Gay-Lussac's reference to the equality of the expansion coefficients suggests that he was assuming, with Haüy and probably Laplace,‡ that the amount of latent heat that was necessary simply to bring about expansion was proportional to that required to effect the same degree of expansion accompanied by heating at constant pressure. It followed that the quantity of heat absorbed as latent heat for a given increase in volume was proportional to the volume heat capacity of the gas under examination, and Gay-Lussac duly concluded that the volume specific heats were greatest for gases of low density, i.e. for those gases in which the cooling effect was largest. One other consequence, derived by the same reasoning, was that the volume specific heat of any gas decreased as it was rarefied. In fact, if my reconstruction of Gay-Lussac's argument is correct, it must be assumed that he supposed the volume specific heat of any one gas to be proportional to its density. This, however, was not the case for variations between different gases. On this point Gay-Lussac concluded: 'The capacities for caloric of equal volumes of different gases are greatest for those gases which have the smallest density.'§ It is true that according to François Delaroche and Jacques Étienne Bérard, the winners of the Institute's

† Gay-Lussac, op. cit., pp. 200–1.
‡ See p. 160.
§ Gay-Lussac, op. cit., p. 202.

prize competition of 1812, Gay-Lussac thought that the volume specific heats of all gases under similar conditions of temperature and pressure were inversely proportional to their densities.† Now Delaroche and Bérard, both of whom were familiar with the members of the Arcueil circle, may conceivably have learnt something of Gay-Lussac's opinions privately, but there is certainly nothing in the 1806 paper to justify their assertion. Moreover, both in the 1806 paper and again in 1812 Gay-Lussac explicitly stated that the exact form of the relationship between specific heat and density was unknown to him.‡

Even Gay-Lussac must have seen his investigation as no more than a modest contribution to a familiar problem. It was work that was hardly worthy of him, and the modern reader is struck not only by the unconvincing nature of the argument he used but also by his failure to remark on the fact, which now seems so significant, that no overall absorption of heat takes place when a gas expands into a vacuum. It was five years before Gay-Lussac returned to the study of the specific heats of gases, but when he described his recent experiments on the subject to the Institute on 20 January 1812,§ he showed as much interest as ever. The simple method of mixtures that he had used was one of striking crudity, for all the confidence that he placed in it. Equal volumes of two gases, one at a high and the other at a low temperature, were allowed to mingle and their final temperature was taken. No serious attempt appears to have been made to eliminate heat losses and it is not surprising that the results were inaccurate. Detecting no difference between the capacities, Gay-Lussac concluded that the volume specific heats not only of the gases examined but possibly of all gases were identical. The result flatly contradicted his earlier conclusion, yet such was his confidence in its correctness that he now (wisely) chose to reject his theoretical interpretation of the 1806 experiments.

However, it was not long before his opinion changed yet again. In July 1812 he inserted a short article in the *Annales de chimie* in which, as a result of further experiments with a larger version of his method of mixtures, he accepted that the volume

† Delaroche and Bérard, *Annls Chim.* **85** (1813), 79.
‡ Gay-Lussac, op. cit., p. 201, and *Annls Chim.* **81** (1812), 103.
§ Gay-Lussac, 'Extrait d'un mémoire sur la capacité des gaz pour le calorique', *Annls Chim.* **81** (1812), 98–108.

specific heats of gases were far from equal.† But the subject now held less interest for him since he had learned that two of his 'friends' were tackling it by what he acknowledged to be a far superior method. There can be little doubt that the 'friends' referred to were Delaroche and Bérard rather than their rivals in the Institute's 1812 competition, Clément and Desormes, although the latter pair did claim friendship with Gay-Lussac.‡ The reason for this conclusion lies, above all, in the fact that Delaroche and Bérard conducted their experiments at Arcueil,§ where Gay-Lussac was a prominent member of Berthollet's circle. Moreover, Delaroche and Bérard had submitted a preliminary report on their work to the Institute on 3 February 1812,¶ of which Gay-Lussac, both as a friend and as one of the judges for the competition, would presumably have known.

Although the competition thus put an end to Gay-Lussac's immediate interest in the problem, as well as to his intention of presenting a much fuller treatment,‖ his article does throw light on the sort of questions that it was thought necessary to answer in 1812. Prominent among these was the dependence of specific heat on temperature, and Gay-Lussac duly showed that the capacities of equal weights of air at -20 and $52\ °C$ were in the ratio 1 to 1·206. Further experiments, in which air was mixed first with hydrogen and then with carbon dioxide at various temperatures, not only showed that the volume specific heats of these gases were far from equal but also confirmed that the variations with temperature were considerable. It is worth pointing out here that his view of the dependence of the volume specific heats of the various gases on their densities had been quite reversed since 1806, for he now found that they were greatest for those gases with the largest density.

Gay-Lussac's results were soon to be superseded, as he himself realized. It is conceivable that they were seen by the contestants in the Institute's competition as an indication of the sort of answers that were expected, but despite Gay-Lussac's eminence

† Gay-Lussac, 'Note sur la capacité des fluides élastiques pour le calorique', *Annls Chim.* **83** (1812), 106–8.
‡ Clément and Desormes, *J. Phys.* **89** (1819), 428, 436, and 443.
§ Delaroche and Bérard, op. cit., p. 110 n.
¶ Delaroche and Bérard, op. cit., p. 81.
‖ Gay-Lussac, *Mém. Phys. Chim. Soc. Arcueil* **1** (1807), 201, and *Annls Chim.* **81** (1812), 106.

it seems rather more likely that he was himself influenced by Delaroche and Bérard, probably through a contact at Arcueil. Nevertheless, there can be no denying that his theoretical views on caloric, which owed much to his early experiments, were to exert a crucial influence on the course of the competition.

The 1812 competition

On 7 January 1811 the First Class of the Institute announced the following subject for its next prize competition in physics.

Determine the specific heats of gases, in particular those of oxygen, hydrogen, azote, and some compound gases, comparing them with the specific heat of water. Determine, at least approximately, the change in specific heat that is produced when the gases expand. Competitors are invited to indicate the chief consequences that these new measurements will have for physical theories.†

Entries were to be submitted by 1 October 1812 and the name of the winner of the 3000-franc prize would be declared in January 1813.

The announcement left the motives for setting this subject in little doubt. After pointing out the great discrepancies between existing values for the specific heats of gases, it continued:

The Institute's Class of mathematical and physical science calls the attention of physicists to this problem, the importance of which is easily appreciated. Indeed, for as long as the specific heats of gases remain undetermined, no exact research can be carried out on the amount of heat given out in various chemical reactions nor on that produced by animals. It may be hoped that the determination of the specific heats of gases will lead to the solution of the outstanding problem of whether some caloric exists in substances in a combined state or whether all the heat given out in a reaction is due to a change in the specific heat of the reacting substance.

Gay-Lussac was predictably enough a member of the Class's committee which had been elected in December 1810 to choose the subject.‡ His interest in specific heat was already well known and his recent work on combining volumes had not only given him a special familiarity with the heat evolved in reactions between gases, but had also convinced him that the laws of nature

† *Mém. Sci. math. phys. Inst. Fr.* **11** (1811), p. xcv.
‡ *Procès-verbaux*, vol. 4, p. 399 (3 December 1810). The other members were Desfontaines, Cuvier, Berthollet, and Bosc.

were to be observed in their simplest form in gaseous matter.†
The changes in capacity that were supposed to accompany reactions were thus an obvious subject for investigation. It was, moreover, a subject that was not of interest simply to Irvinists. In January 1812 even Gay-Lussac himself applied the principle that the thermal capacity of a compound formed in an exothermic reaction must be less than the total capacity of its constituent elements in order to determine a rough value for the specific heat of oxygen.‡

Nevertheless, in the experimental investigation proposed by the Institute it was obviously Irvine's theory and Crawford's explanations of animal heat and heats of reaction that were vulnerable, and the competition must therefore be seen not merely as an attempt to provoke a confrontation between the two main versions of caloric theory (a purpose that is clear enough from the second of the two passages just quoted) but rather as a method of settling the controversy in favour of Lavoisier and Laplace. Thanks to the debate that had taken place during the first decade of the century—and here we must make special mention of the restatement of Irvine's views by his son and the 1793 paper by Lavoisier and Laplace, both of which appeared in 1805§—the issues were now clear and in most cases sides had been taken. In France, of course, the climate of opinion was such that the Irvinists could have little prospect of success, but the existing experimental evidence was so inadequate that the need for a really reliable decisive experiment was felt even there.

The determination of the specific heats of gases presented notoriously difficult problems, but others besides Gay-Lussac were showing an interest. In August 1812, for example, Joseph Mollet communicated a letter to the Institute which was never published but in which he appears to have concluded that the volume specific heats of all gases were equal.¶ And in a description of a new calorimeter which he read to the Institute in

---

† See p. 78.
‡ Gay-Lussac, *Annls Chim.* **81** (1812), 104–6.
§ Irvine's views appeared in *Essays . . . by the late William Irvine*, especially pp. 1–159. The paper by Lavoisier and Laplace appeared in Lavoisier's *Mémoires de chimie*, vol. 1, pp. 121–47, published by his widow.
¶ *Procès-verbaux*, vol. 5, p. 85 (10 August 1812). The nature of Mollet's conclusion is inferred from Mollet, *J. Phys.* **90** (1820), 115–17.

February of that year Count Rumford had emphasized how his apparatus might be used for measuring the specific heats of gases.† Yet, despite this interest, the competition attracted only two entries.

Following a custom that appears to have grown in popularity since the successful collaboration of Lavoisier and Laplace, the contestants worked in pairs. All four contestants would have been well known to the committee of judges, which consisted of Berthollet, Haüy, Charles, Thenard, and Gay-Lussac,‡ not least because three of them, François Delaroche, Nicolas Clément, and Charles Bernard Desormes—all men in their thirties by 1812—had already been unsuccessful candidates for admission to the Institute.§ The fourth, Jacques Étienne Bérard, a chemist from Montpellier, was working at Arcueil by 1808,¶ when he was not yet 20, and was a member of the circle there by 1816, being held in particular affection by Berthollet, who wanted him to be his heir.‖ Bérard in fact declined and even by 1813 he had returned to reside in Montpellier, where he eventually became professor of chemistry in the Faculty of Medicine. Rather less is known of his partner Delaroche, who died in 1813, soon after the competition. However, there is no doubt that he was a native of Geneva, that he was friendly with at least two members of the Arcueil circle, Biot and A. P. de Candolle, and also that his main professional interests were medical.†† He had published a number of papers on physiology and on animal and radiant heat since 1806, when he had graduated as a doctor of medicine in Paris, and one of these had been reported on very

† Rumford, *Nicholson's Journal*, **32** (1812), 113.
‡ *Procès-verbaux*, vol. 5, p. 105 (12 October 1812).
§ *Procès-verbaux*, vol. 4, p. 131 (24 October 1808), when Delaroche was a candidate for a place in the botany section; vol. 4, p. 143 (5 December 1808), when Clément was a candidate for a place as corresponding member of the chemistry section; vol. 4, p. 315 (22 January 1810), when Desormes was a candidate for a place in the chemistry section. Of these three only Desormes ever gained membership; he was elected a corresponding member of the chemistry section on 5 July 1819. Bérard became a corresponding member of the same section on 20 December 1819.
¶ See Gay-Lussac, *Mém. Phys. Chim. Soc. Arcueil* **2** (1809), 216.
‖ *Mém. Phys. Chim. Soc. Arcueil* **3** (1817), p. vii. For evidence of Berthollet's affection for Bérard see the latter's obituary in the Montpellier newspaper *Union nationale*, 13 June 1869, p. 4.
†† Biographical details of Delaroche are hard to find but see especially A. de Montet, *Dictionnaire biographique des Genevois et Vaudois* (2 vols., Lausanne, 1877–8), vol. 2, p. 386.

favourably by Gay-Lussac, Berthollet, and Biot in 1810.†
Equally glowing judgement had been passed on a paper by
Bérard and on two written jointly by Clément and Desormes,
and in each of these three cases also Gay-Lussac had been one
of the two referees, with Chaptal, Berthollet, and Thenard
respectively.‡

Although their work had won praise, Clément and Desormes
were probably rather less well known in the high scientific
circles of Paris than were Delaroche and Bérard. Clément's
interest was industrial chemistry and, with Desormes, whose
daughter he married,§ he ran a factory for the manufacture of
alum at Verberie (Oise). In November 1819 he became the first
professor of applied chemistry at the Conservatoire des Arts et
Métiers in Paris and it was while teaching there that he became
a good and, as I believe, significant friend of Sadi Carnot.¶
Desormes, Clément's closest friend and father-in-law, was an
early product of the École Polytechnique and about 1800 he had
worked as an assistant in Guyton de Morveau's chemical laboratory. He wrote numerous papers on chemistry jointly with
Clément and in July 1819 he finally secured election as a corresponding member in the chemistry section of the Académie des
Sciences. But his interests, like those of his partner, were mainly
industrial, although in later life he earned some distinction for
his political activities as a liberal in the Department of the Oise.

The memoirs of both pairs of contestants‖ were of a remarkably high standard, but the outcome could hardly have been in
doubt. Clément and Desormes used so many concepts that were

† *Procès-verbaux*, vol. 4, pp. 299–301 (15 June 1810). See also p. 274 of the same volume.

‡ *Procès-verbaux*, vol. 4, pp. 321–3 (26 February 1810), pp. 481–2 (27 May 1811), and pp. 508–9 (29 July 1811).

§ Clément's adoption of the surname Clément-Desormes after this marriage (from the mid 1820s) has led to some confusion. In several sources it has caused Clément and Desormes to be treated as a single person. See the comments in note 13 of my paper in *Notes Rec. R. Soc. Lond.* **24** (1969), 248; also J. Payen, *Archs int. Hist. Sci.* **21** (1968), 17.   ¶ See pp. 179–83.

‖ Delaroche and Bérard, 'Mémoire sur la détermination de la chaleur spécifique des différens gaz', *Annls Chim.* **85** (1813), 72–110 and 113–82; Desormes and Clément, 'Détermination expérimentale du zéro absolu de la chaleur et du calorique spécifique des gaz', *J. Phys.* **89** (1819), 321–46. The original of the latter paper, entitled simply 'Détermination expérimentale du zéro absolu de la chaleur', is in the archives of the Académie des Sciences, Paris. It differs only slightly from the published version (though see p. 145 n. †).

quite alien to the caloric theory as it was taught in France at this time, that they must have been fortunate to receive even a 'special mention' (*une mention très particulière*) when the result was announced in December 1812.† There was surely no question of unfairness, although it has recently been maintained that Clément and Desormes did feel aggrieved.‡ It was simply unwise to contradict basic principles that Gay-Lussac, by his work in the preceding years, could only consider as established; and this Clément and Desormes did repeatedly in their memoir, with what were surely inevitable consequences.

Although Delaroche and Bérard's theoretical conclusions were the more acceptable, there was little to choose between the contestants as far as their values for the specific heats of different gases were concerned. Moreover, their results were in remarkable agreement with those accepted today, especially in the case of Delaroche and Bérard, and both papers confirmed one point, namely that the volume specific heats of all gases were not in fact equal.

Yet these were not the real issues. Far more significant both in ensuring their victory and for the future development of caloric theory was that Delaroche and Bérard appeared to have put the opinions of Lavoisier and Laplace on the state of heat in bodies beyond all reasonable doubt. The doctrines of Irvine and Crawford, some of which were retained in a restricted form by Clément and Desormes, were never again a serious rival. The victory for the French orthodoxy was crushing and decisive.

The constant-flow method that Delaroche and Bérard used (see Plate 6) was based on unexceptionable principles. Not surprisingly, the authors were critical of all existing data, but their reason for rejecting Crawford's work is particularly interesting, since it was undoubtedly an important factor in the design of their apparatus. The objection went beyond the mere lack of sensitivity of Crawford's method, as we see in this passage.

This procedure [i.e. Crawford's], apart from its lack of precision, had the disadvantage of not yielding the specific heat in the sense in which we have just defined the term, since the gases, being enclosed as they were, could neither expand nor contract.§

† *Procès-verbaux*, vol. 5, p. 130 (21 December 1812).
‡ B. S. Finn, *Isis* **55** (1964), 12.
§ Delaroche and Bérard, op. cit., p. 84 n.

PLATE 6

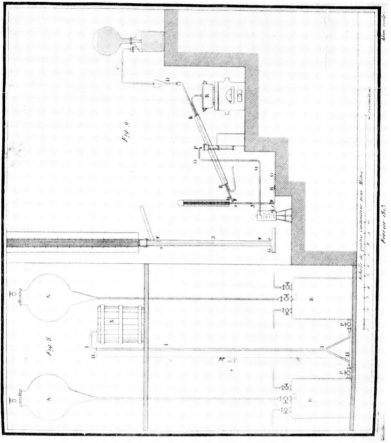

Apparatus used by Delaroche and Bérard in their entry for the Institute's prize-competition of 1812. From *Annls Chim.* **85** (1813), facing p. 224. The main part of the apparatus is on the right of this plate, although here water, rather than a gas, is being passed through the spiral tube in the calorimeter in an attempt to determine the specific heat of air relative to that of water. The apparatus on the left was used in the important supplementary experiment to raise the pressure of the incoming air to 100·58 cm of mercury (or 344 cm of water above atmospheric pressure). Photograph by courtesy of the Manchester University Library

THE RIVAL CALORIC THEORIES

Their opinion that the specific heats of gases should normally be measured at constant pressure rather than at constant volume was probably widely accepted, except by Irvinists, by 1812,† but this was the first occasion on which it had been taken into account explicitly in the design of an experiment. A constant-flow method was the obvious answer and Delaroche and Bérard devised two experiments of this type.

In the first experiment the gas under examination was allowed to flow through a spiral tube inside a cylindrical copper calorimeter filled with water. When equilibrium conditions were established, the gas, which had been heated before entry and which cooled during its passage through the tube, left the apparatus at the same temperature as the water in the calorimeter. It could then be assumed that the heat lost by the gas in cooling equalled the heat given up by the calorimeter to its surroundings, and hence a simple calculation yielded the specific heat. For the second experiment Delaroche and Bérard acknowledged their debt to a suggestion which Rumford had made in his paper of February 1812.‡ In this they did not wait for the establishment of equilibrium conditions, but simply compared the temperature rises caused by the flow of different hot gases when the water in the calorimeter was initially cold. The striking agreement between the results obtained by the two methods inevitably provided strong evidence for their accuracy.

More than a straightforward determination of specific heats was required, however. Of the supplementary problems perhaps the most important was that concerning the dependence of the specific heat of a gas on its density. In order to investigate this, Delaroche and Bérard modified their first experiment so as to allow air to pass through the apparatus at a pressure of 100·58 cm of mercury, as well as at atmospheric pressure, the initial temperature being very nearly the same in both cases. The result, that the volume specific heat at this higher pressure was to that at the ordinary pressure of 74·05 cm as 1·2396 to 1, was decisively, though not seriously, in error. Since the ratio of the pressures was 1·3583:1·0000, it followed that the specific heat by weight had decreased in the ratio 0·9126:1·0000 as a result of the pressure increase. We know now, of course, that no variation at all should have been observed in the specific heat by

† See p. 160 on the view of Haüy. ‡ See p. 135–6.

weight, but in 1812 a decrease in specific heat with increasing pressure was expected and the quite unfounded confidence which Delaroche and Bérard placed in their result almost certainly owed a great deal to this fact. They based their conclusion on only two experiments conducted on air at the single higher pressure of 100·58 cm, the steady variation of pressure being impossible with their apparatus. The discrepancy between the volume specific heats deduced from the two experiments (1·2127 and 1·2665, of which 1·2396 was the mean) should in itself have made them suspicious, but without further examination they proceeded to extrapolate the results to other pressures and confidently assumed that they applied equally well to all gases. Their error, although of less than 10 per cent, was to prove one of the most influential in the whole history of the study of heat. Backed by the prestige associated with victory in the Institute's competition, the result quickly became standard and, as we shall see in later chapters, was to mislead many calorists.

Delaroche and Bérard never lost sight of the central issue in the competition, which was the determination of the form that caloric theory should take, and their attack on the Irvinists' principles was both elaborate and effective. Even a result that at first sight appeared to be wholly consistent with Irvine's views—the one that has just been described concerning the variations in the specific heat of a gas with changes in its density —was interpreted in such a way that it told against Irvine. To make this interpretation, Delaroche and Bérard first calculated the effect of compressing air adiabatically on the basis of Poisson's conclusion that a decrease in volume by 1/116 would raise its temperature by 1 °C. Then, arguing in a perfectly Irvinist fashion, they related the heating predicted in this way to the change in specific heat which, according to their experiments, accompanied the compression. Now it was a necessary consequence of this calculation that the absolute zero of temperature was at $-306\cdot 4$ °C, appreciably higher than any figure previously suggested. That this figure happens to be quite close to the correct one of -273 °C is an interesting coincidence, but it does not alter the fact that an absolute zero at -306 °C was almost unthinkable in 1812 and that the result was put forward as evidence against the Irvinist principles from which it was derived.

THE RIVAL CALORIC THEORIES

However, the whole basis of this argument was so unsubstantial in their own opinion that Delaroche and Bérard set little store by it and instead they looked to two other pieces of evidence that they considered to be far more harmful to the Irvinist cause. The first of these was obtained in a highly suspect determination of the specific heat of steam. The need for a comparison between the specific heats of the same substance in the liquid and vapour states had long been recognized by all those who were interested in Irvine's ideas, whether as critics or supporters, but the high temperatures that it was necessary to maintain in work on vapours had proved a serious obstacle. Crawford's crude calculation had shown, naturally enough, that the specific heat of steam was greater than that of water, but from a comparison of the heating effects of damp and dry air Delaroche and Bérard concluded, though with justified reserve, that the specific heats in the two states were as $0{\cdot}8470{:}1{\cdot}00$, a result which, once established as accurate, would itself overthrow the Irvinist theory of heat.

Their second piece of evidence, arising from their examination of the effect of chemical reaction on specific heat, was rather more convincing. According to Irvine's argument it followed that, in any reaction in which no temperature change occurred, the specific heat of the resulting compound would be $\{ac+(1-a)d\}$, where the masses of the reactants were a and $(1-a)$ and their specific heats by weight were c and d respectively. In any exothermic reaction, however, the predicted specific heat would be below this value, and Delaroche and Bérard's results concerning the influence of changes of density on specific heat suggested that any decrease in volume during a reaction would also serve to reduce the specific heat. Hence in no case where heat was emitted or where contraction occurred could a specific heat greater than that given by the above expression be accounted for on Irvinist principles. In fact the experimental values for certain compounds formed in exothermic reactions were less than the calculated figures, but for red lead oxide and, even more strikingly, for water, in the formation of which much heat was liberated and a large decrease in volume occurred, the specific heat *exceeded* the calculated figure, as Delaroche and Bérard demonstrated. The evidence, in their eyes, was 'insurmountable', for no conceivable error in the specific heats of oxygen and hydrogen could account for the discrepancy. While they did not exclude completely the possibility that the specific heat of a compound might be related to those of its constituents,† there could be only one conclusion. They concluded their paper with the comment:

... so if we are not mistaken, we must reject once and for all the hypothesis whereby the release of heat which occurs when substances combine is attributed solely to a change in the capacity of these substances for caloric, and consequently we must accept, with Blake [sic], MM. Lavoisier and De Laplace and a large number of physicists, that caloric exists in a combined state in bodies. A knowledge of the specific heat of oxygen alone is enough to lead us to take this view. For it is so small that it is almost impossible to account for the large quantity of heat which is released in the combustion of most substances if we do not assume that the heat in question was previously present in the substances in a combined state; if the opposite view is adopted, one has to ascribe to this gas a specific heat at least fifteen times greater than its true value.‡

For Delaroche and Bérard, as we see, the Irvinist case was quite indefensible.

The contrast with the approach of Clément and Desormes was a marked one. Whereas Delaroche and Bérard conducted a simple, straightforward determination of specific heats, based on principles and coming to conclusions that could have aroused no objection among the judges, Clément and Desormes used an indirect method, the theory of which could only have met with disapproval, especially from Gay-Lussac. Moreover, the two perfectly fair attempts at direct determinations which they did make were reduced in their paper to the status of digressions, for in their view of the problem such results could not be regarded as being in any sense final. A far more important determination for them and the one to which they gave the greatest attention was of the heat capacity of the vacuum. That the void did in fact possess a definite capacity for heat was, as usual, based chiefly on the ease with which it accounted for the phenomena accompanying the entry of gas into an evacuated container. In particular, as Clément and Desormes later pointed out, it answered one problem that may well have caused more concern

† See the passage quoted on p. 203.
‡ Delaroche and Bérard, op. cit., pp. 174–5.

PLATE 7

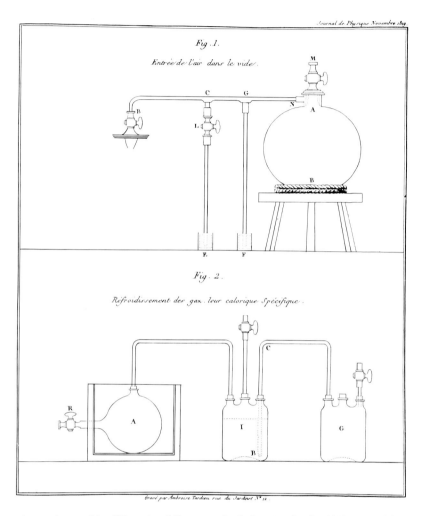

Apparatus used by Clément and Desormes in their entry for the 1812 competition.
From *J. Phys.* **89** (1819).

to calorists than the literature suggests, namely the puzzling fact that some cooling was not observed when a gas first expanded into a vacuum.† Not until 1822 do we find another reference to the problem, this time by Gay-Lussac and J. J. Welter, who based their solution on their observation that air rushing from a vessel where it had been kept under pressure experienced no fall in temperature.‡ They interpreted this as indicating that the passage of air through a narrow aperture caused heating that counteracted the normal cooling effect, and they argued that this was precisely what happened when a vacuum was suddenly opened. Almost inevitably, a year later, came the best of all solutions, when Auguste de la Rive and François Marcet gave an independent experimental demonstration that the slight initial cooling did in fact occur!§

The principle of Clément and Desormes's method for the measurement of the heat capacity of the void was to determine experimentally the amounts of caloric present in a vacuum of fixed volume at two different temperatures and hence to find the amount of caloric required to effect the rise from one temperature to the other. Since they supposed that air entering an evacuated chamber was heated because it absorbed all the caloric previously contained in the void, this could be achieved in an ideal experiment simply by measuring the heating effects obtained at each of the two chosen temperatures. But such factors as heat loss and the imperfection of the vacuum made any experiment far from ideal, and it was to removing, or at least to allowing for, such imperfections that Clément and Desormes devoted a large part of their memoir. They sought to lessen heat loss by using a large spherical glass vessel of 28·4 litres capacity which they only partially evacuated (see Plate 7, Fig. 1). Above all, this provision reduced the time for which air was actually entering and ensured that the rise in temperature and hence also the rate of loss of heat were small.

For their analysis of the effect of using a partial rather than a perfect vacuum, Clément and Desormes looked to Gay-Lussac's work of 1806 on adiabatic expansion. They argued that since the temperature changes in Gay-Lussac's two vessels had

† Desormes and Clément, *J. Phys.* **89** (1819), 436–8.
‡ Gay-Lussac and Welter, *Annls Chim. Phys.* **19** (1822), 437.
§ De la Rive and Marcet, *Bibliothèque universelle* **12** (1823), 273–9.

been equal and opposite, the total quantity of caloric present was unchanged by the expansion. It followed that any volume of air at half atmospheric pressure, for example, was exactly equivalent to half that volume of air at 1 atmosphere together with an equal volume of evacuated space. A similar argument could be applied to air at any other pressure below atmospheric. Thus, since Clément and Desormes found it most satisfactory to reduce the pressure in their experiment by only about 14 mm of mercury, the heat produced on admitting air to their 28-litre vessel was to be attributed to the destruction of only approximately half a litre of vacuum. Temperature measurement was effected by determining the momentary increase in pressure (caused by heating) that accompanied the entry of air into the partial vacuum. Although the ingenious method that they used is now generally associated with their name in physics textbooks, it is worth bearing in mind that the principle had been used some years earlier for the same purpose by Dalton.[†] However, Clément and Desormes used a rather more sensitive, conventional type of manometer fitted externally to the vessel instead of the capillary tube that Dalton suspended inside.

Clément and Desormes repeated the procedure with carbon dioxide instead of air in an attempt to correct for the remaining heat losses (by using a gas of higher capacity) and then concluded that at 12·5 °C the total quantity of caloric present in a vacuum was sufficient to raise the temperature of an equal volume of air measured at atmospheric pressure by 114 °C. Understandably, they were not slow to point out the agreement between this figure and the heating predicted by Poisson in 1807.[‡] From their opinion that the heat evolved in the mechanical compression of a gas resulted from the destruction of a certain volume of void it followed that the effect of compressing atmospheric air by 1/114 of its original volume was essentially to destroy this same volume of evacuated space, producing a temperature rise in the air of approximately 1 °C. Thus Clément and Desormes were simply substituting the figure 114 for Poisson's 116.

It now remained only to discover the quantity of caloric present in the same volume of evacuated space at a higher temperature, but in this Clément and Desormes encountered

† See p. 87–8. ‡ See p. 86.

the problem that led them to their only direct determinations of specific heats. It was clearly essential to know the specific heat of air, which was the thermometric fluid in their experiments, and in particular to ascertain how it was affected by temperature changes. Of the two types of experiment which they devised the first was a constant-flow method in which the heat given up by the gas in cooling was measured by the melting of ice;† but unsuitable winters prevented any extensive use of the apparatus, or of Lavoisier and Laplace's ice-calorimeter with which they began their work. Their second method (see Plate 7, Fig. 2) was the more successful. In this, air at various pressures and a number of other gases were heated in a closed spherical vessel surrounded by a water bath and fitted with an external manometer which was used to measure the pressure and hence also the temperature of the enclosed gas. A comparison of the rates of heating (as also, in a supplementary experiment, of cooling) yielded a simple means of calculating the volume specific heats. That they were here measuring specific heat at an approximately constant volume, whereas in their flow method they would have found a constant-pressure figure, does not appear to have concerned them, but their comment that the expansion occurring in their second method was so small as to necessitate no 'correction'‡ suggests that it was the constant-volume value that they saw as the 'true' specific heat.

Using a slightly modified apparatus, Clément and Desormes repeated their experiments on air entering a partial vacuum at 18 and 98 °C. After corrections for the effect of temperature on the specific heat of the air, made by applying their experimental results on the assumption that isothermal expansion and expansion accompanied by heating influenced the specific heat in exactly the same manner, they concluded that the quantities of caloric contained in equal volumes of vacuum at 18 °C and at 98 °C would raise the temperature of an equal volume of air at atmospheric pressure by 102 and 132·24 °C respectively. Hence it followed that the caloric that raised the temperature of the void by 80 °C would heat the air by only 30·24 °C, so that the relative volume capacities of air and vacuum were as 80:30·24 or 1000:377.

---

† The description of this method was omitted from the printed version of the paper.   ‡ Desormes and Clément, *J. Phys.* **89** (1819), 337.

Two completely fortuitous agreements confirmed Clément and Desormes's confidence in their analysis of adiabatic heating. First, their conception of a partial vacuum as a mixture of a perfect vacuum and of air at normal pressure led directly to a simple relationship giving the volume specific heat of air at any pressure,

$$C' = \frac{NC+N'c}{N+N'}, \qquad (2)$$

where C', C, and c were the volume specific heats of air at low pressure, air at atmospheric pressure, and the void respectively. N and N' were the volumes of air at normal pressure and of vacuum which were supposed to be equivalent to volume $(N+N')$ of low-pressure air. On inserting in eqn (2) the specific heats that they had determined experimentally at three pressures below atmospheric, Clément and Desormes obtained three remarkably similar values for c/C and a mean of 410/1000. Even in this calculation, however, the consistency was obtained purely by chance and could not have occurred if their experimental data had been accurate. Yet in their eyes it was convincing, and the point could not now be missed that this second value for c/C was very close to the first figure of 377/1000, although it was arrived at by a completely different method. The capacity of the void was duly set at 400, where that of an equal volume of air at atmospheric pressure was 1000.

The other fortuitous agreement which appeared to confirm the correctness of Clément and Desormes's treatment was between two values for the absolute zero of temperature deduced from two quite unrelated principles. The first method of calculation depended on their estimate that the quantity of caloric in a vacuum at 18 °C was to that at 98 °C as 102:132·24. Arguing on Irvinist lines and assuming that the volume specific heat of the void was independent of temperature, Clément and Desormes concluded that, where x was the absolute temperature at 18 °C, $x:80::102:30\cdot24$ and hence that the absolute zero was at $-251\cdot8$ °C. A modified procedure, also involving the capacity of the void, gave $-267\cdot5$ °C. Their second, quite different method of calculation was essentially the one originally suggested by Amontons and since used by Martine, Lambert, and Volta. Naturally they used Gay-Lussac's result for the expansion coefficient, but only after checking it themselves, and in this way they were led to place the zero at −266·66 °C. The calculation is notable for its correctness, since of earlier nineteenth-century writers only Dalton and J. T. Mayer had adopted anything resembling Amontons's definition of the zero, and Dalton, as we have seen, had applied it in a most unconvincing fashion.† This neglect seems particularly surprising when we recall that the common eighteenth-century belief that gases expanded to widely differing extents when heated, which presumably served to divert attention from Amontons's definition, had been convincingly rejected as early as 1802. For Clément and Desormes, however, the important point was the agreement between the results of two completely different methods of procedure. Once again, it seemed, their basic principles had been more than adequately confirmed.

To what extent Clément and Desormes's theoretical ideas influenced their selection of what they considered to be the true readings we cannot know. As it stands, however, their paper is a thoroughly consistent document, and for the most part a quite plausible one, in which experimental results strongly vindicate their theoretical analysis of adiabatic heating. It is incidentally an indication of the authors' great confidence in their analysis that the specific heats listed in the final table were given not merely for equal volumes, as we should expect, but relative to that of an equal volume of vacuum. Moreover, the variations in specific heat with density were determined not experimentally but by calculation with eqn (2). Only in fixing the specific heats of different gases relative to air were experimental results used directly.

On 21 December 1812 Delaroche and Bérard's victory was announced to the First Class of the Institute. The report of the committee was unfortunately never published, despite a request that it should be,‡ and it has not survived. Yet the reasons for the decision are clear enough. Although it would be wrong to see the two memoirs as simply representing the rival versions of caloric theory and the result as a straightforward victory for Lavoisier and Laplace over Irvine and Crawford, there is

---

† On Dalton see pp. 73–7. For Mayer's work see his 'Commentatio de lege vis elasticae vaporum', *Commentationes societatis regiae scientiarum Gottingensis recentiores* **1**, commentationes mathematicae (1811), 28–9.

‡ *Procès-verbaux*, vol. 5, p. 130.

something of this in the situation. Clément and Desormes had never denied the possibility that some caloric might exist in a combined state in bodies† but, on the other hand, they had never given any support to this view. Instead they sought to avoid the issue by working in terms of the caloric of the void, in which there could be no question of latent caloric. Irvine's principles might reasonably be expected to hold here and, as we have seen, they were applied, if only in a restricted form. The idea that a vacuum could have a definite capacity for heat also had Irvinist overtones, not least in the mind of Gay-Lussac who, in trying to discredit the idea in 1806, had directed his attack particularly at Dalton, the best-known Irvinist of the day. Moreover, Clément and Desormes themselves intimated that their own analysis of the evacuated receiver experiment was identical to Dalton's,‡ although in this they were largely mistaken, as should now be clear. Dalton's explanation of the heating effect that accompanied the rapid entry of air into a vacuum, we remember, was based on the belief that the capacity of the void was *greater* than, and not *less* than, that of an equal volume of air. Of course, the results of both pairs of contestants in the competition supported Dalton's position to the extent that, when extrapolated to pressures greater than atmospheric, they predicted a decrease in specific heat by weight with increasing pressure. In fact the agreement between the two memoirs on this point was extremely close, for while Delaroche and Bérard observed an increase in volume capacity of 22·3 per cent when the pressure increased from 75·80 to 100·58 cm of mercury, Clément and Desormes predicted 21·5 per cent for the same increment.§

It certainly appeared that the theory of heat had received important clarification as a result of the competition, but problems remained. In particular, the effect of temperature and pressure on specific heat had not been adequately investigated

† Desormes and Clément, op. cit., pp. 326–7. In later years, however, they apparently thought the existence of latent caloric 'a useless hypothesis'; see L. J. Thenard, *Traité de chimie* (4th edn, 1824), vol. 1, p. 84.

‡ Desormes and Clément, op. cit., p. 436.

§ Desormes and Clément, op. cit., p. 429. On the scale where the volume specific heat of air at 75·80 cm of mercury is 1000, the specific heat at 1000·58 cm, as given by Delaroche and Bérard's results, should not be 1·2396, as Desormes and Clément wrote, but 1·223. This correction brings the figures of the contestants into even better agreement.

and there was a growing realization that the choice between constant-pressure and constant-volume measurements might be an important one. On 21 December 1812 Gay-Lussac, speaking on behalf of a committee consisting of Berthollet, Haüy, Cuvier, J. N. Hallé, and himself, appeared before the First Class to announce the subject for the next competition, the result of which was to be declared in January 1815.† It was as follows:

Determine the specific heats of the elastic fluids at twenty degree intervals, between the temperature of melting ice and that of boiling water. The measurements should be made at two different pressures, one being twice the other; the gases may be maintained at a constant volume or allowed to expand freely under the action of heat.‡

The setting of this subject was in no sense a criticism of Delaroche and Bérard's highly esteemed results, but merely an extension based on problems arising out of the previous competition.

The new competition was not a success and little record of it has survived.§ On 5 December 1814 Berthollet announced to the Class that no paper of sufficient merit had been received and he recommended that the subject be withdrawn.¶ If we are to judge by the one entry that has come to light the decision was fully justified. This memoir came from one Vincent Frédéric Olmi, a teacher of natural science at the *lycée* in Sorèze (Tarn), who tried to extend the well-known dependence of the expansion coefficient of a gas on humidity to specific heat, claiming that variations in specific heat were caused simply by the different quantities of water vapour present in the various

† *Procès-verbaux*, vol. 5, p. 130. On the slight doubt concerning the date of the declaration of the result see p. 150 n. †.
‡ From p. 1 of the official pamphlet announcing the subjects for prizes to be awarded in 1816 and 1817 by the First Class of the Institute. The pamphlet, which was printed for the public meeting of the Class held on 9 January 1815, recorded that the subject quoted here had been withdrawn. The subject was also given on p. 3 of the corresponding pamphlet for the public meeting of 4 January 1813, when the competition was first announced.
§ It is not mentioned, for example, in E. Maindron, *Les fondations des prix à l'Académie des Sciences* (1881), or in any of the annual reports of the work of the Class.
¶ *Procès-verbaux*, vol. 5, p. 435. The other referees appointed with Berthollet on 17 October 1814 were Gay-Lussac, Thenard, Charles, and Poisson.

gases.† In fact he did not tackle the problem seriously and gave none of the results that he may, or may not, have obtained with his '*hygro-calorimètre-pneumatique*', an instrument designed to determine simultaneously the expansibility, specific heat, and humidity of a gas.

As will be seen in Chapter 7, only a few years passed before serious doubt began to be cast on the accuracy of Delaroche and Bérard's results. For the moment, however, they won general acceptance and they continued to exert a powerful influence on calorists until the 1830s. But, in conclusion, it cannot be emphasized too strongly that of far more immediate and decisive importance for the future of caloric theory was the convincing way in which Delaroche and Bérard had refuted the Irvinist doctrines and, by their victory, given official sanction to the rival version associated principally with the names of Lavoisier and Laplace.

After 1812

For the calorists, especially those in France, something of a mopping-up operation followed the 1812 competition. The basic principles of caloric theory were now established beyond reasonable doubt, and the problems that remained, such as those set in the competition of 1815, although important, were not of a fundamental nature. Thus, despite the disappointing outcome of the latter competition, Gay-Lussac could declare with every confidence in 1816: '... we can now affirm that the capacity of a compound does not depend on the aboslute quantity (*quantité absolue*) of heat that it contains'.‡

The impression that the pattern for the future development of caloric theory had been settled was further strengthened by the failure of Clément and Desormes to publish their paper until 1819, by which time they could have had even less hope of

† Olmi's paper, entitled 'Mémoire sur la détermination de la chaleur spécifique des fluides élastiques', is in the archives of the Académie des Sciences. The paper, which is profusely dedicated to Laplace, is dated 5 October 1813 and curiously refers on the title-page to the fact that the prize was to be awarded in January 1814. Why Olmi should have made this mistake is not clear, although it is possible that he was misled by a misprint in the official announcement, which at one point gave the date for the declaration of the result as the first Monday in January 1814. The correct date of 1815 was given elsewhere in the announcement.

‡ Gay-Lussac, *Annls Chim. Phys.* **1** (1816), 216.

success than in 1812. By 1819 the intensity of the debate over the state of caloric in bodies had died down. New and even more fundamental issues, touching on the very existence of the fluids of heat and light, had now arisen, while for those who remained calorists Delaroche and Bérard's results and opinions were proving so successful as to require no modification. Yet, despite this quietening of the debate, it was a paper interpreted by Clément and Desormes as a major attack on their theory of heat that seems to have encouraged them to publish their work. The paper, by Gay-Lussac, appeared in November 1818.† It advocated the application of the cooling associated with the rapid expansion of a gas to the attainment of low temperatures, a proposal that was to be taken up in later years, of course. The ignition of tinder by the compression of air to one-fifth of its original volume in a fire piston, which Gay-Lussac saw as an indication that a temperature of at least 300 °C had been attained, showed that the heating effect might even exceed that predicted by Poisson in 1807. So there was surely no reason to doubt that much higher temperature changes would occur if the air was compressed still more strongly and rapidly and, conversely, that equivalent degrees of cold might be attained if air at very high pressure was suddenly allowed to expand. There might be no limit to this cold, Gay-Lussac argued, and if this were so, 'the determination of the absolute zero of heat must appear an utterly fanciful question'.‡ It is interesting to note that earlier in the same year, in their prize-winning memoir on thermometry and the laws of cooling, A. T. Petit and P. L. Dulong had come to a very similar conclusion.§

Gay-Lussac's view could hardly fail to arouse Clément and Desormes, even though they were not mentioned in the 1818

---

† Gay-Lussac, 'Sur le froid produit par la dilatation des gaz', *Annls Chim. Phys.* **9** (1818), 303–10. Desormes and Clément never stated that Gay-Lussac's comments had been responsible for their decision to publish, but they declared that criticism had been voiced in the Institute and they even cited Gay-Lussac's paper of 1818 as a 'violent attack' on them; see *J. Phys.* **89** (1819), 428 and 444. In his paper 'Le "calorique du vide" de Desormes et Clément', *Archs. int. Hist. Sci.* **21** (1968), 6–7, P. Costabel agrees that it was criticism by Gay-Lussac that provoked Desormes and Clément to publish, but he argues that the immediate incentive was not the paper of 1818 but a privately circulated copy of a paper that Gay-Lussac eventually published in March 1820 (see p. 153 n. ‡ for reference).

‡ Gay-Lussac, *Annls Chim. Phys.* **9** (1818), 310.   § See p. 239.

paper. In successive issues of the *Journal de physique*, in November and December 1819, they published first the paper that they had entered for the competition and then a detailed reply to the supposed attack.† Their observation, made at the beginning of this reply, that the nature of heat was still unknown suggests that they had made some concession to the changing climate of opinion, possibly to Gay-Lussac himself, who was to make the same point in 1820.‡ But their argument remained essentially unaltered and they made no attempt to reconcile it with any but the material theory. Indeed, much of the reply was taken up with a vigorous reaffirmation of their belief in the capacity of the void. Gay-Lussac, they maintained, had drawn the wrong conclusion from his demonstration that changes in the volume of a vacuum did not cause heating. The difference when a gas was present was simply that the gas presented an obstacle to the escape of heat through the walls of the container. In a vacuum, where there was no such obstacle, the heat escaped without being detected by the thermometer, and this was what had happened in Gay-Lussac's experiment. Moreover, if the idea of the capacity of the void was abandoned, how else was the heating that accompanied the entry of air into an evacuated receiver to be explained ? Rejecting all the other possibilities, Clément and Desormes repeated their own explanation, which had now received further support from a recent application of Bréguet's metallic thermometer to the problem.§ Experiments with this highly sensitive instrument, conducted by the Bréguets, had shown that when air entered a vacuum its temperature rose by 70 °C, the highest figure yet obtained by direct measurement on a thermometer and one in fair agreement with their own predicted figure, which, as will be recalled, was of the order of 100 °C.

Turning to Gay-Lussac's more recent strictures on the absolute zero, a concept that they regarded as a no less essential

† For details of the original paper see p. 137 n. ||. The supplementary paper appeared on pp. 428–55 of the same volume (vol. 89) of the *Journal de physique*. The title was as for the published version of the original paper.

‡ Desormes and Clément, op. cit., p. 428 n. Gay-Lussac's statement on the nature of heat appeared in *Annls Chim. Phys.* **13** (1820), 304. As was pointed out on p. 151 n. †, P. Costabel has suggested that the paper that contained this statement was known to Desormes and Clément by 1819.

§ Desormes and Clément, op. cit., pp. 440–1. The experiments with the Bréguet thermometer were described in *Annls Chim. Phys.* **5** (1817), 312–15.

part of their theory of heat, Clément and Desormes claimed that two errors had been made. First, since tinder ignited at a lower temperature in high-pressure air, Gay-Lussac had overestimated the temperature changes occurring in the fire piston. More seriously, he had neglected the dependence of the specific heat of a gas on its pressure, so that there was no reason to suppose that the abstraction of a given quantity of heat at low pressure would produce the same cooling effect as on the identical gas at high pressure. In fact the temperature changes would be expected to be less in the former case, which led to the prophetic conclusion: 'The attainment of the absolute zero of heat is no doubt as impossible as the creation of a perfect vacuum.'† The zero was none the less real, and its existence was given further plausibility by the two additional determinations of its value that Clément and Desormes now described. The first determination, based on the increase in the pressure of a gas with temperature, did not differ essentially from the one based on the use of the expansion coefficient, which they had conducted in their entry for the competition; as before, it yielded −266·66 °C. In their second determination they located the zero from a knowledge of the change in capacity accompanying the conversion of ice to water and of the latent heat absorbed in the process. The method was Irvine's, but the insertion of their own experimental value of 0·72 for the specific heat of ice, instead of the usual 0·9, gave yet another figure in close agreement with all their others.

Gay-Lussac was quick to rise in his own defence, though again without mentioning Clément and Desormes. In a paper that appeared in 1820,‡ he recalled how, about the time of the 1812 competition, he had repeated his earlier experiments on the compression and expansion of a vacuum on a considerably larger scale. The results had been unchanged and had served merely to confirm his views, no doubt to the detriment of Clément and Desormes's chances. The whole question of the amounts of heat contained in various bodies was dismissed by Gay-Lussac, not as unsolvable, but as one of little importance, which it most certainly was in any but Irvine's theory of heat.

† Desormes and Clément, op. cit., p. 445.
‡ Gay-Lussac, 'Sur le calorique du vide', *Annls Chim. Phys.* **13** (1820), 304–8.

In the face of this attack such Irvinists as remained were noticeably silent. There were still those, Mollet, Navier, and de la Rive and Marcet, for example, who advocated the application of evacuated receiver experiments to the determination of specific heats,† but there is no evidence that any of them proposed the retention of the concept of the capacity of the void or, indeed, of any principles that might be considered Irvinist. Clément and Desormes themselves did not reply to Gay-Lussac, but in a lecture that he gave at the Conservatoire des Arts et Métiers in 1825 we find Clément stating as obviously quite familiar facts that the quantity of heat in a body was proportional to its thermal capacity and that the absorption of latent heat had the effect of increasing capacity during a change of state.‡ Since he made no mention of the capacity of the void, it seems possible that he had been at least partially persuaded by Gay-Lussac's arguments.

Even Dalton himself had to make concessions. Just how quickly he did so, however, is not clear, since his views on Delaroche and Bérard's results did not appear until 1827.§ Even then he was reluctant to admit defeat and he had accepted Delaroche and Bérard's results as irrefutable only after a thorough repetition of their experiments, which had proved to his own satisfaction that the specific heat of air was indeed much lower than Crawford (and he) had earlier supposed. It was in his interpretation of the results, however, that Dalton still held some reservations and, incidentally, showed the weakness of his position. His claim that even Crawford had all along believed caloric to be combined with ponderable matter and that therefore Delaroche and Bérard had said nothing new must appear either as a particularly weak attempt to justify his own change of heart or as an indication of some serious misunderstanding of the essential points at issue. After all, Crawford had at no time suggested that heat entering into combination thereby became latent, a point that Dalton appears to have understood

† Mollet, *J. Phys.* **90** (1820), 113–30; Navier, *Bull. Soc. philomath. Paris*, N.S. **7** (1820), 101; de la Rive and Marcet, *Bibliothèque universelle* **22** (1823), 280.

‡ In a lecture given on 19 December 1825 (Library of the Conservatoire National des Arts et Métiers, Paris, MS. 8° Fa 40 (2), pp. 58–60 of the second of the two notebooks bound in the second voulme of the MS.). See p. 180 n. † for detailed reference.

§ Dalton, *New system* (Manchester, 1827), vol. 2, pp. 268–71 and 282–8.

perfectly well in 1803.† On one point, however, there could be no concealing the fact that Dalton's opinion had changed, for he now had to admit that Delaroche and Bérard's experiments had shown conclusively that the total quantities of heat in gases were not proportional to their specific heats. But even this concession was hedged about with reservations and Dalton was in fact retreating less than might appear at first sight. The reason for the breakdown of proportionality, he argued, was merely that the specific heats of gases varied with temperature, so that in the case of solids and liquids, where this variation did not seem to occur, there was as yet no reason to abandon Irvine's principle. If the Irvinist method of determining the absolute zero had to be abandoned, as now seemed likely, it was for this and not for any more fundamental reason.

In his reaction to Dulong and Petit's discovery of 1819 that the atomic heats of a number of metal elements were identical Dalton was still less gracious,‡ and this despite the support which the law that was based on the discovery gave to the atomic theory. The essence of his criticism was to cast doubt on the data used and hence on the constancy of the atomic heats adopted by Dulong and Petit; yet he was quick to point out that the relationship might yet be proved to hold for gases and so, he supposed, be reconciled with those views on caloric theory and gases which he had expounded in 1808 in the first volume of the *New system of chemical philosophy*. This claim that he was in some sense a precursor of Dulong and Petit hardly bears examination, although it has been taken seriously by one zealous biographer.§ Its falsity should be obvious to anyone who reads Dulong and Petit's highly critical opinion of the Irvinist theory and in particular of Dalton's interpretation of it.¶ In the minds of Dulong and Petit the very existence of caloric was open to serious doubt, so that they were unlikely to find much of value in Dalton's highly speculative views on the sizes of gas particles and on caloric. These views, we should remember, had not been taken seriously, even by calorists. Moreover, in predicting that the 'atoms' of all gases had

† See Dalton, *Nicholson's Journal*, 2nd ser. **5** (1803), 36.
‡ Dalton, *New system*, vol. 2, pp. 280–1 and 293–7.
§ W. C. Henry, *Memoirs of the life and scientific researches of John Dalton* (1854), p. 68; also in the *D.N.B.* article on Dalton.
¶ See Petit and Dulong, *Annls Chim. Phys.* **10** (1819), 398–9 and 408.

156 THE RIVAL CALORIC THEORIES

identical capacities for heat Dalton had made no attempt to distinguish between elements and compounds, whereas Dulong and Petit were quite sure that the law for compounds, when found, would be different from that for elementary atoms. It was clearly not in Dalton's interest to emphasize such differences between his own work and that of Dulong and Petit, but it also seems possible that he did not fully understand the argument of the French scientists. At one point, for example, we find him in the ludicrous situation of criticizing them for adopting the Irvinist view that the specific heat of a substance was proportional to the quantity of heat that it contained!†

Such muddled thinking reflects the hopelessness of the Irvinist position in the 1820s. When its leading exponent could produce nothing better than this confused defence, the cause was surely lost. We shall find certain elements of Irvine's theory surviving, notably in the work of Amedeo Avogadro discussed in Chapter 6, but the general pattern for the future development of caloric theory was now firmly based on quite different principles, principles that from the early 1820s became increasingly associated with the name of Laplace. It was Laplace who gave caloric theory its most sophisticated form, and it is in his work that we see the natural culmination of the important developments in early nineteenth-century physics that have been examined in the last two chapters. With the special status of gases beyond all doubt, with the Irvinists virtually routed in the 1812 competition, and with experimental data that gave every appearance of reliability at last available, Laplace could confidently proceed to apply his powerful mathematical techniques to the elaboration of a caloric theory which, in terms of elegance and completeness, bore little resemblance to the versions that we have encountered so far.

† Dalton, op. cit., vol. 2, p. 296.

5

TWO GREAT CALORISTS: LAPLACE AND CARNOT

LAPLACE made the most lasting of his contributions to the study of heat in a short paper that he read to the Académie des Sciences in December 1816.† In it he first showed that the effect of taking into account the small temperature changes accompanying the passage of a sound-wave was to multiply Newton's expression for the velocity of sound by a factor $\sqrt{(c_p/c_v)}$. He then used quite independent data to obtain a value for the correction factor that brought his own theoretical figure for the velocity into good agreement with the experimental value, and hence he not only established the correctness of his treatment but also put the need for a clear distinction between c_p and c_v beyond doubt. The latter was no small achievement, since scholars for some thirty years had been groping to the realization that c_p and c_v were in some respect different without ever confirming, let alone quantifying, their belief. It seems appropriate, therefore, that we should preface our account of Laplace's paper with a brief review of this earlier work.

c_p or c_v?

In Chapter 2 we saw how, in 1788 in the second edition of *Animal heat*, Crawford used the grossly inaccurate experiment data then available to him to determine a value for the ratio c_p/c_v.‡ Although his figure of 1·03 (for oxygen) had the merit of being the first ever obtained, its smallness inevitably led him to dismiss the distinction between the two possible definitions of specific heat as unimportant and so to refer to it in only the briefest fashion. Of course, this cursory treatment by such a prominent authority could serve only to mislead, and the fact must clearly be taken into account in attempting to understand

---

† Laplace, 'Sur la vitesse du son dans l'air et dans l'eau', *Annls Chim. Phys.* 3 (1816), 238–41.   ‡ See pp. 37–8.

why Dalton alone of the later Irvinists was to return to the problem. If the misleading nature of Crawford's comments is ignored, it becomes even more difficult to explain why anyone who used changes in capacity to account for the heat evolved in the compression of a gas should have failed to realize that c_p must be greater than c_v. De Luc, Leslie, and Murray were among those who fell into this category† but who made no reference to the distinction between the two quantities.

There were others, besides the Irvinists, who could easily have made the discovery quite independently of Crawford and even without any knowledge of adiabatic phenomena. Lavoisier is the obvious example. Not only did he believe that volume was one of the factors that determined specific heat‡ but, more importantly, with Laplace he had also clearly distinguished the heat that caused expansion from that which caused heating.§ Did it not follow from this that the amount of heat required to effect any temperature change was greater when expansion as well as heating occurred than when the volume of the heated substance remained constant? The deduction seems a simple one, but there is no evidence that Lavoisier and Laplace made it or ever appreciated that in their experiments on gases they had been measuring a quantity quite different from the one determined by Crawford. Unfortunately their views on the twofold function of specific heat were largely ignored, with only De Luc commenting on them explicitly and then merely to reject them.¶ It was all too easy to accept, with De Luc, that the specific heat added to a substance always brought about both heating *and* expansion simultaneously, and this may well account for the fact that some authorities who shared this opinion|| did not think it worthy of detailed comment. That Crawford's experiments provided an exception was not pointed out until 1812.††

The ways in which the two main versions of caloric theory could have led to an appreciation of the difference between c_p

† See pp. 52–3 and 107–9.
‡ Lavoisier, *Traité élémentaire de chimie* (1789), vol. 1, pp. 17–19. The other important factor was the magnitude of the attractive force of cohesion between the particles of a substance. § See p. 31.
¶ De Luc, *Idées sur la météorologie* (London, 1786), vol. 1, pp. 187–9. See also his comments in his *Introduction à la physique terrestre* (1803), vol. 1, p. 243.
|| See, for example, M. Landriani, *J. Phys.* 26 (1785), 89–90, and G. Adams, *Lectures on natural and experimental philosophy* (1794), vol. 1, p. 220.
†† By Delaroche and Bérard; see p. 138.

and c_v are well illustrated by the treatments of Dalton and Haüy. Writing, as they did, soon after the turn of the century, both men benefited from a full acquaintance with the adiabatic phenomena, which they invoked in support of their arguments.† The knowledge was particularly valuable to Dalton who, following Crawford, used it as evidence that specific heat was dependent upon volume. His comments date from 1808 and occur in the section of his *New system of chemical philosophy* which deals with thermometry.‡ Here he was concerned to show not so much that c_p for a gas was greater than c_v, a fact that he evidently expected to be readily accepted, but rather that c_p (unlike c_v) increased steadily with temperature, i.e. as the gas expanded. If a standard of specific heat was to be obtained, therefore, it was necessary to choose a body whose volume could be kept constant, and in this respect, of course, only gases were practicable. This was the only point that Dalton was trying to make and there was obviously no need for him to estimate the difference between c_p and c_v, as Crawford had done. Instead Dalton merely cited his own experiments with the air-pump receiver as evidence that the variations in specific heat with volume were far from negligible.

When he wrote the second edition of his *Traité élémentaire de physique* in 1806, Haüy too had little to say on the matter of quantification but he paid rather more attention than Dalton to showing that c_p must be greater than c_v.§ Although his view of the function of caloric in effecting expansion was precisely that proposed by Lavoisier and Laplace (with latent caloric causing the increase in volume and sensible caloric causing heating), he had an important advantage over them in possessing independent evidence, in the form of the adiabatic phenomena, that even the isothermal expansion of a gas necessitated the addition of heat. According to Haüy this heat, which was the latent caloric (*calorique latent*) required solely to effect expansion, was equal to the difference between c_p and c_v.

Like Dalton, Haüy believed that the ratio c_p/c_v normally varied with temperature and that this was a consideration of

† Haüy, *Traité élémentaire de physique* (2nd edn, 1806), vol. 1, p. 88, and Dalton, *A new system of chemical philosophy* (Manchester, 1808), vol. 1, part 1, pp. 6–7.
‡ Dalton, op. cit., vol. 1, part 1, pp. 4–6.
§ Haüy, op. cit., vol. 1, pp. 86–8.

some importance in thermometry. The ideal thermometric substance, he pointed out, was one in which the increase in its volume was proportional to that in its true temperature, i.e. in the 'tension' of the caloric that it contained. But this condition could not hold unless the relative quantities of sensible and latent caloric that went to make up the specific heat were themselves in proportion to each other. Arguing in a way that probably owed a great deal to Laplace,† Haüy maintained that only air fulfilled these requirements and so should be adopted as the standard thermometric substance.‡ It should be noted that Haüy, unlike Dalton, had no objection to the use of a constant-pressure instrument as a standard. For someone who was not an Irvinist, after all, it was not necessary to suppose that expansion brought about any increase in specific heat, and in his references to 'specific caloric' Haüy evidently understood c_p rather than c_v. This is clear from his reference to the two distinct contributions which made up the total specific heat c_p, where he wrote:

> Experiment cannot distinguish these from each other, and so whenever we define specific caloric—the caloric which is needed to raise the temperature of a substance by a given number of degrees—there is included, by our definition, that part of the whole whose sole function is to bring about expansion, this latter effect being a necessary accompaniment of a rise in temperature.§

So by 1808 there was no reason why any calorist, whichever version of the theory he adopted, should have ignored the distinction between c_p and c_v. Both pairs of contestants in the Institute's prize competition of 1812 were aware of it, and, as has been shown, Delaroche and Bérard at least saw it as an important factor in the design of their apparatus, although they made no attempt to estimate the relative magnitudes of the two quantities.¶ This latter problem was to remain unsolved for yet another four years during which the only contribution seems to have been an unsuccessful attempt by Dulong to measure c_v

---

† There is a good deal in Haüy's argument that is reminiscent of Laplace's comments on thermometry which had recently appeared in the fourth volume of the *Traité de mécanique céleste*; see p. 73.
‡ Haüy, op. cit., vol. 1, pp. 88, 160, and 167–8.
§ Haüy, op. cit., vol. 1, p. 118.
¶ See pp. 138–9. On the views of Clément and Desormes see p. 145.

for vapours experimentally.† The practical difficulties facing Dulong in this work must have been daunting, and it is significant that when the solution eventually did come, from Laplace in 1816, no new experiments were called for.

Laplace's chief concern in the paper of 1816 was to develop his now familiar explanation of the error in Newton's expression for the velocity of sound. Just how he arrived at the correction factor $\sqrt{(c_p/c_v)}$ was not described, but if we are to judge from a paper that he wrote in 1821,‡ it seems likely that the first part of his argument was essentially that given in present-day textbooks.§ Thus he probably began by showing that in a gas of pressure P and density ρ the velocity was given by $v = \sqrt{(\mathrm{d}P/\mathrm{d}\rho)}$. It was Laplace's great achievement, of course, to identify Newton's error at this stage and to show that $\mathrm{d}P/\mathrm{d}\rho$ in a sound-wave was not P/ρ but $(c_p/c_v)(P/\rho)$, so that v became $\sqrt{\{(c_p/c_v)(P/\rho)\}}$. In deriving this relationship he assumed not only that the difference between c_p and c_v represented the latent heat that was required solely to bring about expansion but also that it was this same heat which, by losing its latent state and becoming sensible, caused heating when the gas was rapidly compressed. By this argument a decrease in the volume of unit mass of gas from V_0 to $(V_0 - \Delta V)$ would convert an amount of heat $(\Delta V/\alpha V_0)(c_p - c_v)$, from its latent form and, if all this heat was absorbed by the gas, would cause a rise in temperature of $\Delta V/\alpha V_0\{(c_p-c_v)/c_v\}$, where α was the temperature coefficient of expansion (and also the coefficient for pressure increments). The effect of this rise in temperature would be to increase the pressure of the compressed gas by $P_0(\Delta V/V_0)\{(c_p-c_v)/c_v\}$, in addition to the pressure increment that would be expected for an isothermal compression, so that the total increase in pressure, from P_0 to $(P_0+\Delta P_a)$, the subscript a indicating an adiabatic process, was given by

$$\left(\frac{\Delta P}{\Delta V}\right)_a = \frac{P_0}{V_0} + \frac{P_0}{V_0}\left(\frac{c_p}{c_v}-1\right) = \frac{c_p}{c_v}\left(\frac{\Delta P}{\Delta V}\right)_i,$$

where $(\Delta P/\Delta V)_i$ represented the pressure increment that would

† See pp. 204–5.
‡ Laplace, *Connaissance des tems* . . . *pour l'an 1825* (1822), pp. 224–7.
§ For example, in C. A. Coulson, *Waves* (7th edn, Edinburgh and London, 1955), pp. 88–9. I must point out that my reconstruction of the argument of the 1816 paper differs appreciably from that given by Dr B. S. Finn in his paper on 'Laplace and the speed of sound' in *Isis* **55** (1964), 15.

have been obtained under isothermal conditions. Since $(dP/dV) \propto (dP/d\rho)$, it followed simply from this that the velocity of sound was $\sqrt{\{(c_p/c_v)P/\rho\}}$. It was clearly essential to suppose that all the heat that was converted from the latent state in compression did in fact go to heating the gas, and there was no proof of this; but Laplace was confident that the vibrations constituting the sound-wave occurred so quickly that there was no time for this heat to escape before it was reabsorbed, as latent heat, in the succeeding period of rarefaction.

The next stage was to calculate c_p/c_v. In the absence of any reliable experimental data for c_v, Laplace used an ingenious though fallacious argument to deduce the ratio from Delaroche and Bérard's results. Again his account of his reasoning was cryptic, although he admitted to making one additional assumption, namely that the total amount of heat in any given mass of air at a constant pressure was always proportional to the volume of the air at the different temperatures. Despite the lack of information in Laplace's paper, it is not difficult to make a plausible reconstruction of his argument.

It seems that he considered a fixed mass of air undergoing, first, a process of cooling at constant pressure and, secondly, one of heating at constant volume, the temperature at the end of the two processes being equal to the initial temperature. As is indicated in the diagram of the processes (Fig. 1), the effect of allowing the air to cool at pressure P_0, so that its volume decreased from V_0 to V_1, was that it lost an amount of heat $(Q_0 - Q_1)$ or, in accordance with Laplace's additional assumption referred to above, $Q_0\{1-(V_1/V_0)\}$, where Q_0 and Q_1 represented the total heat content of the air before and after the cooling. In the second process, in which the air was supposed to be heated at constant volume until the temperature returned to its original value and the pressure increased to P_1, where $P_1 = P_0 V_0/V_1$ by Boyle's law, an amount of heat $(Q_2 - Q_1)$ was added to the gas. Now Laplace could show, at least to his own satisfaction, that the ratio Q_2/Q_1 was equal to $c_{p_1}(v)/c_{p_0}(v)$, the ratio between the volume specific heats of the air (at constant pressure) after and before the heating (respectively $c_{p_1}(v)$ and $c_{p_0}(v)$). To see how he did this, imagine the air initially at pressure P_0 and volume V_1 being heated at constant pressure through a small temperature increment of δt. Then imagine the same sample of air with initial volume V_1 but with pressure P_1 being heated, also at constant pressure, through precisely the same small increment in temperature. If it was assumed that the gas was perfect, the

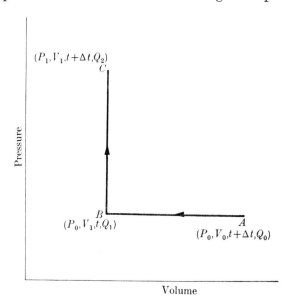

FIG. 1. Heat lost by the air in process $AB = C_p \triangle t = Q_0 - Q_1$
Heat gained by the air in process $BC = C_v \triangle t = Q_2 - Q_1$

expansion occurring in each of these two processes would be the same, say δV, so that the ratios

$$\left\{ \frac{Q_2 + V_1 c_{p_1}(v) \delta t}{Q_2} \right\} \quad \text{and} \quad \left\{ \frac{Q_1 + V_1 c_{p_0}(v) \delta t}{Q_1} \right\}$$

could be equated, since both were equal to $(V_1 + \delta V)/V_1$, in accordance with Laplace's assumption that the heat content and volume of a gas expanding at constant pressure were proportional to each other. Thus

$$\left\{ \frac{Q_1 + V_1 c_{p_0}(v) \delta t}{Q_1} \right\} = \left\{ \frac{Q_2 + V_1 c_{p_1}(v) \delta t}{Q_2} \right\}$$

and
$$\frac{Q_2}{Q_1} = \frac{c_{p_1}(v)}{c_{p_0}(v)}.$$

To return to the main argument of Fig. 1, it could now be

seen that the heat gained by the gas in the constant-volume process was

$$Q_1\left\{\frac{c_{p_1}(v)}{c_{p_0}(v)}-1\right\} \quad \text{or} \quad Q_0 \cdot \frac{V_1}{V_0}\left\{\frac{c_{p_1}(v)}{c_{p_0}(v)}-1\right\},$$

while that lost in the constant-pressure process was $Q_0\{1-(V_1/V_0)\}$, as stated above. Since the temperature changes in the two processes were identical, say Δt, it was a simple matter to incorporate the quantities $c_{p_1}(v)$ and $c_{p_0}(v)$, for which experimental data were available, in an expression for (c_p/c_v). For the ratio between the heat lost in the constant-pressure process $(c_p \Delta t)$ and the heat gained in the constant-volume process $(c_v \Delta t)$ became

$$\frac{c_p \Delta t}{c_v \Delta t} = \frac{Q_0(1-V_1/V_0)}{Q_0 \cdot (V_1/V_0)\{c_{p_1}(v)/c_{p_0}(v)-1\}}$$

$$= \frac{V_0/V_1 - 1}{\{c_{p_1}(v)/c_{p_0}(v)-1\}}$$

$$= \frac{P_1/P_0 - 1}{\{c_{p_1}(v)/c_{p_0}(v)-1\}}, \tag{1}$$

which, on the insertion of Delaroche and Bérard's results, yielded

$$\frac{(1\cdot 36/1\cdot 00)-1}{(1\cdot 24/1\cdot 00)-1}, \quad \text{or} \quad \frac{3}{2}, \quad \text{for} \quad \frac{c_p}{c_v}.$$

It is not at all difficult to find weaknesses in this argument. One of the most obvious is Laplace's failure to observe that the pressure varied during the rise in temperature in the second stage of his thought experiment, so that c_v in his final expression was really a constant-volume specific heat for a mean pressure between P_0 and P_1, while his c_p was a constant-pressure specific heat for heating at P_0. Moreover, his justification for using Delaroche and Bérard's results in eqn (1) was extremely doubtful since the ratio of $1\cdot 24:1\cdot 00$ between the specific heats of air at $1\cdot 36$ and $1\cdot 00$ atmospheres had been determined for equal volumes of gas over the *same* range of temperature. These conditions did not of course hold for the pressure increment from P_0 to P_1, which was the result of an increase in temperature and not of density.

The remarkable agreement between Laplace's figure for c_p/c_v and the present accepted value of $1\cdot 40$ was the result of pure coincidence, errors in his argument being largely compensated for by those in Delaroche and Bérard's data. Yet Laplace's reasoning, although faulty, was not haphazard. As we have seen, he made just one not unreasonable assumption, and it seems unnecessary to suppose, as one recent commentator has done,† that he also assumed the quantity of caloric in unit volume of a gas to be proportional to its pressure. That it was proportional to $c_p(v)$, on the other hand, certainly was important, as we have seen, but this was a simple deduction from his initial premise and not just another assumption.

Laplace must have been highly satisfied, for the velocity of sound calculated with $c_p/c_v = 1\cdot 5$, 345·35 m/s, was in striking agreement with the value of 337·18 m/s that was accepted at that time. Moreover, the agreement had been attained with the minimum of speculation concerning the nature and properties of heat. Indeed, it might well seem that caloric theory had not been furthered by Laplace's paper, but an important foundation had been laid for the detailed elaboration of the theory that he was to begin some five years later.

Laplace

There is no difficulty in following the final stages in the development of Laplace's caloric theory of gases. Between September 1821 and April 1823 Laplace inserted accounts of his work on the subject in many of the leading French scientific journals, especially in the *Connaissance des tems*, as well as in the fifth volume of the *Traité de mécanique céleste*. An examination of these papers will show that, despite the shortness of this period of activity, the theory was still undergoing important changes.

In analysing the phenomena of gases Laplace leant heavily on his earlier work on gravitation that had appeared in the first volume of the *Mécanique céleste* as early as 1799.‡ In a paper that he read to Académie des Sciences on 10 September 1821 he applied his already familiar expressions for the gravitational forces between spherical bodies to the repulsion that he supposed to exist between the particles of a gas.§ For the one case, of

† Finn, op. cit., p. 15.
‡ See especially Laplace, *Traité de mécanique céleste* (1799), vol. 1, pp. 47–9 and 138–44.
§ Laplace, 'Sur l'attraction des sphères, et sur la répulsion des fluides élastiques', *Connaissance des tems . . . pour l'an 1824* (1821), pp. 328–43.

course, the force was attractive, falling off in inverse proportion to the square of the distance r, while in the other it was repulsive and decreased as $1/r$, in accordance with Newton's theoretical treatment of gas structure; but the necessary changes were easily made. One further modification, that the repulsive force was only effective over a very short range, was also required if the predicted behaviour of gases was to agree with that observed, but here again there was little difficulty. The assumption had the backing of Newton himself and such short-range forces had already been successfully applied by Laplace in his work on capillary action,† as well as by his pupil Biot in the study of refraction.‡ Indeed, a passage in one of Laplace's papers dating from either December 1821 or possibly early in 1822§ suggests a clear intellectual lineage for his theory of heat and brings out its relationship to a far more general scheme of scientific investigation, a Laplacian programme for physics, which sought to interpret all physical phenomena in terms of short-range forces. Referring to the laws and phenomena of gases, he wrote:

> These laws and phenomena are thus explained in terms of attractive and repulsive forces which act only over insensible distances (*distances imperceptibles*). In my theory of capillary action I related all the effects of capillarity to such forces. I dare to hope that this new extension of the theory of these forces will be of interest to geometers. Nearly all terrestrial phenomena depend on them, just as celestial phenomena depend on universal gravitation. It seems to me

† See the two supplements (to Book X) that were added to the fourth volume of the *Traité de mécanique céleste* (1805).

‡ A study which Biot carried out with Arago and which both men described in March 1806 in a long paper to the First Class of the Institute; see *Mém Sci. math. phys. Inst. Fr.* **7** (1806), 301–87. The passage from the paper that is quoted on p. 200 is especially relevant to the interest in short-range forces. Étienne Malus, the discoverer of the polarization of light, was another good Laplacian who took an interest in short-range forces for the explanation of optical phenomena. See, for example, his prize-winning memoir of 1809 on double refraction in *Mém. prés. div. Sav. Acad. Sci. Inst. Fr.* **2** (1811), 489–96.

§ The doubt about the dating arises from the fact that the paper in *Connaissance des tems . . . pour l'an 1825* in which the comment appeared was not published until 1822. The comment was made in one of two undated continuations of a paper that bears the date 12 December 1821, so that it may itself date from that time or have been added subsequently. The fact that it also appeared in the February 1822 issue of the *Journal de physique* (see p. 167 n. ‡) confirms that it had been written by the early part of 1822. On p. 171 n. ¶ and p. 172 n. § I argue that the continuations had not been written, or at least completed, by December 1821.

that this theory should now be the chief goal of our research in mathematical philosophy; I believe that it would be useful to introduce it even in proofs in mechanics, laying aside abstract considerations of flexible or inflexible lines without mass.† A number of trials have shown me that by coming closer to nature in this way one can make these proofs no less simple and far more lucid than by the methods used hitherto.‡

By applying these principles to the consideration of the equilibrium of a spherical shell taken at random inside a gas, Laplace showed, in the paper of September 1821, that the pressure P of the gas must be proportional to $\rho^2 c^2$, where ρ represented its density and c the quantity of heat contained in each of its particles.§ There were several grounds for objection in the argument. In particular, the all important assumption which he made at the outset, that the repulsive force between any two particles was proportional to c^2, was presented as self-evident, although in fact it was quite novel and without independent justification. A second, equally original result followed from the same basic physical picture.¶ Constancy of temperature, Laplace maintained, was the result of a dynamic equilibrium, with the particles of a gas both radiating and absorbing caloric at an equal rate. Postulating the simplest of mechanisms for this process, he argued that radiation from any given particle was not spontaneous but resulted from the detachment of some of the particle's own caloric by incident radiant caloric (*rayons caloriques*), the density of which in a gas, $\Pi(t)$, was a function,

† Dr J. R. Ravetz has pointed out to me that here Laplace was almost certainly referring to the self-taught mathematician Sophie Germain who used 'abstract considerations' of the type mentioned; see especially her paper 'Recherches sur la théorie des surfaces élastiques', published and presented to the Académie des Sciences in 1821. Laplace would have known of this work in 1821, since he, Prony, and Poisson acted as referees for the Académie. Sophie Germain's work was greatly admired by Fourier and Legendre. See H. Stupuy's biographical sketch of her in *Œuvres philosophiques de Sophie Germain* (1879), pp. 47–9; also the letters from Fourier reproduced on pp. 350–1 and 357–69 of the *Œuvres*.

‡ Laplace, op. cit., p. 323; also in *J. Phys.* **94** (1822), 90. The statement of the Laplacian programme given in April 1823 in the *Traité de mécanique céleste*, vol. 5, p. 99, differed from the statement quoted here in certain minor respects. For example, the sentence beginning 'I dare to hope ...' was omitted, as was the word 'Nearly' at the beginning of the next sentence. Also the words 'and of perfectly hard bodies' appeared at the end of the penultimate sentence, following the words 'without mass'.

§ Laplace, *Connaissance des tems* ... *pour l'an 1824* (1821), pp. 328–34.

¶ Laplace, op. cit., pp. 334–5.

and hence also a true measure, of temperature. Of the incident caloric a fraction q, determined by the nature of the gas, would be absorbed, while the quantity detached would be proportional both to c and to the total density of caloric in the gas, ρc.† From these quite gratuitous assumptions it followed simply that

$$\rho c^2 = q\Pi(t). \qquad (2)$$

Most consequences of the treatment only served to confirm its correctness. The laws of Boyle and Gay-Lussac, and Dalton's law of partial pressures could be readily predicted. It could be shown also that the volume of a gas expanding at constant pressure must be proportional to $\Pi(t)$,‡ a clear endorsement of Laplace's earlier support for the air thermometer, even though the obvious next step, of assuming $\Pi(t)$ to be proportional to the (absolute) temperature u, was not taken in this paper; it appeared very soon afterwards.§ Yet there was one deduction that must have caused concern. From eqn (2) it followed that a doubling of density at constant temperature would cause c to decrease by a factor $\sqrt{2}$,¶ so that heating was inevitable whenever compression took place in a container impermeable to heat, i.e. when this decrease could not take place. Although such a conclusion was perfectly acceptable in qualitative terms, it was clear to Laplace, for reasons that are examined below, that the predicted decrease by a factor $\sqrt{2}$ in an isothermal compression was too great to account accurately for the error in Newton's expression for the velocity of sound.‖ His arbitrary solution, that the alternate compressions and rarefactions occurred slowly enough for there to be some heat exchange with the surroundings, could hardly have convinced even Laplace himself, and three months later he had abandoned it. By December 1821, therefore, he was arguing that there was in fact no heat exchange during the passage of a sound-wave and that the heat 'expelled' in excess of that required to reconcile the theoretical and experimental figures for the velocity of sound remained attached to the particles of

† For Laplace ρc was an accurate measure of the total amount of caloric in a gas, since the quantity of detached, radiant caloric was thought to be negligible by comparison with that which was present in the gas particles.

‡ Laplace, op. cit., p. 338.

§ In a paper in the November 1821 issue of *Annls Chim. Phys.* 18 (1821), 274.

¶ Cf. the view on this matter which Laplace had held in 1803.

‖ Laplace, *Connaissance des tems . . . pour l'an 1824* (1821), p. 336.

the gas, but merely became latent and thereby ceased to contribute to the inter-particle force.† No modification in the expressions for P and $\Pi(t)$ was required, although c now represented not the total quantity of heat in a particle of gas but only that part of it that was 'free'. A new term i had to be introduced to represent the quantity of latent heat. The form of the expression for the velocity of sound, i.e. for $\sqrt{(\mathrm{d}P/\mathrm{d}\rho)}$, which he now derived,‡ $\sqrt{\{(2P/\rho)(1-C)\}}$, where P and ρ represented the pressure and density of the undisturbed gas and $C = -\rho/c(\mathrm{d}c/\mathrm{d}\rho)$, was similarly influenced in no way by the change. That this expression, when $\mathrm{d}c/\mathrm{d}\rho$ and hence also C are zero (i.e. under the condition that the heat in the compressed gas neither escapes nor becomes latent), yields far too high a value for the velocity is obvious§ and, as was mentioned above, this point was evidently recognized by Laplace even in his paper in September.

The precise determination of C, which was so essential if Laplace's analysis was to be proved correct, was impossible by any known direct method, but a simple argument showed that $(1-C) = \tfrac{1}{2}c_p/c_v$ and so reduced the expression $\sqrt{\{(2P/\rho)(1-C)\}}$ to the $\sqrt{\{(c_p/c_v)(P/\rho)\}}$ already derived in 1816.¶ To make this simplification Laplace assumed that the total heat content of a particle, $(c+i)$ or Q, say, was a function of any two of P (i.e. of $\rho^2 c^2$), ρ, and the absolute temperature (u) of the gas. From the supposition that Q remained constant, it followed that in a sound-wave

$$\mathrm{d}P\left(\frac{\partial Q}{\partial P}\right)_\rho + \mathrm{d}\rho\left(\frac{\partial Q}{\partial \rho}\right)_P = 0, \qquad (3)$$

where partial differentials are introduced for clarity, although they were not in use when Laplace wrote. Hence $(1-C)$, or $(\rho/2P)(\mathrm{d}P/\mathrm{d}\rho)$, became equal to $-(\rho/2P)\{(\partial Q/\partial \rho)_P/(\partial Q/\partial P)_\rho\}$, which to anyone unfamiliar with the interconvertibility of heat and work was itself equal to $\tfrac{1}{2}c_p/c_v$ or, in the customary modern notation, $\tfrac{1}{2}\gamma$. In this way the determination of $(1-C)$, a quantity

† Laplace, 'Développement de la théorie des fluides élastiques et application de cette théorie à la vitesse du son', *Connaissance des tems* ... *pour l'an 1825* (1822), p. 223. See also his comments in the continuation of this paper on pp. 302–5. ‡ Laplace, op. cit., pp. 224–6.
§ It yields a velocity of $2P$, instead of the correct value, $1\cdot 40P$.
¶ Laplace, op. cit., pp. 302–4.

intimately related to Laplace's speculations on the state of caloric in bodies, was reduced to that of γ, a quantity which in 1821–2 it was possible to measure by experiment far more reliably than in 1816 and quite independently of any physical explanations of pressure and radiation.

The necessary experimental advance had been made, albeit unwittingly, by Clément and Desormes, for although they themselves had not been concerned to determine γ, their work provided all the data required for the calculation. Naturally Laplace could not accept Clément and Desormes's explanation of the temperature changes occurring in their experiments. Instead he argued that opening the stopcock of a vessel in which the pressure had been slightly reduced caused the gas already in the vessel to be compressed and so heated.† In fact, provided the initial pressure is only slightly below that of the gas entering, this explanation, which is widely used today,‡ is a good approximation to the truth, and the necessary conditions were certainly fulfilled in the experiments of Clément and Desormes. Laplace's expression for γ in terms of quantities given by Clément and Desormes was therefore correct, within these limitations, and his result of 1·35 was close to the true figure for air of 1·40. Further similar experiments conducted by Gay-Lussac and the chemist and industrialist Jean Joseph Welter during 1822, probably at Laplace's instigation,§ only served to confirm the essential correctness of the method. The new experiments were preferable in several respects but above all in the reduction that was effected in the time for which the stopcock was open and so also in heat loss. Hence it is not surprising that they yielded values of 1·37244 and, a little later, of 1·3748 which brought the predicted figures for the velocity of sound (335·2 m/s and 337·14 m/s respectively) into rather better agreement with experiment. That the agreement was still not perfect was almost certainly the motive for the experimental redetermination of the velocity that Laplace persuaded the Bureau des Longi-

† Laplace, op. cit., pp. 304–6.
‡ See, for example, G. R. Noakes, *A text-book of heat* (1945), pp. 262–7, or virtually any textbook on heat.
§ See Laplace, op. cit., p. 306; also, on the dating of the experiments, p. 372, where the work was said to be in progress at the time of writing (i.e. in 1822). See also Laplace, *J. Phys.* **94** (1822), 89, and the comments of Gay-Lussac and Welter themselves in *Annls Chim. Phys.* **19** (1822), 436–7.

tudes to undertake, but, as it happened, the new measurements, made in June 1822 by a distinguished group under Arago,† only served to make the discrepancy slightly greater.

The additional observation by Gay-Lussac and Welter that γ remained very nearly constant over a wide range of temperature and pressure was an important one. It led to the conclusion, for example, that the velocity of sound should be independent of pressure, a fact that measurements made at the high altitude of Quito by a group of French and Spanish academicians in 1740 had already shown to be true. Unfortunately, it also encouraged Laplace to make one quite erroneous prediction.‡ When the condition $\gamma = $ constant was introduced into the relationship $\gamma = -\{(\partial Q/\partial \rho)_P/(\partial Q/\partial P)_\rho\}(\rho/P)$ and integration performed, it followed that Q must be a function of $P^{1/\gamma}/\rho$ or, where $P = k\rho u$, of $kuP^{(1/\gamma - 1)}$. k was here an unknown constant. By postulating the simplest possible relationship between Q and $P^{(1/\gamma - 1)}$, viz. proportionality, Laplace could show that

$$Q = HkuP^{(1/\gamma - 1)}, \S \qquad (4)$$

where H was another constant, and hence also that Q was proportional to u for heating at constant pressure. This last result he saw as strong confirmation for his work, for it was essentially no more than he had assumed to be true in 1816, and in a similar type of argument conducted before the results of Gay-Lussac and Welter became available he had even begun with it as his basic assumption.¶

It was now a simple step from eqn (4) to the prediction that the ratio between the volume specific heats of any gas measured at

† Arago, 'Résultats des expériences faites par ordre du Bureau des Longitudes, pour la détermination de la vitesse du son dans l'atmosphère', *Annls Chim. Phys.* **20** (1822), 210–23.

‡ Laplace, *Connaissance des tems* . . . *pour l'an 1825*, pp. 307–8.

§ A slightly different form of this equation appeared in April 1823 in the *Traité de mécanique céleste*, vol. 5, p. 128.

¶ Laplace, *Bull. Soc. philomath. Paris*, N.S. **8** (1821), 170–1. The paper in which the argument appears bears the same title as the papers in the *Connaissance des tems* for 1825 and, like the first of those papers, it is dated 12 December 1821. However, it differs in certain respects from the three *Connaissance des tems* papers, notably in that no mention is made in it of the work of Gay-Lussac and Welter. This fact supports my view that at least the second and third of the *Connaissance des tems* papers had not been written by December 1821 (see p. 166 n. § and p. 172 n. §).

two different pressures P_0 and P_1, but at the same temperature, was given by

$$\frac{c_{p_1}(v)}{c_{p_0}(v)} = \left(\frac{P_1}{P_0}\right)^{1/\gamma}. \tag{5}$$

Unfortunately Delaroche and Bérard's results only confirmed this false relationship. Indeed, their insertion in eqn (5) yielded a good value of 1·425 for γ, different, it should be noted, from the value which Laplace had derived from the same experimental data in 1816.[†] In fact it is clear that now (in 1822) he had abandoned this earlier method of calculation, although he had adopted an argument very similar to it in a paper published in the *Bulletin des sciences* of the Société Philomathique late in 1821.[‡] In the latter paper, as in that of 1816, he had used the argument to show $\gamma = 1{\cdot}5$, and it may well have been this same method that he was still using in December 1821 when, without explanation, he gave an approximate figure of 0·8 for $(1-C)$, i.e. for $\tfrac{1}{2}c_p/c_v$.[§] Not the least reason for supposing that this may have been the case is that otherwise a somewhat complex reconstruction of Laplace's reasoning[¶] seems to be necessary. However, the point is not a crucial one, for, whatever Laplace's attitude to his earliest method for calculating γ may have been in the period up to December 1821, there can be no doubt that he had rejected it by the early part of the following year and that he then placed far greater confidence in experiments of the Clément and Desormes type than in those of Delaroche and Bérard.[||]

The definitive version of Laplace's views appeared as Book XII of the fifth volume of the *Mécanique céleste*, published in April 1823.[††] It is unnecessary to examine the work in detail, since it consisted almost entirely of the papers that had already been

[†] See pp. 161–4.
[‡] Cited on p. 171 n. ¶.
[§] Laplace, *Connaissance des tems* . . . *pour l'an 1825*, p. 226. The change in Laplace's value for γ is, of course, further evidence that the second and third of the three papers on elastic fluids in the *Connaissance des tems* for 1825 were written some time after the first. Laplace's ideas were clearly changing very rapidly about this time.
[¶] Such as that given in Finn, *Isis* **55** (1964), 16.
[||] For evidence of his scepticism towards Delaroche and Bérard's results see Laplace, op. cit., p. 226.
[††] Laplace, *Traité de mécanique céleste*, vol. 5, pp. 87–144. Although the volume as a whole bears the date 1825, the books were published and dated individually.

published in the *Connaissance des tems*.† There were minor modifications of course. For example, it was now assumed from the outset that the caloric of a body existed in two states, and no mention was made of the application of Delaroche and Bérard's results to the determination of γ. Such changes inevitably gave the caloric theory of gases the appearance of even greater consistency and rigour, so that the question of why this work, backed by the authority of Laplace and published in his greatest book, exerted so little influence becomes all the more interesting. There can be no doubt that any explanation must take account of the more general decline of Laplacian science which is examined in Chapter 7, but some attempt to assess the merits of the theory considered in isolation will not be out of place here.

It was clearly a notable achievement to have devised a physical picture of the state of caloric in gases from which the laws of Boyle, Gay-Lussac, and Dalton could be predicted, but Laplace's theory left much to be desired. For example, even his initial assumptions concerning the dependence of P and $\Pi(t)$ on ρ and c were by no means beyond objection. The precise form that they took had clearly been determined by the requirement that the deductions made from them should agree with the observed laws, and they had no basis either in Laplace's earlier thought or in the work of any contemporary calorist. The *ad hoc* nature of the assumptions is brought out well in an important modification that Laplace made early in the development of his theory. In his first paper on the subject, in September 1821,‡ he had shown that the main gas laws could be predicted if it was assumed that there was an attractive force between the ponderable 'molecule' at the centre of any one gas particle and the caloric attached to other molecules. This attractive force was supposed to be far smaller than the inter-particle repulsion and so to be overcome by it, but it seemed reasonable to expect that it existed and to take it into account in calculation. Only some three months later, however, after detecting an error in his earlier argument and so realizing that the gas laws were not in fact predicted on this basis, he denied outright the existence

† His short papers in *Connaissance des tems* . . . *pour l'an 1825*, pp. 371–2 and 386–7, did not appear, however.

‡ Laplace, *Connaissance des tems* . . . *pour l'an 1824*, pp. 338–42.

of any attractive force.† In short, there was nothing inviolate about his initial premises. They were designed first and foremost to 'save the phenomena' and only by the false relationship expressed in eqn (5) was anything resembling truly independent confirmation possible.‡

That the same initial premises also led, or were made to appear to lead, to the correction factor $\sqrt{(c_p/c_v)}$ in the expression for the velocity of sound was perhaps their most remarkable property. But here again Laplace's argument was open to criticism, for it depended not on the more speculative parts of the theory but only on the belief, which was then virtually beyond dispute, that Q was a function of P and ρ only. Nevertheless, writing in 1822, Laplace used the experimental determinations of γ that were now available to extend his knowledge of the state of caloric in bodies.§ Was not $(1-C)$ equal to $\frac{1}{2}\gamma$ and so also to 0·6874 (by the experiments of Gay-Lussac and Welter)? And did not this in turn give a definite figure for C and hence for $(\rho/c)(\mathrm{d}c/\mathrm{d}\rho)$? For Laplace the fact that C was a positive quantity was sufficient proof that the quantity c in a gas undergoing adiabatic compression decreased with increasing density. The question whether the caloric thus 'expelled' was radiated or became latent was essentially a separate issue, although Laplace now had no hesitation in claiming that his determination of C proved not only that $(\mathrm{d}c/\mathrm{d}\rho) < 0$ but also, since $\mathrm{d}c = -\mathrm{d}i$ in conditions such that there was no heat loss, that $\mathrm{d}i$ too was finite. He concluded: 'Hence the existence of latent heat (i) and the fact that its quantity is increased by pressure are the results of observation.'¶ The statement was bold, to say the least, although the presence of latent as well as sensible heat in gases did have the additional support of Delaroche and Bérard's paper of 1812, as Laplace pointed out.

How the mechanisms that Laplace devised to account for the behaviour of gases were viewed by his contemporaries is difficult to ascertain, but the very scarcity of comment‖ is surely signifi-

† Laplace, *Connaissance des tems* . . . *pour l'an 1825*, p. 223 (in the paper dated 12 December 1821).

‡ It is hardly necessary to stress that the confirmation was quite illusory, with the errors in Delaroche and Bérard's data compensating the error in eqn (5).   § Laplace, op. cit., pp. 308–9 and 318–19.

¶ Laplace, op. cit., p. 319.

‖ A particularly detailed criticism by John Herapath appeared in *Phil. Mag.*

TWO GREAT CALORISTS

cant. Even such a close disciple as Poisson avoided using the mechanisms when, in the spring of 1823, he took up some of the problems that Laplace had raised.† Like his master, Poisson used Clément and Desormes's data to determine the correction factor to be applied to Newton's expression for the velocity of sound, inserting them in a result that he had already given in 1807,‡ namely

$$v = \sqrt{\left[gHn\left\{1 + \frac{\alpha.\Delta t}{(1+\alpha t)(\Delta V/V)}\right\}\right]}.$$

In the first place he used the data simply to calculate the increment in temperature Δt caused by a fractional decrease in volume $\Delta V/V$, but he also showed that his correction factor to the Newtonian expression $\sqrt{(gHn)}$ was equal to $\sqrt{(c_p/c_v)}$. The argument was simple. He imagined a gas being heated at constant pressure from a temperature $(t-\Delta t')$ to t and then being compressed, without heat loss and with a further rise in temperature of Δt, back to its original volume. Finally, in a third stage, the gas regained its initial conditions by cooling at constant volume from its temperature after compression, $(t+\Delta t)$, to $(t-\Delta t')$. Since the initial and final conditions were the same and since also heat was neither added to nor abstracted from the gas during the second stage, Poisson could equate the amounts of heat gained and lost in the first and third stages respectively. Hence he put

$$c_p \Delta t' = c_v(\Delta t' + \Delta t)$$

and, where $\Delta t' = \dfrac{(1+\alpha t)\Delta V/V}{\alpha}$, he arrived at

$$\frac{c_p}{c_v} = \frac{\Delta t' + \Delta t}{\Delta t'} = 1 + \frac{\alpha.\Delta t}{(1+\alpha t)\Delta V/V}.$$

By substituting $d\rho/\rho$ for $\Delta V/V$ and dt for Δt, it was now a simple

---

62 (1823), 61–6 and 136–9. In the same volume of the *Philosophical Magazine* (pp. 329–38) Herapath also attacked Poisson in the footnotes to his translation of some of Poisson's work on the caloric theory. For another critical comment see W. Whewell, *A history of the inductive sciences* (1837), vol. 2, pp. 530–3.

† Poisson, 'Sur la vitesse du son', *Connaissance des tems . . . pour l'an 1826* (1823), pp. 257–77. A shortened version appeared in *Annls Chim. Phys.* **23** (1823), 5–16.

‡ Poisson, *J. Éc. polytech.* **7**, cahier 18 (1808), 360–4.

matter to derive the now familiar relationships for adiabatic volume change,

$$t+266 \cdot 7 = 266 \cdot 7 \left(\frac{V_0}{V}\right)^{\gamma-1}, \qquad (6)$$

where V_0 and V were the volumes of a gas at 0 °C and t °C respectively and where the only error arose from Gay-Lussac's figure (1/266·7 per °C) for the expansion coefficient of gases. From this the other relationships

$$PV^\gamma = \text{constant} \qquad (7)$$

and $$T^\gamma P^{1-\gamma} = \text{constant} \qquad (8)$$

follow immediately.† It is hard to believe that Laplace himself would not have seen how these important equations could be derived from eqn (3),‡ but he never gave them explicitly and Poisson must therefore retain at least some credit for their discovery.

Later in the same year, in August, Poisson published a further extension of Laplace's views on heat.§ Again he avoided detailed discussion of the state of caloric in bodies, merely assuming that the amount of heat in a gas in excess of that present at some standard temperature and pressure was given by $q = f(P, \rho)$. This assumption, similar to that already made by Laplace, led first to an expression identical in form to eqn (3), though with Poisson's q replacing Q, to eqns (5), (6), and (7), and then to the relationship

$$q = A + B(266 \cdot 7 + t) P^{(1/\gamma - 1)},¶$$

where A and B were constants. Expressions for the specific heats by weight, $BP^{(1/\gamma-1)}$ and $(1/\gamma)BP^{(1/\gamma-1)}$ for c_p and c_v respectively, were also given and the magnitude of B was determined by inserting Delaroche and Bérard's data.

Throughout these two papers there is strong evidence that Poisson leant heavily on Laplace. Most of his results had already been obtained by Laplace or could have been derived by him

---

† Although only eqns (6) and (7) were given by Poisson.
‡ In fact he came very close to them in *Connaissance des tems ... pour l'an 1825*, p. 307.
§ Poisson, 'Sur la chaleur des gaz et des vapeurs', *Annls Chim. Phys.* 23 (1823), 337–52.
¶ Cf. the expression in Laplace, *Traité de mécanique céleste*, vol. 5, p. 128.

without difficulty. Even Poisson's treatment of the quantity of heat in water vapour, which led to the erroneous result that γ for steam was 1·073,† was an obvious extension of a passage that had appeared in the *Mécanique céleste*.‡ Poisson's greatest contribution, then, was rather to free Laplace's work of its more suspect elements, merely by picking up the argument at the stage $q = f(P, \rho)$. In doing so he was showing in a most effective manner just how irrelevant much of Laplace's theory was, so that even to a reader convinced of the physical reality of caloric Poisson's must have seemed undeniably the more fruitful approach.

Sadi Carnot

Any addition to the mass of literature relating to Sadi Carnot must necessarily be made with caution. Carnot's contributions to the theory of heat engines and to theoretical physics have received the well-deserved attention not only of numerous historians but also of the writers of our modern textbooks on thermodynamics. Yet problems remain and it is in an attempt to remove certain specific and limited gaps in our understanding of Carnot that this account will seek to relate his contributions in the field of physical theory to the caloric theory of gases, to the work of those of his contemporaries who accepted the theory, and to the experimental data that supported it.

Recent studies have confirmed beyond all reasonable doubt that Carnot owed far more to his contact with the world of power engineering than he did to the physics of his day.§ It is now clear that the precedents for his innovations are to be found not so much in the scientific traditions and textbooks of the

† Poisson, op. cit., pp. 342–8.
‡ Laplace, op. cit., vol. 5, pp. 139–40.
§ See especially T. S. Kuhn, 'Engineering precedent for the work of Sadi Carnot', *Actes du IXe Congrès International d'Histoire des Sciences. Barcelona—Madrid, 1959* (Barcelona and Paris, 1960), pp. 530–5, and D. S. L. Cardwell, 'Power technologies and the advance of science, 1700–1825', *Technology Cult.* 6 (1965), 188–207. The view that Carnot's *problem* in the *Réflexions sur la puissance motrice du feu* was one of engineering, and he was there addressing engineers rather than physicists, is now, of course, a commonplace. See, for example, J. T. Merz, *A history of European thought in the nineteenth century* (4 vols., Edinburgh and London, 1896–1914), vol. 2, p. 217; L. Rosenfeld, 'La genèse des principes de la thermodynamique', *Bull. Soc. R. Sci. Liège* 10 (1941), 197–212; S. Lilley, 'Social aspects of the history of science', *Archs int. Hist. Sci.* 2 (1948–9), 392–4.

early nineteenth century as in the practices and beliefs that were current among engineers and were described in such treatises as Lazare Carnot's *Principes fondamentaux de l'équilibre et du mouvement* (1805), A. Guenyveau's *Essai sur la science des machines* (1810), J. N. P. Hachette's *Traité élémentaire des machines* (1st edn, 1811; 2nd edn, 1819), and Héron de Villefosse's three-volume *De la richesse minérale* (1810–19).

More precisely, I believe, Carnot's work can be seen, in many respects, as a natural if unusually brilliant product of a period when power technology in general and steam engines in particular were arousing unprecedented interest in France. The appearance of this new interest can be dated to 1815, when high-pressure engines employing Watt's expansive principle were first introduced into France from England by Humphrey Edwards, a former associate of the great Cornish engineer Arthur Woolf.[†] Edwards's engines, which were of the Woolf type (i.e. operating at roughly 2 atmospheres and with one small and one large cylinder) were immediately successful—not surprisingly in a country where even the low-pressure Watt engine had never become widely used but where the virtues of steam power were at long last being recognized. By November 1817 Edwards had already erected fifteen of his engines in France; by 1824 this number had risen to 300,[‡] and some of the most talented minds of the day had already begun to take an interest in the theoretical problems raised by the new machines. Probably the best treatment of the theory of the Woolf engine was that given in 1818 by Hachette, the former professor of geometry and mechanics and the successor of Monge at the École Polytechnique;[§] but it was a paper published in the same year by the gifted young physicist A. T. Petit,[¶] who, although he was not yet 27, had held the chair of physics at the École Polytechnique for nearly three years,

[†] On the arrival of the high-pressure Woolf engine in France in 1815 and on its great influence on French power technology see R. Fox, 'Watt's expansive principle in the work of Sadi Carnot and Nicolas Clément', *Notes Rec. R. Soc. Lond.* **24** (1969), 238–40.

[‡] R. Jenkins, 'A Cornish engineer: Arthur Woolf, 1766–1837', *Trans. Newcomen Soc.* **10** (1932–3), 59 and 61.

[§] Hachette, *Bull. Soc. Encour. Ind. natn.*, XVIIe année (1818), pp. 169–74. This was reproduced in the following year in Hachette's *Traité élémentaire des machines*, pp. 210–16.

[¶] Petit, 'Sur l'emploi des forces vives dans le calcul de l'effet des machines', *Annls Chim. Phys.* **8** (1818), 287–305.

which attracted the greatest attention. Petit's paper was not concerned exclusively with the Woolf engine but more generally with the expansive use of steam. It showed, quite erroneously of course,† that in an ideal heat engine operating expansively the use of air as the working substance in place of steam would bring about a fourfold increase in the effect obtainable from a given quantity of heat. Other writers were unconvinced and in August 1819 Clément and Desormes, the unsuccessful entrants in the Institute's prize competition of 1812, presented a paper before the Académie des Sciences in which they appear to have pointed out certain errors in Petit's treatment,‡ though not the most important error, which was that Petit, along with virtually all his contemporaries (Carnot is the only known exception), had considered the work done in the expansion stroke only and not the work done in a complete cycle of operations.

Of all the papers on the theory of the steam-engine that were written in the 1810s there can be no doubt that Clément and Desormes's was the one that had the greatest influence on Carnot, and it is unfortunate, therefore, that this paper was evidently thought unworthy of publication by the referees, Fourier, Arago, Thenard, and Gay-Lussac. No trace of it has been found and all we have is a brief summary that appeared in the *Bulletin* of the Société Philomathique for August 1819.§ We know, however, that Carnot read the paper in its full manuscript form, and in the *Réflexions sur la puissance motrice du feu*,¶ as also in a recently published manuscript paper,‖ he made an elaborate acknowledgement to Clément both for the general style of his approach to the problem and for a new 'law' that was announced for the first time in the 1819 paper. Elsewhere I have examined at some length the nature and extent of Carnot's debt to Clément and have tried to show that it was indeed very

† For a recent comment on the errors in Petit's paper see E. Mendoza, 'Contributions to the study of Sadi Carnot', *Archs int. Hist. Sci.* **12** (1959), 393–5.
‡ The reading of the paper, by Clément, and the selection of referees are recorded in *Procès-verbaux*, vol. 6, pp. 480 and 481 (16 and 23 August 1819).
§ Desormes and Clément, 'Mémoire sur la théorie des machines à feu (extrait)', *Bull. Soc. philomath. Paris*, N.S. **6** (1819), 115–18. In the *Procès-verbaux* the reference is to 'un mémoire sur les machines à vapeur', no fuller title being given.
¶ Carnot, *Réflexions sur la puissance motrice du feu* (1824), p. 98 n.
‖ W. A. Gabbey and J. W. Herivel, 'Un manuscrit inédit de Sadi Carnot', *Revue Hist. Sci. Applic.* **19** (1966), 151–66.

considerable.† Hence further detailed discussion seems unnecessary, yet the matter is briefly introduced here because it does throw some light on the relationship between Carnot's work and contemporary knowledge concerning the thermal properties of saturated vapours.

The chief problem which Clément and Desormes tackled in their paper was the familiar one of determining the maximum effect that could be obtained from a given quantity of working substance, which could be a vapour or a gas, under various conditions of temperature and pressure. The fact that their paper was never published in full makes complete reconstruction somewhat difficult, but references in two sets of notes that were taken a few years later at lectures given by Clément‡ leave the general principles of their solution, at least as far as steam was concerned, in little doubt. The basis of the argument was the simple thought experiment shown in Plate 8. In this Clément and Desormes imagined steam being introduced, in the form of a bubble, at the bottom of a tall vessel that was filled to the brim with water. Hence, as the steam was introduced and the bubble grew bigger, water necessarily overflowed (at D in the figure), indicating that work had been performed or, to use the terminology of the paper, that a certain quantity of 'mechanical power' (*puissance mécanique*) had been expended. This, however, was not the only effect that could be obtained from the working substance, as Clément and Desormes pointed out and as any power engineer of the day who was familiar with the expansive use of steam would also have argued. For if the bubble was now allowed to rise in the vessel, it would expand still further, so causing more water to overflow and more work to be performed.

† In the paper cited on p. 178 n. †. See also 'The intellectual environment of Sadi Carnot: a new look', a paper which I read in August 1968 at the XIIe Congrès International d'Histoire des Sciences in Paris and which is to be published shortly in the *Actes* of the congress.

‡ The more extensive of the two sets of notes was taken at Clément's lectures at the Conservatoire des Arts et Métiers by J. M. Baudot between 1824 and 1828. The notes are bound in three volumes and are now MS. 8° Fa 40 (2) in the library of the Conservatoire. The relevant notes were taken early in 1825 and appear in the first volume. The other set of notes was taken, also at the Conservatoire but in 1823–4, by the mathematician L. B. Francoeur. It is now MS. 407 in the library of the École Nationale Supérieure des Beaux-Arts, Paris. I am grateful to Professor E. Mendoza for drawing my attention to these notes.

PLATE 8

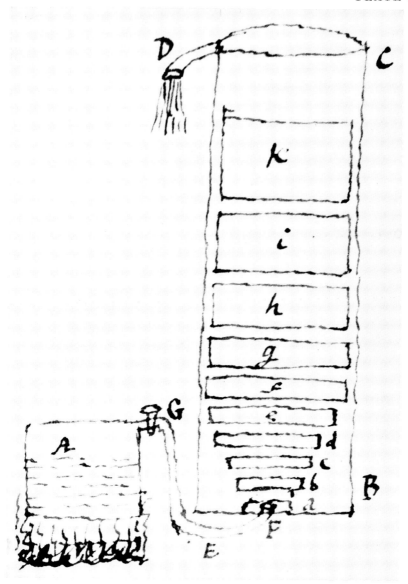

Thought experiment devised by Clément and Desormes to illustrate their views on the production of 'mechanical power' in a heat engine operating expansively. The rectangles $a, b, c, d, \ldots$ represent the volume of the working substance (normally steam) at successive points during its ascent under adiabatic conditions through the water filling the vessel. The volume at cut off, just before the expansive phase, is represented by a. This sketch appears in a set of notes taken by Louis Benjamin Francoeur at the lectures which Clément gave at the Conservatoire des Arts et Métiers in 1823–4. Photograph by courtesy of the Director of the École Nationale Supérieure des Beaux-Arts, Paris.

The mechanical power expended by the working substance in the first stage of this experiment, a stage that Clément later termed '*production*', was easily calculated from the readily available tables of vapour pressure.† But the corresponding calculation for the second stage, the 'expansion' (or *détente*), required the acceptance of the new 'law' to which reference has already been made and according to which the quantity of heat in unit mass of any given saturated vapour was always the same, being independent both of its temperature and pressure. Now Clément and Desormes had laid down that the bubble in their thought experiment was impervious not only to water but also to heat, so that it followed that the expansion of the working substance in the second stage took place adiabatically. Hence, by the new 'law', steam that was introduced at the bottom of the vessel in a saturated state would remain saturated as it rose, expanded, and cooled, and its pressure and temperature at any point would therefore constantly be related in accordance with the vapour-pressure tables.‡

In discussing the extent of Clément and Desormes's influence on Carnot, it is important to assess the originality of the thought experiment just described. In the first place, the distinction between the production and expansion stages, which corresponded clearly to the two stages in the expansion stroke of an expansive engine, and the belief that maximum effect was obtained only if the working substance was allowed to expand freely after the cut-off of supply, were far from novel in 1819. In these respects the treatment simply reflected the current interest in the expansive use of steam.§ But the view that the expansion after cut-off should take place adiabatically certainly was original, for until Carnot's work was rediscovered about the middle of the nineteenth century it was almost axiomatic, at

---

† It was, of course, simply the product of the vapour pressure of the saturated steam being introduced and its total volume.

‡ For a detailed account of the calculation see Fox, *Notes Rec. R. Soc. Lond.* **24** (1969), 240–5. Clément's table of results giving the maximum quantities of mechanical power that could be obtained from a steam-engine operating expansively was published separately in 1826 and reprinted in F. P. C. Dupin, *Géométrie et mécanique des arts et métiers et beaux-arts* (Brussels, 1826), vol. 3, p. 378. It demonstrates clearly the advantages to be gained by using steam expansively and at high pressure.

§ See pp. 178–9.

least for the purposes of calculation, that the expansion was isothermal.†

Inevitably, as a result of this difference, Clément and Desormes's treatment departed markedly from others that were available at the time in the method of calculating the effect obtained in the second part of the expansion stroke (i.e. in the process of *détente*). While Petit, for example, used Boyle's law to show that the total effect after cut-off was proportional to $\ln(V_2/V_1)$ for an expansion from volume V_1 to V_2, and while Hachette similarly made use of Boyle's law in his treatment of the Woolf engine,‡ Clément and Desormes used their 'law', and the assumption that the expansion took place adiabatically, in order to derive a quite different result. In doing this they made the additional assumption that the saturated steam could be treated as a perfect gas from the point of view of its obedience to the laws of Gay-Lussac and Boyle. Hence, knowing the pressure and so also the temperature of the steam at any point as it rose in the vessel, they were able to determine the changes in volume that occurred. This made the calculation of the mechanical power a simple matter, for by considering the rise of the bubble of steam over a short distance only, they could put the mechanical power equal to the product of the mean pressure during that part of the expansion and the increase in volume. By a somewhat tedious process of addition it was then possible to determine (numerically) the total effect obtained for any stated change in pressure or temperature.

The evidence for believing that Carnot was influenced by this treatment is strong, if not absolutely conclusive. It is well known, for example, that in Carnot's cycle of operations for an ideal heat engine the second part of the expansion stroke was treated as a case of adiabatic expansion, as it was in Clément and Desormes's thought experiment but nowhere else.§ Moreover, from a manuscript paper by Carnot, which was probably written

† To cite just one particularly noted example, the assumption was still made by Victor Regnault in his monumental writings on steam-engines in the 1840s; see *Mém. Acad. Sci. Inst. Fr.* **21** (1847), 6–11.

‡ Petit, *Annls Chim. Phys.* **8** (1818), 291–2, and Hachette, *Traité élémentaire des machines* (2nd edn, 1819), pp. 212–16. On Hachette's reservations concerning the use of Boyle's law here see the next note.

§ Although certain engineers were aware that *some* cooling might well occur as a result of the expansion. See, for example, Hachette, op. cit., pp. 198 and 214–15, and D. Gilbert, *Phil. Trans. R. Soc.* **117** (1827), 34.

about 1827,† we know that he was using a method for the calculation of the effect obtained after cut-off which was basically identical to Clément and Desormes's although it was somewhat more elegant and led to a general expression rather than to numerical results for particular instances. It is conceivable, of course, that Carnot arrived at his analysis of the expansion stroke quite independently of Clément and Desormes, but two considerations make this unlikely. First, to repeat a point already made, it cannot be emphasized too strongly that in 1819 the thought experiment described above was quite unique in a number of important respects, but most notably in the belief that maximum effect could be obtained only if the expansion after cut-off took place adiabatically and also in the use of Clément and Desormes's 'law' for the purposes of calculation. And secondly, we must point to the very relevant fact that Carnot and Clément are known to have been good friends. This is vouched for explicitly by Hippolyte Carnot, Sadi's younger brother and his best-known biographer, who records that in the late 1820s Carnot was a frequent visitor to the Conservatoire des Arts et Métiers, where Clément held the chair of industrial chemistry from 1819 until his death in 1841.‡ Admittedly Hippolyte associates the visits particularly with the *late* 1820s but since Sadi is known to have attended at least one of Clément's lectures in January 1825 and since the account of the visits was not given until 1878 we can hardly rule out the possibility that the friendship in question was already a close one *before* the publication of the *Réflexions*. And in any case, whether or not the friendship was a close one, we know that *some* contact had been made by 1824, since in the *Réflexions* Carnot acknowledged the kindness of 'the author' in allowing him to examine the joint paper of 1819.§

It must be emphasized in conclusion that Carnot did not owe

† Gabbey and Herivel, loc. cit. On the dating of the paper see Fox, op. cit., pp. 246–7. The title of the paper, which is now in the library of the École Polytechnique, is 'Recherche d'une formule à représenter la puissance motrice de la vapeur d'eau'.

‡ See Hippolyte's biographical sketch appended to the Gauthier Villars edition of the *Réflexions sur la puissance motrice du feu* (1878), p. 77. Reference to the friendship was also made in a short biography written by Sadi Carnot, Hippolyte's son, and communicated to the Italian Count Paolo di San Roberto; see *Atti Accad. Sci., Torino* **4** (1868), 157.

§ Carnot, *Réflexions* (1st edn.), p. 98 n. The 'author' referred to was Clément.

everything to his association with Clément, even though this may well have provided the greatest single stimulus for his achievements. For example, both the idea of a closed cycle of operations and the fruitful analogy between the fall of water in a water-powered engine and the 'fall' of caloric in a heat engine appear to have their precedents, such as they are, in quite different aspects of the engineering tradition of the early nineteenth century.† But the association seems scarcely less important for that. It adds weight to the already widespread belief that Carnot's true intellectual milieu was among the engineers and that his relationship with the scientific 'establishment' of Paris, centred on the Académie des Sciences and the École Polytechnique, was of the most tenuous kind.

Yet he did have certain problems in common with a number of the great men of Parisian science and, perhaps because of his detachment from the traditions in which they were working, he was able to view some of these problems in a refreshingly original manner. For our purpose, of course, it is of special interest to examine the way in which Carnot applied the fundamental new concepts of the *Réflexions* to the theoretical study of the thermal properties of gases. Here the closed cycle of operations proved a powerful tool, although errors inevitably arose from Carnot's assumption that the motive power produced in the cycle resulted from the 'fall' of caloric from the temperature of the boiler to that of the condenser (rather than from its conversion into work). Among those of his results that were correct, at least for a perfect gas, one of the most important was the conclusion that the difference between the volume specific heats at constant pressure and at constant volume must be the same for all gases, under similar conditions of pressure, etc. The demonstration‡ depended entirely on the principle, which was established early in the *Réflexions*, that the motive power produced in a Carnot cycle was not affected by the nature of

† See especially Kuhn, *Actes du IXe Congrès International d'Histoire des Sciences* (*1959*), p. 532, and Cardwell, *Technology Cult.* **6** (1965), 200–5. I use the term water-powered engine here, instead of the more common 'waterwheel', in the light of Dr Cardwell's comments in his paper. Dr Cardwell argues that in drawing the analogy between a head of water and a 'head' of caloric as sources of power, Carnot probably had in mind the column-of-water engine rather than the water-wheel.

‡ Carnot, op. cit., pp. 41–8.

the working substance. Hence if equal volumes of two gases at the same temperature and pressure were put through the same cycle of operations, they would yield identical quantities of power. If it was now supposed that the cycle operated between two temperatures which were so nearly equal that the contribution of the two adiabatic stages to the total output could be neglected, it followed, since Carnot's was the most efficient possible cycle, that equal volumes of all gases would absorb the

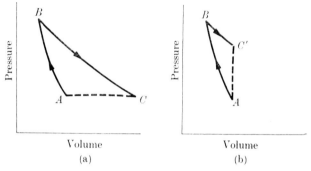

FIG. 2. (a) Heating at constant pressure. (b) Heating at constant volume.

same quantity of caloric when they expanded isothermally by the same amount and at the same temperature. Since this quantity of caloric was seen to represent the part of $c_p(v)$ that went solely to effect the volume change during heating, Carnot could then argue that $\{c_p(v)-c_v(v)\}$ must be the same for all gases. He could even quantify the difference between the two specific heats by considering the heating of two identical samples of air, in one case at constant pressure and in the other at constant volume. Temperature increments of one degree, he argued, could be effected either directly, by adding the appropriate quantities of heat, which would be proportional respectively to c_p and c_v, or in each case by a different two-stage process. For the heating at constant pressure (see Fig. 2 (a)) the volume of the air could first be reduced adiabatically from V to $\{V-(V/116)\}$ and hence, as Poisson had shown,† heated by 1 °C; this was process AB. It could then be expanded isothermally (along BC), absorbing heat until its volume became $\{V+(V/267)\}$, the value which it would have had if it had been

† See p. 86.

heated directly. For the temperature change at constant volume (see Fig. 2 (b)) the first stage of the indirect process (AB) would be the same but in the second stage the isothermal expansion (BC') would be from $\{V-(V/116)\}$ to V. Now it was fundamental to any calorist that the quantity of heat absorbed by the air should be the same whether the simple direct process or the indirect, two-stage process was followed. Therefore Carnot naturally, though falsely, assumed that since no change in heat content occurred during the adiabatic processes AB, the quantities of heat absorbed during the stages of isothermal expansion (BC and BC') must be respectively c_p and c_v in the two cases. By making the further assumption that these quantities of heat were proportional to the changes in volume, he could then show that for air c_p/c_v, or $\{c_p(v)\}/\{c_v(v)\}$, was equal to $\{(1/267+1/116)/1/116\}$ or approximately $1\cdot 0/0\cdot 7$, and consequently that $c_v(v)$ for any gas could be found by subtracting $0\cdot 3$ from Delaroche and Bérard's values of $c_p(v)$, on the scale where $c_p(v)$ for air was unity.

In this case Carnot was fortunate that his erroneous assumptions had not led him too far astray. But elsewhere in the *Réflexions* the inadequacy both of the caloric theory and of the experimental data on which he based his argument become very apparent to the modern reader. There are two obvious illustrations. In one, viz. his proof that the difference between the specific heats of a gas measured at any given temperature but at two different densities ρ_1 and ρ_2 was proportional to $\ln(\rho_1/\rho_2)$,† Carnot was misled entirely by the natural calorist assumption that the heat gained or lost by a gas in passing from one state to another was independent of the path taken. Unfortunately, of course, the belief that variations with density occurred at all was supported by Delaroche and Bérard's results, and Carnot was even able to draw up a table of specific heats at different densities on the basis of their single observation on the matter.‡ In the second illustration, his investigation of the dependence on temperature of the motive power produced by a given 'fall' of caloric,§ he relied still more heavily on Delaroche and Bérard, despite his obvious doubts concerning the dependability of their work.¶ He considered equal masses of air being taken through

† Carnot, op. cit., pp. 56–62. ‡ Carnot, op. cit., p. 61.
§ Carnot, op. cit., pp. 68–79. ¶ Carnot, op. cit., p. 73.

two Carnot cycles operating between temperatures 100 °C and (100−Δt) °C and between 1 °C and (1−Δt) °C respectively. By applying the limitation that Δt was very small he was able to ignore the stages of adiabatic expansion and compression, so that the net gain in motive power became simply the difference between that produced by the gas during isothermal expansion and that expended in compressing it isothermally back to its original volume. Provided the processes of expansion were conducted between the same limits of volume in each of the two cycles, the quantities of motive power produced would thus be independent of the temperature at which the cycle operated.† This was not the case, however, for the quantities of caloric which in each case underwent the 'fall' in temperature Δt, as Carnot proceeded to show. He argued that the temperature of a sample of air at 1 °C could be raised to 100 °C with any desired accompanying increase in volume in two ways. It could be heated at 100 °C at constant volume (along AB in Fig. 3) and then expanded isothermally (BC) or, alternatively, the isothermal expansion could be effected at 1 °C (AB') and the air could *then* be heated at constant volume to 100 °C ($B'C$). Once again he now made the crucial mistake of supposing that the total quantities of heat absorbed in the two processes were equal. But, by accepting Delaroche and Bérard's data as evidence that the quantity of heat required to bring about the rise in temperature increased with decreasing pressure, he arrived at the perfectly correct conclusion that the quantity of heat absorbed during an isothermal expansion between any two given volumes was greater at higher temperatures. Hence, in accordance with his theory of the heat engine, it followed simply that, as he put it: 'The fall of caloric produces more motive power at low temperatures than it does at high ones.'‡

To the extent that the Carnot cycle is indeed more efficient (for a given fall in temperature) at low temperatures, the result was a true one, but it cannot be emphasized too strongly that Carnot's analysis of steam-engine operation was faulty and that he had obtained his result only by chance after a quite false argument.

† They would both be equal to the product of the change in volume and the increment in pressure caused by the temperature change Δt.
‡ Carnot, op. cit., p. 72.

Of course, Carnot was not always so fortunate as to arrive at a correct result. We have already noted his mistaken conclusion concerning the dependence of variations in specific heat on changes in density, and it was in part this error that led him to yet another wrong relationship, this time between the volume and temperature of a gas undergoing adiabatic volume changes.† Here again we see how he was misled by the caloric theory, for

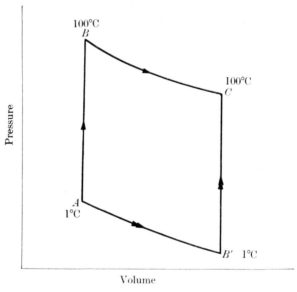

Fig. 3.

he argued that the rise in temperature during an adiabatic compression could be determined by considering the process in two stages, isothermal compression followed by the absorption by the gas of the heat expelled in the first stage, the volume of the gas in this second stage being kept constant. He had already shown, by an unexceptionable argument that involved none of the misleading tenets of caloric theory, that the amount of heat evolved during the isothermal compression of a gas from volume V_1 to V_2 at any given temperature was $\{X + Y \ln(V_2/V_1)\}$, where X and Y were unknown constants.‡ The expression, as

† Carnot, op. cit., pp. 62–8.
‡ Carnot, op. cit., pp. 52–63. See also pp. 73–6 n.

we see, reduces easily to the correct one, $K \ln(V_2/V_1)$, with K here being a function of temperature only.† By now using this expression in conjunction with that for the specific heat of the gas at volume V_2, which also had the form $\{X'+Y'\ln(V_2/V_1)\}$,‡ Carnot could deduce that the rise in temperature during an adiabatic compression from V_1 to V_2 was

$$\left(\frac{X+Y\ln V_2/V_1}{X'+Y'\ln V_2/V_1}\right),$$

an expression differing considerably from the one that Poisson had given in 1823.§ Unfortunately Carnot made no comment on the discrepancy and it is not absolutely certain that in 1824 he even knew of Poisson's derivation, although he appears to have read another of his papers in the same volume of the *Annales de chimie et de physique* in which the $TV^{\gamma-1} = $ constant relationship was announced.¶

Carnot's manuscript notes, most of which were published posthumously by his brother Hippolyte, indicate that he had abandoned the caloric theory before his death in 1832 and that he had adopted the belief that heat was simply motion of the particles of ordinary ponderable matter, motive power being gained only if a corresponding amount of heat was consumed.‖ Unfortunately we have no evidence that Carnot ever investigated the full implications of his new theory, and the only complete paper that was written by him after 1824 was the one discussed earlier in this section, which I have dated to about 1827. In this he sought to derive an expression for the motive power produced by steam in an ideal expansive engine operating between temperatures that differed by an appreciable amount, as opposed to the quite unrealistic 'fall' of only 1 °C used in the *Réflexions*. Evidently he was trying to remedy one of the most obvious of the defects in his book.

Few people were impressed by Carnot and after a single (highly favourable) review of the *Réflexions*, which appeared

† As we now know, of course, K is proportional to the absolute temperature.
‡ Derived from the result on p. 188.
§ See pp. 175–6. Cf. also Navier's expression in *Bull. Soc. philomath. Paris*, N.S. 7 (1820), 97–101.
¶ Carnot, op. cit., p. 43 n.
‖ *Sadi Carnot, biographie et manuscrit*, ed. É. Picard (1927), pp. 76–85.

in the pro-republican journal *Revue encyclopédique*,† ten years elapsed before the French engineer Émile Clapeyron again drew attention to his work. In a lengthy paper published in 1834 in the *Journal de l' École Polytechnique*‡ Clapeyron sought to give Carnot's results the analytical form which they did not have in the *Réflexions*, and he also introduced the now familiar indicator diagram to illustrate the argument.§ Throughout he adhered to the basic principles of caloric theory. For him, as for Laplace, Poisson, and Carnot, the amount of caloric in a gas (Q) was a function only of pressure and volume, so that the heat gained or lost in any change of conditions was given by

$$dQ = \left(\frac{\partial Q}{\partial V}\right)_P dV + \left(\frac{\partial Q}{\partial P}\right)_V dP. \P \qquad (9)$$

From this Clapeyron was able to derive the equation that was fundamental to the whole paper, namely that for a perfect gas

$$Q = R(B - C \ln P), \| \qquad (10)$$

where B and C were unknown functions of temperature and R was the quantity $\{PV/(267+t)\}$. Invoking unspecified experimental evidence, presumably Dulong's results on the specific heats of gases,†† Clapeyron deduced that B was probably the same for all elementary gases at least. With regard to C, on the other hand, there was no doubt that this quantity was completely independent of the nature of the gas, for it could be shown from the Carnot cycle that the greatest quantity of motive power obtainable from a given quantity of heat 'falling' through Δt degrees was $\Delta t/C$ for all gases. C was in fact proportional to the absolute temperature T, but Clapeyron knew only that it was some function of temperature, although he was able to show, by an argument that involved the acceptance of

† *Revue encyclopédique*, **23** (1824), 411–14. The reviewer was the engineer and veteran of the Egyptian campaign P. S. Girard.

‡ Clapeyron, 'Mémoire sur la puissance motrice de la chaleur', *J. Éc. polytech.* **14**, cahier 23 (1834), 153–90.

§ The indicator diagram was not used by Carnot himself. Indeed, despite the fact that it had been used by Boulton and Watt, it had been kept a closely guarded secret by them and was still very little known in Clapeyron's day. I owe my information on the early history of the indicator diagram to Dr D. S. L. Cardwell and Dr A. J. Pacey.

¶ Clapeyron, op. cit., p. 166.

‖ Clapeyron, op. cit., p. 167   †† See pp. 253–5 and Tables A and B.

Delaroche and Bérard's results, that it increased very slowly as a gas was heated.† So once again the derivation of this perfectly correct result depended principally on the false belief, and evidence, that the specific heat of a gas was influenced by changes in its volume.‡ The rest of the paper was devoted largely to showing how Carnot's results could be derived from eqn (10), but there are other points of interest, notably the first announcement of the famous Clapeyron (or Clausius–Clapeyron) equation.§

It should now be clear that Carnot occupies a somewhat isolated place in our story. Certainly the basic principles of his caloric theory and the experimental evidence that he used in expounding it would have been accepted by all calorists in the 1820s, but there is too much in Carnot's work that distinguishes him from Laplace and Poisson for us to accept him as a true Laplacian. His background, as has been pointed out, was firmly in the traditions of power technology and it was in this tradition that he found both his initial problems and his main inspiration. Despite the contributions that he made to the study of the thermal properties of gases, Carnot had no close contact with any major theoretical school, and therefore he stands virtually independent both of the great calorists and of those, such as Dulong, who were already beginning to attack the caloric theory.

Other work

With Carnot and Clapeyron we come to the end of our consideration of the work of the great French calorists. It is true that even after Laplace's death Poisson continued to work in the context of caloric theory. In a lengthy paper, which he read before the Académie des Sciences in October 1829 and published sixteen months later in the *Journal de l'École Polytechnique*¶ he repeated in some detail all the basic principles of Laplace's caloric theory. But by this time, and still more so by 1835, when his *Théorie mathématique de la chaleur* was published, Poisson's work was a relic of a bygone age and it was seen as such by his contemporaries.|| In the 1830s the prevailing attitude towards

† Clapeyron, op. cit., p. 182.
‡ Although Clapeyron did give other supporting evidence (on pp. 183–5).
§ Clapeyron, op. cit., pp. 171–3.
¶ Poisson, 'Mémoire sur les équations générales de l'équilibre et du mouvement des corps solides élastiques et des fluides', *J. Éc. polytech.* **13**, cahier 20 (1831), 1–174; see especially pp. 4–8 and 85–90. || See p. 274.

the nature of heat was one of agnosticism, with caloric no more than, at best, a useful aid to thought. Indeed, when we consider the isolation in which Laplace and Poisson worked even in the early 1820s, it seems likely that this same attitude was widespread in the previous decade also. There were no Laplacian disciples to develop the theory further and most of those who did pursue the theoretical study of heat in terms of caloric were, like Carnot, of little standing in the scientific community.

A notable exception in this respect was the engineer C. L. M. H. Navier, but his paper on the temperature variations accompanying rapid changes in the volume of a gas, first published in July 1820,† was hardly worthy of his already considerable reputation. Taking it as a basic experimental fact that the specific heat of a gas varied with its density and making certain other assumptions that were still less justified, Navier showed that for any given gas the rise in temperature Δt occurring during adiabatic compression was proportional to $\ln(c_1/c_2)$, where c_1 and c_2 represented the gas's specific heat before and after the compression. His examination of the results both of Clément and Desormes and of Delaroche and Bérard led him also to the empirical relationship for the specific heat of air at any pressure h:

$$c = 0 \cdot 76 \bigg/ \sqrt{\left(\frac{1}{h} + 0 \cdot 42\right)},$$

where c was unity for $h = 0 \cdot 76$ m of mercury, and hence to the conclusion that, even for $h = \infty$, Δt could not exceed 360 °C, a result that few, and certainly not Gay-Lussac for example,‡ would have accepted. There was no limit, on the other hand, to the degree of cold obtainable, as Gay-Lussac had recently been suggesting.‡ The eclecticism of Navier's attitude to the existing experimental evidence and theories emphasizes his independence of the leading calorists. In the study of heat Navier was no more than an interested amateur and he can hardly have been surprised, or greatly distressed, that his results were ignored, despite his attempt to apply them to the familiar problem of the economy of heat engines. Laplace, after all, was the acknowledged authority in the field of caloric theory and even he hardly obtained a hearing.

† Navier, *Bull. Soc. philomath. Paris,* N.S. **7** (1820), 97–101.
‡ See p. 151.

If we were to assess the merits of scientific contributions in terms of their bulk alone, there could be no doubt that two of the most important writers on heat in the 1820s were the Scotsmen James (later Sir James) Ivory and Henry Meikle. Ivory's concern with the subject arose from his work on astronomical refraction, for his acceptance of an explanation for the decrease in atmospheric temperature with altitude similar to Erasmus Darwin's led him naturally to an interest in the relationship between the temperature, volume, and pressure of air expanding adiabatically.† When Poisson's expressions became known, Ivory adopted them readily, first with $\gamma = \frac{4}{3}$‡ and later with $\gamma = \frac{11}{8}$.§ His numerous papers on this and related topics which followed contained little that was original, although the clarity with which he expounded Laplace's distinction between the latent heat required to effect expansion and the sensible or free heat that went to increase temperature is noteworthy.¶

Between 1827 and 1829 Ivory's view on heat brought him into an acrimonious dispute with Meikle.‖ Like Ivory, Meikle accepted Poisson's expressions relating to adiabatic volume change, although as a result of experiments of his own,†† he supposed γ to be 4/3. Yet there was much in the earlier calorists' work which he rejected. He could not agree, for example, that the addition of equal quantities of heat to an air thermometer caused equal increments of volume,‡‡ and he showed by a perfectly sound argument that a necessary consequence of Gay-Lussac and Welter's demonstration that γ was independent of pressure and temperature was that the specific heat of a gas must be unaffected by changes in volume, although he did not deny the variation with temperature.§§

One of the few points of agreement between Ivory and Meikle

† Ivory, *Phil. Mag.* **58** (1821), 25.
‡ Ivory, *Phil. Mag.* **66** (1825), 9.
§ Ivory, *Phil. Mag.*, 2nd ser. **1** (1827), 93.
¶ See, for example, *Phil. Mag.* **66** (1825), 4–6; 2nd ser. **1** (1827), 90–3 and 168–70; 2nd ser. **4** (1828), 322–6; 2nd ser. **5** (1829), 105–6.
‖ See Meikle, *Edinb. new phil. J.* **3** (1827), 149–57; *Q. Jl Sci.*, 2nd ser. **4** (1828), 124–35 and 315–19; also 2nd ser. **5** (1829), 57 n., 58–9, 62, and 109–13. Ivory's answers to Meikle appeared in *phil. Mag.*, 2nd ser. **4** (1828), 321–6, and 2nd ser. **5** (1829), 104–6.
†† Meikle, *Edinb. new phil. J.* **6** (1828), 28.
‡‡ See especially Meikle, *Edinb. new phil. J.* **1** (1826), 332–6.
§§ See Meikle, *Edinb. new phil. J.* **1** (1826), 339–40; *Ann. Phil.*, 2nd ser. **12** (1826), 367–9; *Q. Jl Sci.*, 2nd ser. **5** (1829), 65–7.

is found in their acceptance of $\sqrt{\gamma}$ as the correction factor to be applied to Newton's expression for the velocity of sound. In this they were merely reflecting the mood of the time, for Laplace's method of removing the discrepancy between theory and experiment had won general approval. However, there were still a few critics. Notable among these was John Herapath, the early exponent of the kinetic theory of gases, who in 1830, following a forceful attack on the work of Laplace and Poisson on the caloric theory of gases that he delivered in 1823†, managed to remove the discrepancy by the simple device of inserting an unexplained factor $\sqrt{2}$ under the square root sign in Newton's expression.‡ Herapath's views, not surprisingly, attracted little attention and nearly twenty years passed before Laplace's identification of Newton's error once again became an important issue. The men responsible for this revival of interest were James Challis, Plumian professor of astronomy and experimental philosophy at Cambridge, and Richard Potter, professor of natural philosophy and astronomy at University College in London, both of whom put forward the extraordinary criticism that, since rarefaction as well as compression occurred in a sound-wave, a diminution in the velocity of sound was quite as likely by Laplace's argument as an increment.§ The response to this naïve suggestion and to the alternative solutions which the two men proposed were swift and crushing, with the Astronomer Royal, G. B. Airy, replying to Challis in 1848¶ and Professor G. G. Stokes, William Rankine, the Revd. Samuel Haughton (a Fellow of Trinity College, Dublin), and Auguste Bravais, professor of physics at the École Polytechnique, all rising in immediate defence of Laplace when Potter's views were published in 1851.‖

The authority of this response about the middle of the century only emphasizes how widely Laplace's work on sound had become accepted and how successfully it survived the early

† See p. 174 n. ‖.
‡ Herapath, 'On the velocity of sound and variation of temperature and pressure in the atmosphere', *Q. Jl Sci.*, 2nd ser. **7** (1830), 167–75. The factor was explained later in Herapath's *Mathematical physics* (1847), vol. 2, p. 63.
§ Challis, *Phil. Mag.*, 3rd ser. **32** (1848), 283–4, and Potter, *Phil. Mag.*, 4th ser. **1** (1851), 101.
¶ Airy, *Phil. Mag.*, 3rd ser. **32** (1848), 339–43.
‖ Stokes, *Phil. Mag.*, 4th ser. **1** (1851), 305–17; Rankine, *Phil. Mag.*, 4th ser. **1** (1851), 225–7; Haughton, *Phil. Mag.*, 4th ser. **1** (1851), 332–4; Bravais, *Annls Chim. Phys.*, 3rd ser. **34** (1852), 82–9.

development of the new science of thermodynamics and the final, definitive rejection of caloric. That the $\sqrt{\gamma}$ factor should have been generally accepted by calorists and non-calorists alike is not surprising, since Laplace's derivation of it in 1816 had involved only limited assumptions concerning the properties of caloric and they were ones to which no one could seriously have objected, whether or not he was familiar with the inter-convertibility of heat and work. The really suspect parts of Laplace's caloric theory, on the other hand—the 1816 determination of γ and the extravagant treatment developed after 1821—could be, and for the most part were, rejected quite independently. With the advent of thermodynamics this situation underwent little change. The cause ascribed to the production of heat in the compression of a gas was now different, of course, but the fact that it did occur and that it affected the velocity of sound was not in serious doubt and even the correction factor was seen to be the right one.† Laplace was fortunate that the erroneous view of the adiabatic compression process as equivalent to an isothermal compression followed by heating at constant volume, which he adopted in 1816, was a good approximation to the truth for the small changes in density that occur in sound-waves, and in this respect he must be contrasted with Carnot, who was led into serious error by similar basic assumptions. The correction factor has therefore undergone no modification since it was first derived by Laplace and it remains as a solitary reminder in modern physics textbooks that the theory of heat must be numbered among the most important interests of this great mathematical physicist.

† A point made as early as 1847 in J. P. Joule's paper 'On the theoretical velocity of sound', *Phil. Mag.*, 3rd ser. **31** (1847), 114–15.

6

THE CALORIC THEORY OF AMEDEO AVOGADRO

IF we now went straight on to examine the process of the decline of caloric theory our narrative would have continuity and every appearance of completeness. However, in doing so we should not only be ignoring one of the most elaborate of all the treatments of the theory but we should also be missing a much-needed opportunity of increasing our understanding of its author, the Italian Amedeo Avogadro. As the originator of the celebrated hypothesis that, under similar conditions of temperature and pressure, equal volumes of all gases contain equal numbers of molecules, Avogadro is familiar enough, but of his other activities all too little is known. In his work on caloric theory in particular he appears in a somewhat unaccustomed light, not as the chemist of profound insight whose views were misunderstood in his lifetime, but as a physicist clinging to a discredited theory of heat. His voluminous writings on the theory, which are examined in this chapter, can have done little to enhance his stature in the scientific community, and it seems that this factor should at least be borne in mind in any attempt to account for the neglect of his gas hypothesis between 1811, when it was first proposed, and 1860, when it was finally acclaimed at the Karlsruhe congress, four years after his death.

It is in order to illustrate the outdated character of much of Avogadro's thought that we begin with an attempt to identify the research tradition in which he was working.

Affinity for light

In Chapter 5 we saw that one of the main objects of Laplace's work in physics was the extension of the concepts established in his treatment of gravitational forces from the celestial to the molecular scale. In chemistry too a similar approach was being attempted in the first few years of the nineteenth century by Laplace's close friend Berthollet. Since Avogadro's early scientific work was conducted in the years when these two men were at the height of their powers and when French influence in Italy, if only for political reasons, was considerable, it is not surprising to find that his treatment of specific heats as being determined by a characteristic attractive power between a particle of ponderable matter and the surrounding caloric owed much, both in detail and general conception, to the views on affinity developed in Berthollet's *Essai de statique chimique* of 1803.†

One of Berthollet's greatest innovations in the study of affinities had been to take specific account of the masses of the reactants in a chemical reaction. He had emphasized his belief in the gravitational nature of chemical forces by arguing that the 'chemical action' of a substance, i.e. the measure of its tendency to combine chemically with another given substance, was proportional to its own mass multiplied by its characteristic affinity for the second substance. Although earlier writers, notably Bergman,‡ had already suggested that the mechanism of chemical affinity was gravitational, the explicit distinction between the effects of mass and affinity was quite original, and its influence was soon to be felt not only in chemistry but also in attempts to extend the new principles to embrace imponderable matter, in the first instance the particles constituting light. This important extension was first made by Biot and Arago in their experimental work on the refractive powers of gases which they announced to the Institute in 1806.§ Since it is not difficult to find close parallels with Avogadro's work in this paper and since here also the debt to Berthollet and Laplace was made explicit by the authors' profuse acknowledgements,¶ it is a significant document for our purpose. Avogadro was familiar with it‖ and he almost certainly owed a great deal to the treatment of the forces between ponderable and imponderable matter that it contained.

The application of what were essentially Newtonian principles

† See especially Berthollet, *Essai de statique chimique*, vol. 1, pp. 1–22.
‡ Bergman, *Opuscula physica et chemica* (Uppsala, 1783), vol. 3, pp. 291–4.
§ Biot and Arago, 'Mémoire sur les affinités des corps pour la lumière et particulièrement sur les forces régringentes des différens gaz', *Mém. Sci. math. phys. Inst. Fr.* **7** (1806), 301–87.
¶ Biot and Arago, op. cit., p. 304.   ‖ See p. 213.

to light was not new. In the *Opticks* Newton himself treated refraction as the result of a 'refracting force' between the transparent material and light,† and this view of refraction had continued to attract attention and support throughout the eighteenth century, not least among chemists. Indeed, in 1758 Georges Louis Le Sage of Geneva had even advocated the inclusion of light in a table of chemical affinities.‡ To early nineteenth-century writers, however, at least for those in France, it must have been the 1806 paper by Biot and Arago that represented the modern approach to the problem, especially since it had the backing of some very great names.

The instigator of Biot and Arago's experiments was Laplace, who had elaborated Newton's treatment of refraction in terms of attractive forces in the fourth volume of his *Traité de mécanique céleste*, published in 1805.§ It was he who persuaded the Institute to organize the experimental work and he was no doubt instrumental in arranging for the appointment of his protégé Biot, who in turn engaged Arago as his assistant. As would be expected at a time when the materiality of light was in little doubt, least of all for such a close disciple of Laplace as Biot, the results were given in terms of refractive power (p) and not of refractive index (μ). Laplace had retained Newton's concept of a refracting force, equal to (μ^2-1), as the measure of the force of attraction that a body exerted on incident particles of light, the magnitude of the force being proportional to the increase in the square of the velocity of these particles, an increase that was supposed to occur in accordance with the normal laws of dynamics.¶ This force, conceived as a gravitational one, was thought to vary in proportion to the density of the material, and it was the refractive power, i.e. the refracting force divided by the density, ($\mu^2-1)/\rho$, which, according to both Newton and Laplace, yielded the quantity more closely related to the true nature of the substance and characteristic of it alone. The distinction between the effects of mass and innate affinity was thus already implied in this procedure. Newton himself had

† Newton, *Opticks* (4th edn, 1730), pp. 245–51.
‡ Le Sage, *Essai de chymie méchanique* (Geneva, [1758]), pp. 50–2. For an account of other work on similar lines see H. Metzger, *Newton, Stahl, Boerhaave et la doctrine chimique* (1930), pp. 68–82.
§ Laplace, *Traité de mécanique céleste*, vol. 4, pp. 231–7.
¶ For Newton's argument on this point see Newton, op. cit., pp. 245–7.

obtained a continuous range of refractive powers for some twenty different substances and had tentatively attributed the variations to varying quantities of what he termed 'sulphureous oily particles', but the powers of gases other than air had not been investigated. By 1806 the need to rectify this omission had become pressing. Quite apart from the fact that the experimental re-examination of refraction in gases would provide data for Laplace's immediate problem, atmospheric refraction, and even for the study of optical dispersion, the study of gases in general had now acquired a new significance, as was pointed out in Chapter 3. It is certain that Biot and Arago fully shared the prevailing contemporary belief that in gases the laws of nature would appear in their simplest form,† and there can be little doubt that this provided an important incentive for their work.

But the chief significance of the paper lies in the obvious debt that Biot and Arago owed to the principles of chemical affinity recently established by Berthollet. For anyone working under the influence of Berthollet's chemistry, as virtually all French chemists were in the first decade of the nineteenth century, tables of affinity must have represented the most promising means of accounting for and systematizing the properties of matter. The new tables that Berthollet envisaged were intended to go much further than their eighteenth-century counterparts, which provided merely a rule-of-thumb method of predicting the course of a reaction and lacked any sound theoretical basis.‡ Berthollet considered that chemical affinity for a substance such as oxygen, what he termed a 'dominant affinity', would account for and allow prediction of a far wider range of properties than those governing the immediate chemical reaction.§ And 'affinity' for light might be even more revealing, as Biot and

† See Biot and Arago, *Mém. Sci. math. phys. Inst. Fr.* **7** (1806), 301–2. Also see pp. 77–8.

‡ However, the growing interest in affinity tables after about 1750 does seem to reflect the more general acceptance of a Newtonian-style chemistry in which attractive forces between the particles of matter played a crucial role. The point is well made in A. W. Thackray, 'Quantified chemistry—the Newtonian dream', in D. S. L. Cardwell (ed.), *John Dalton & the progress of science* (Manchester, 1968), pp. 101–5. For a general account of affinity tables in the eighteenth century see A. M. Duncan, 'Some theoretical aspects of eighteenth-century tables of affinity', *Ann. Sci.* **18** (1962), 177–94 and 217–32.

§ Berthollet, *Essai de statique chimique*, vol. 1, p. 61, and vol. 2, pp. 3–6.

Arago commented in a passage that emphasizes their interest in what were essentially chemical problems. They wrote:

... as the action of bodies upon light is effective only at very small distances, the intensity of this action is necessarily dependent on the nature and arrangement of the particles composing the bodies or, in other words, on their most intimate properties (*leurs propriétés les plus intimes*). Therefore the physicist who observes the refracting forces of substances and compares them with one another is proceeding in exactly the same way as the chemist who causes a given base to react in turn with all the acids (or who alternatively causes a given acid to react with all the alkalis) in order to determine in each case the tendency to combine and the degree of saturation. In our experiments the substance to which all bodies are exposed is light and we measure the action of these bodies upon light in terms of their refracting force or, in other words, of the increase in *vis viva* (*force vive*) which the light acquires by the action of their particles. ... In this way the different velocities acquired by the light, as indicated by the differing refracting forces, allow us to draw up an extensive series in which all bodies can be arranged at intervals and which thus can serve either as a means of distinguishing and characterizing the bodies or as a means of detecting them and of recognizing their presence in compounds.†

The mere fact that the eight gases examined by Biot and Arago were found to have widely differing refractive powers was sufficient evidence that the first of the two aims mentioned in the last sentence of this quotation, namely the acquisition of a method of classification, had been attained. The effect of a chemical reaction on the characteristic affinities for light was more problematical, however, but it seemed likely that there was some connection between the high refractive power of ammonia, for example, and the fact that one of its constituent elements was hydrogen, which had a power higher than that of any other gas. Berthollet, whose work had been purely qualitative, had offered no solution of the analogous problem for ponderable matter, except in so far as he believed it reasonable to suppose that a property so closely linked to the essential nature of a substance as chemical affinity was somehow conserved‡.Yet the substance would retain only that part of its

† Biot and Arago, op. cit., pp. 326–8.
‡ Berthollet, op. cit., vol. 1, pp. 308–32. See also his *Recherches sur les lois de l'affinité* (1801), pp. 62–9.

affinity, and hence of its characteristic properties, which was not neutralized by the effect of the other reactant. This was as far as Berthollet could go, since there were too many unaccountable factors, such as the effect of the change in volume during a reaction, for the matter to be treated quantitatively. Affinities for light, on the other hand, were subject to no such difficulties, at least in the opinion of Biot and Arago. They maintained that the smallness (*ténuité*) of the light particles by comparison with the distance between the particles of ponderable matter would diminish any effect that a change in volume during a reaction might have, so that the refractive power of a compound would be determined in a wholly predictable way by those of its constituents, except in cases where very great decreases in volume occurred.† Moreover, they had no qualms about extending Berthollet's principles of affinity to imponderable matter, although they did not hesitate to make slight modifications where necessary. It was assumed, for example, that each substance had a characteristic affinity for light that was quite independent of its affinity for other substances, so that the contributions of the various elements making up a compound would be unaffected by the chemical reaction. In other words, neutralization of affinity did not occur in the case of light, as it did when ordinary ponderable matter entered into chemical combination. Also, since it was axiomatic (for good Laplacians) that the attractive force exerted by any given body on the particles of light was proportional to its mass, it followed that the respective contributions of the reactants to the refractive power of the resulting compound would be proportional to the weights in which they combined as well as to their affinities for light. On this basis it was possible to establish a conservation law in which the refractive powers of reactants and compound were intimately related. The law‡ was expressed as follows:

$$p'a'+p''a''+p'''a'''+ \ldots = p, \qquad (1)$$

where a', a'', a''' ... were the masses of the various elements of refractive power p', p'', p''' ... which combined to form unit mass of a compound of refractive power p. In the case of air, a mixture, the refractive power calculated from those of its

† Biot and Arago, op. cit., pp. 329–30.
‡ Biot and Arago, op. cit., p. 330 n.

constituent gases was within 0·6 per cent of the observed value. The calculated and experimental values for ammonia, a compound in the formation of which a comparatively small decrease in volume occurred, showed even closer agreement, so close in fact as to suggest that observations of refractive properties might be used in the determination of the composition of compounds by weight. In the case of water, the only other compound for which a test was possible, the discrepancy of over 10 per cent was serious, but this was readily dismissed as the effect of the large decrease in volume, an explanation that was sanctioned by Berthollet's own views on the effect of volume changes on affinity.†

Largely because of the lack of precise data concerning the composition of compounds, Biot and Arago's treatment, supported by results for only one gaseous compound (ammonia), was far from vindicated and, although J. B. J. Delambre's contribution to the History of the First Class of the Institute for 1806 contained a glowing account of their work and of its importance in many branches of science,‡ the promise of the optical method of investigating the properties of matter was never fulfilled. When Arago and Petit re-examined the matter in 1815, they could find no justification for a simple relationship of the type expressed in eqn (1) and, largely as a result of this conclusion, they cast grave doubts on the corpuscular theory of light as a whole.§ It is hardly necessary to add that thereafter the growing popularity of the wave theory, with the supporters of Fresnel gradually gaining the ascendancy over the Laplacians,¶ ensured that the study of ponderable matter based on the search for characteristic affinities for light could never again become an important research tradition in science.

Affinity for heat

The situation regarding the theory of heat was somewhat different. For all its apparent inadequacy on certain points, there was no obvious alternative to caloric theory that had the stature of the wave theory of light, and it is therefore not

† Given in Berthollet, *Essai de statique chimique*, vol. 1, p. 312.
‡ Delambre, 'Analyse des travaux de la Classe . . . partie mathématique', pp. 9–11, in *Mém. Sci. math. phys. Inst. Fr.* **7**, part 1 (1806).
§ Arago and Petit, *Annls Chim. Phys.* **1** (1816), 1–9.
¶ On this see pp. 233–5.

surprising to find that problems analogous to those treated by Biot and Arago were being tackled in terms of caloric even after 1815, with specific heat replacing refractive index as the significant measurable quantity. We have already seen that in principle the prediction of the specific heat of a compound from those of its constituent elements was possible in the Irvinist version of caloric theory,† but even here the solution was not simple, if only because it required a knowledge of the heat evolved in the relevant reaction. For someone holding the rival view, according to which both latent and sensible caloric was present in bodies, the problem was even more complex, of course, but the conviction that the specific heat of a compound and those of its constituents were somehow related was present among these calorists also. Delaroche and Bérard, for example, concluded their prize-winning memoir of 1812 with the words:

> However, it must not be thought that there is no relationship between the specific heat of a compound and those of the elements composing it; the existence of such a relationship is supported by too many facts for it to be denied. In this regard, water shows the greatest deviation for which really reliable evidence is available, and yet this deviation amounts to no more than one third of the actual specific heat. Generally it may be said that the constituents contribute their own specific heat to that of the compound which they form. This is observed very clearly in compounds containing hydrogen, the substance with the largest of all specific heats, for the compounds in which it is present have specific heats far greater than those of other substances—hence the large specific heats of water, olefiant gas [ethylene], and of vegetable and animal substances.‡

But in the absence of any knowledge concerning either the relative quantities of the two types of caloric in a body or of their relationship with the only observable quantity, specific heat, the problem remained unsolved, and even Laplace's definitive version of the caloric theory of gases, expounded in the early 1820s, was to throw no light on the problem.

The deduction of some quantity that would characterize substances and allow a meaningful classification of them was another problem that was of permanent interest, although increasingly the approaches adopted differed from the one that

† See p. 141.
‡ Delaroche and Bérard, *Annls Chim.* 85 (1813), 175–6.

Biot and Arago had adopted in 1806. Indeed, in their paper of 1815 on refraction Arago and Petit, having abandoned the corpuscular theory of light, did not even mention this possible application of their work; and a similar attitude to the earlier tradition was apparent when, in the same year, Ampère pointed out the inadequacy of Berthollet's proposals for a classification of elements based on affinity for oxygen, a quantity that Ampère considered to have no significance.† In rejecting the uniqueness of oxygen, which he believed Lavoisier had over-emphasized, Ampère stressed the need for a 'natural' classification that would bring out analogies by grouping together similar substances, suggest the best procedure for further investigation of their properties, and allow the prediction of the properties of a compound from a knowledge of those of its constituents. Affinities could not provide the answer and Ampère's classification was to be based on purely observable chemical properties only.

In this search for a system of classification, specific heats were not overlooked, but the only reliable data, those of Delaroche and Bérard, appeared to be quite unrelated to such fundamental properties as acidity or basicity. Biot himself seems to have looked in vain for some correlation and he suggested in 1816 that his failure might be due to the fact that specific heats had normally been measured at constant pressure and so included the heat required both to raise the temperature and to cause expansion. He commented:

> So long as no distinction between these two effects is made by our experiments, the measurement of specific heats will yield complex results; and it is perhaps this complexity which hitherto has prevented the discovery of any clear relationship with the chemical nature of bodies.‡

According to Biot, Dulong was at this time engaged on measuring the specific heats of vapours at constant volume§ and it seems at least possible that the experiments were being conducted with the end in view that Biot saw, for he showed great familiarity with Dulong's investigations. But the failure of what must have been some very difficult experimental work was

† Ampère, 'Essai d'une classification naturelle pour les corps simples', *Annls Chim. Phys.* **1** (1816), 295–308 and 373–94, and **2** (1816), 5–32 and 105–25. ‡ Biot, *Traité de physique*, vol. 4, p. 728.
§ Biot, op. cit., vol. 4, pp. 728–30.

only to be expected and the results were never published. In fact, despite this work and despite Crawford's crude attempts some forty years earlier, it was not until 1890 that the direct measurement of specific heats at constant volume became a practical possibility.† Why Biot looked to specific heats for help in this problem is not clear and there is no definite evidence that he saw any special analogy with his work on refractive power. In 1816, however, he was still greatly influenced by Laplace and Berthollet, as his praise for their work in his *Traité de physique*‡ and his dedication of the book to Berthollet emphasize. Moreover, his professed agnosticism towards the nature of heat scarcely conceals a preference for the material theory, which he felt was supported by the analogies between heat and light revealed by recent work on radiation.§ So it is quite conceivable that Biot did expect the correlation with chemical properties which he had observed for refractive powers to be repeated somehow in terms of specific heats. However, in the absence of any developed theory of affinity for caloric, he would certainly have had difficulty in taking the matter further. And this is to say nothing of his own doubts, expressed in the *Traité*, on the question of whether it really was affinity that retained caloric in bodies in any case.¶

Such, then, was the state of the problem when Avogadro first tackled it in 1816. In understanding his approach it seems important to begin by emphasizing his isolation, especially from Paris. He had few contacts, even in Italy, and very rarely travelled abroad.‖ Hence from the start he was not influenced by what I see as a growing crisis in the Laplacian programme for physics (as this programme was expressed in the passage quoted on pp. 166–7) and he had none of the scepticism towards

† See J. Joly, *Proc. R. Soc.* **48** (1890), 440–1.
‡ Biot, op. cit., *passim* but especially vol. 1, pp. i–ii and xxiii–xxiv.
§ Biot, op. cit., vol. 4, pp. 611–17. In two pieces published in the *Mercure de France* in 1809 Biot had been even more adamant in his agnosticism with regard to the nature of heat. In an otherwise favourable review of a French translation of Jane Marcet's *Conversations on chemistry* (3 vols., Geneva, 1809), he criticized the author for her acceptance of the literal truth of the caloric theory, and later he repeated his scepticism towards caloric (as well as the fluids of electricity and magnetism) in 'Sur l'esprit de système'; see Biot, *Mélanges scientifiques et littéraires* (1858), vol. 2, pp. 102–3 and 114–16. There is evidence of a more favourable attitude to caloric in Biot's obituary notice of Petit: see *Annls Chim. Phys.* **16** (1821), 334. ¶ Biot, op. cit., vol. 4, p. 730.
‖ See I. Guareschi's introduction to *Opere scelte di Amedeo Avogadro* (Turin, 1911), p. vi.

imponderable fluids felt by certain younger French scientists even before 1820.† On the other hand, even though he was a convinced calorist, he was in no way committed to the Laplacian version of the theory, which had still to reach its highest stage of development despite the notable and decisive victory that it had already won over its Irvinist rival. Hence Avogadro was able to devise a theory of heat that borrowed something from Laplacians and Irvinists alike. His isolation was important in other ways also, for it allowed him to continue to use Berthollet's increasingly unpopular theory of affinities and to apply his own famous gas hypothesis effectively and with every confidence. These were opportunities largely denied to contemporary French scientists, working as they were in a more tightly defined context of acceptable theories and results.

Like Dalton, Avogadro believed that gases were composed of particles of ponderable matter each surrounded by a sphere of caloric that was retained by an attractive force between it and the particle.‡ It was this force, he maintained, which determined the quantity of caloric in any given molecule§ and he thus differed from Dalton, for whom the quantity of caloric was the same in all molecules under similar conditions of temperature and pressure, irrespective of the nature of the gas and the magnitude of the force. Also for Dalton, and particularly in his second theory of mixed gases (adopted in 1805), it had been fundamental that the size of a gas molecule was determined by its chemical nature, and here again Avogadro differed. Since he accepted, like Dalton, that adjacent gas molecules were in contact, it followed from his hypothesis that they must be of the same size in all gases, and it was natural, therefore, that he should have shown no interest in Dalton's diffusion problem when he first announced the hypothesis in 1811. The immediate inspiration for his views was Gay-Lussac's law of combining volumes, which Dalton could not accept.¶

† On this criticism of Laplacian physics after 1815 see pp. 233–5 and 238–48.

‡ See, for example, the opening paragraph in the paper of 1811 in which he first announced his gas hypothesis, in *J. Phys.* **73** (1811), 59.

§ Avogadro used the terms *molécule* or *molécule intégrante* for what we should now call a molecule, and the terms *demi-molécule* or, later, *molécule partielle* for an atom. In this he differed from Dalton, who used 'atom' to convey the idea of both a molecule and an atom in the modern sense.

¶ Avogadro, op. cit., especially p. 58. For Dalton s view of Gay-Lussac's law see his *New system*, vol. 1, part 2, pp. 555–60.

It is well known that despite the independent derivation of a somewhat different version of the gas hypothesis by Ampère in 1814,† there were few contemporaries who accepted Avogadro's view of gas structure, and such acceptance as there was was only partial. Berzelius, for example, considered the number of gas molecules per unit volume to be identical in elementary gases only,‡ and the same was almost certainly true of Dulong and Petit.§ And we can be sure that none of these men, all of whom broke with the traditional caloric theory before 1820, would have entertained Avogadro's views on molecule size. This latter comment undoubtedly applies to Joseph Mollet who, in 1817, expounded the view that equal volumes of all gases did contain equal numbers of molecules, as Avogadro had said, but maintained that the sizes of the molecules varied according to the nature of the gas.¶ His molecules, of course, differed from Avogadro's and Dalton's in that they did not touch one another.

So Avogadro's view of gas structure provided him with a new and virtually unique tool for the development of caloric theory. In fact, he did not examine the implications of his hypothesis for the theory of heat in his well known paper of 1811 nor in the supplement to it published in 1814,‖ but once he had examined the implications in conjunction with Delaroche and Bérard's results some very pertinent conclusions emerged. Avogadro described these in the obscure journal *Biblioteca italiana*, published in Milan, in December 1816 and January 1817.†† Here he argued that, by his hypothesis, the volume specific heats of all gases would represent the quantities of heat which identical numbers of molecules must receive in order that their temperature should rise by one degree. The possibility that a readily observable thermal property might thus be used for the examination of matter on the molecular scale was attractive and one that Avogadro's views on caloric and the constitution of gases alone made practicable. By introducing the additional

† Ampère, *Annls Chim.* **90** (1814), 43–86.
‡ Berzelius, *Essai sur la théorie des proportions chimiques* (1819), pp. 52–4.
§ See pp. 216–17; also my comment in *Br. J. Hist. Sci.* **4** (1968–9), 19 n.
¶ Mollet, *J. Phys.* **90** (1820), 113–30.
‖ Avogadro, *J. Phys.* **78** (1814), 131–56.
†† Avogadro, 'Memoria sul calore specifico de' gaz composti parragonato a quello de' loro gaz componenti', *Bibltca ital.* **4** (1816), 478–91, and **5** (1817), 73–87.

assumption that the quantity of caloric required to produce a given temperature increment in unit volume of any gas was determined solely by the characteristic 'attractive power' (*potere attrattivo*) of the molecules, and assuming also that it was this quantity, and not specific heat, that was conserved in a chemical reaction, Avogadro was able to establish a general conservation law governing attractive power. The analogy between this and Biot and Arago's law for refractive power was a close one, as will become clear.

Unfortunately Avogadro's determination of the attractive power from the one observable quantity, specific heat, was open to serious objection. It was based on the quite unjustified assumption that the power of an individual molecule was proportional to some integral power m of the volume (or, for Avogadro, molecular) specific heat. m, of course, was unknown and the small number of the gases whose specific heats had been measured made any determination of its magnitude extremely suspect. Yet Avogadro showed every confidence in his method. By reading 'molecule' for 'volume' in the data concerning the volumes in which gases were known to combine chemically and by using Delaroche and Bérard's results, he was able to derive several equations expressing the conservation of attractive power in a reaction. His first example concerned the formation of one volume of carbon dioxide from one volume of carbon monoxide and half a volume of oxygen. In molecular terms he could write:

$$\text{1 molecule of carbon monoxide} + \tfrac{1}{2} \text{ molecule of oxygen} \longrightarrow \text{1 molecule of carbon dioxide},$$

whence the application of Delaroche and Bérard's results yielded

$$(1{\cdot}0340)^m + \tfrac{1}{2}(0{\cdot}9765)^m = (1{\cdot}2583)^m,$$

giving $m = 1{\cdot}888$.

The only other reaction in which values of specific heat were available for both the reactants and the resulting compound was

$$\text{1 molecule of nitrogen} + \tfrac{1}{2} \text{ molecule of oxygen} \longrightarrow \text{1 molecule of nitrous oxide},$$

giving $(1{\cdot}0058)^m + \tfrac{1}{2}(0{\cdot}9765)^m = (1{\cdot}3503)^m$

and $m = 1{\cdot}333$.

THE CALORIC THEORY OF AMEDEO AVOGADRO

Even in this equation the specific heat for nitrogen was not an experimental one but was estimated from the specific heat of air treated as a mixture of oxygen and nitrogen. The other reactions that were used involved the undetermined volume specific heat of carbon vapour, but Avogadro easily eliminated the unknown quantity and made further determinations of m from the following three equations, all of which describe purely imaginary reactions. x throughout is the unknown specific heat.

$\frac{1}{2}$ volume of carbon vapour $+$ 1 volume of oxygen $\longrightarrow$ 1 volume of carbon dioxide

$$\tfrac{1}{2}x^m + (0\cdot9765)^m = (1\cdot2583)^m \qquad (2)$$

1 volume of carbon vapour $+$ 2 volumes of hydrogen $\longrightarrow$ 1 volume of ethylene

$$x^m + 2(0\cdot9033)^m = (1\cdot5530)^m \qquad (3)$$

$\frac{1}{2}$ volume of carbon vapour $+$ $\frac{1}{2}$ volume of oxygen $\longrightarrow$ 1 volume of carbon monoxide

$$\tfrac{1}{2}x^m + \tfrac{1}{2}(0\cdot9765)^m = (1\cdot0340)^m \qquad (4)$$

By taking in turn equations (2) and (3), (2) and (4), and (3) and (4), values for m of 2·812, 1·888, and 2·331 were obtained. Since now the mean of all his values for m was 2·091 when the repeated figure of 1·888 was neglected, Avogadro confidently and, in view of the wide discrepancies between the values, very rashly assumed that m was exactly 2. Hence the general equation expressing the conservation of attractive power for caloric in a reaction became

$$c^2 = n'c'^2 + n''c''^2 + n'''c'''^2 + \ldots \qquad (5)$$

where c was the volume specific heat of the resulting compound and c', c'', c''' ... those of the constituent gases. n', n'', n''' ... were the respective numbers of molecules of these gases that combined to form one molecule of the compound.

When Delaroche and Bérard's results for oxygen, hydrogen, and carbon monoxide and Avogadro's own estimated value for nitrogen were inserted, eqn (5) yielded theoretical specific heats for a wide range of compounds in the gaseous state. For many of these no experimental data were available, and even where comparisons were possible, agreement between the calculated and observed values was poor, the discrepancy for nitrous oxide being over 10 per cent (see Table A and Table E, column 2). Avogadro nevertheless considered the calculated specific heats to lie within the limits of experimental error.

Avogadro's debt to the work of Laplace and Berthollet and to the important movement in early nineteenth-century French science which sought to extend the principles of gravitational attraction first to chemistry and then to the study of light, is nowhere more apparent than in his interpretation of the quantity that he termed attractive power for caloric. In the paper published in 1816–17 he did not even mention the earlier work in France, but in stating as self-evident that the attractive power of a molecule was 'necessarily' proportional to the product of the mass of the molecule and its characteristic affinity for caloric, Avogadro was evidently assuming a close analogy between the mechanism that retained caloric round a molecule and that which, according to Berthollet, governed the chemical combination between two substances. This extension from the realm of ponderable matter to that of imponderable matter was apparently thought to require no justification. The gas hypothesis made the derivation of the new quantity, affinity, extremely simple, since the mass of a molecule could be taken as proportional to the density of the gas, and division of the attractive powers by the known densities hence yielded the affinities, as listed in Table E, column 7.

The subject of affinities for caloric would never have assumed the importance that it was to have for Avogadro if he had not immediately observed a point of unmistakable significance in the figures obtained. He pointed out that when the affinities were arranged in order, there emerged remarkable analogies with the results of some earlier work of his on chemical affinities. Since this earlier work had been conducted quite independently and without reference to specific heats, it seemed difficult to attribute the analogies to chance, and this Avogadro was certainly not prepared to do.

It was in 1809 that he had published the clearest exposition of his early views on chemical combination.† In this exposition, published in French in the *Journal de physique*, he used the

† Avogadro, 'Idées sur l'acidité et l'alcalinité', *J. Phys.* **69** (1809), 142–8.

concept of an antagonistic force between acids and alkalis to explain the tendency of substances to combine, arguing that alkalinity and acidity were not absolute properties of matter but were relative. Hence in his view the same substance A might be at once acidic with respect to B and alkaline with respect to C. A's behaviour, in fact, could be predicted by means of a single table of affinities in which it would act as the 'acid' component in a reaction with all substances below it but would be alkaline with respect to any above it. In order of decreasing acidity, the table would be headed by oxygen, followed by carbon and the acids, then by water and the neutral salts, and lastly by the alkalis and hydrogen. Avogadro found support for his views, and in particular for his rejection of any idea that chemical affinity was an absolute property, in Davy's recent and much publicized attempt to identify the strength of chemical affinity with the degree of electrical heterogeneity,† although, for his part, he was not fully convinced that the mechanism of affinity was necessarily electrical. Neither Avogadro in 1809 nor Davy in his earlier writings had drawn up complete affinity tables of their own, although Avogadro had commented in the 1809 paper that the best available method of establishing his proposed table was one based on the tables already drawn up by Volta and Pfaff from purely electrical considerations and for a quite different purpose.‡ Volta, for example, working in terms of the contact theory of electricity (which Davy had adopted by 1807), had given a series on which the distance between two metals was a measure of the degree to which they assumed electricity of the opposite sign on contact, the presence of a conducting liquid being necessary without being thought to be the seat of the galvanic phenomenon.§ Davy's references to Volta suggest that he, like Avogadro, would have adopted distance apart on this series as a measure of chemical affinity.

Avogadro's view of the probable connection between chemical affinity and electrical properties was not entirely novel, even in 1809, although he made no mention of his precursors apart

† Davy, 'The Bakerian Lecture, on some chemical agencies of electricity', *Phil. Trans. R. Soc.* **97** (1807), 1–56. This work won one of the annual prizes of 3000 francs for galvanism which had first been offered by Napoleon in 1802.
‡ Avogadro, op. cit., p. 146.
§ Volta, *Neues Journal der Physik* (ed. F. A. C. Gren), **2** (1795), 142.

from Davy.† Already, in 1798, the German chemist Johann Wilhelm Ritter had pointed out the resemblance between Volta's series and Gren's table of affinities for oxygen.‡ Although Avogadro specifically rejected Lavoisier's view of the uniqueness of oxygen as the cause of acidity, maintaining that an even more electropositive substance may yet be found, with respect to which oxygen itself would be alkaline, his proposed series too was to be essentially one of affinity for oxygen. The term 'oxygenicity' (*oxigénicité, ossigenicità*) was introduced in the 1809 paper as a measure of proximity on the series to oxygen, and chemical affinity was to be measured by the difference between the oxygenicities of two reactants.

It was on the similarity between the newly derived values of the affinities for caloric and the degrees of oxygenicity, as inferred from the tables of Volta and Pfaff, that Avogadro rested his case in the paper of 1816–17. He pointed out that the order of increasing affinity for caloric, with one or two exceptions, resembled that of increasing affinity for oxygen, i.e. of decreasing acidity, with the affinity between hydrogen and caloric on the one hand and that between hydrogen and oxygen on the other being greatest of all. Since even oxygen itself had affinity for caloric, it followed, as Avogadro put it in 1825 in a phrase that emphasizes his belief that affinities for oxygen and for caloric were of the same nature, that caloric was the 'sostanza ossigenica per eccelenza'.§

We have already noted the interest that was being shown during the early years of the nineteenth century in the possi-

† By 1813, however, Avogadro was happy to acknowledge that strong support for his views had appeared in the recently published work on electrochemistry by Berzelius; see Avogadro, *Annls Chim.* **87** (1813), 286–92. Avogadro pointed out that he differed from Berzelius on a number of points, notably in that he followed Davy and the exponents of the contact theory of electricity in the meaning that he attached to the terms electropositive and electronegative, although in fact it was in 1813 also that Berzelius adopted Davy's nomenclature and so began to consider oxygen as electropositive rather than electronegative.

‡ Ritter, *Beweiss dass ein beständiger Galvanismus den Lebensprocess in dem Thierreich begleite* (Weimar, 1798), especially pp. 70–1. See also A. Hermann (ed.), *Die Begründung der Elektrochemie und Entdeckung der ultravioletten Strahlen von J. W. Ritter*, Ostwalds Klassiker, N.S. vol. 2 (Frankfurt, 1968), pp. 22–9. Gren's table appeared in his *Systematisches Handbuch der gesammten Chemie* (Halle, 1796), vol. 4, p. 162.

§ Avogadro, *G. Fis.* Decade II, **8** (1825), 9.

bility that at least two of the imponderables, light and caloric, might be essentially identical, being merely modifications of one and the same fluid.† For Avogadro it was an obvious extension of this idea to suppose that affinity for caloric and refractive power might be connected, and in a paper published in 1817 he duly succeeded in establishing a relationship between the two quantities.‡ A comparison of his own values for the affinities with Biot and Arago's refractive powers showed that the relationship was not one of proportionality, but the variations could nevertheless be accounted for on the assumption that light was merely caloric in rapid motion. On this view those molecules that had a large affinity would attract a large quantity of caloric about them, but since the caloric particles were mutually repulsive, the process of accumulation would tend only to diminish the resultant attractive power of the molecules for light particles. Hence the effect of the high affinity would tend to be diminished, though without being entirely counteracted. It followed that affinity for caloric would be expected to show greater variation from substance to substance than would affinity for light, and this agreed with the observations.

From results relating to the five gases for which adequate data were available Avogadro derived the following purely empirical equation involving refractive power p and affinity for caloric A:

$$p = zA+(1-z)\sqrt{A}, \qquad (6)$$

where z, a constant, was found to be $0{\cdot}4193$ by insertion of the relevant figures for hydrogen. The equation thus became

$$p = 0{\cdot}4193A + 0{\cdot}5807\sqrt{A}. \qquad (7)$$

By inserting the known affinities for caloric, other refractive powers could now be predicted and compared with Biot and Arago's experimental results. In the absence of any but the most rudimentary physical picture of the nature and behaviour of the imponderable fluid, Avogadro was obviously under no restriction concerning the form which he adopted for his equation. There was no conservation principle to be expressed

† See pp. 96–8.
‡ Avogadro, 'Memoria sopre la relazione che esiste tra i calori specifici ei poteri refringenti delle sostanze gasose', *Memorie Mat. Fis. Soc. ital. Sci.* **18** (1820), 153–73.

and, with no limit on its complexity, the equation could always be made to accommodate any new data. Such a trial-and-error approach could hardly fail to give good agreement, although ethylene, for which the theoretical and experimental refractive powers would have been in serious disagreement, was significantly omitted.

In 1819 Avogadro was elected a Fellow of the Reale Accademia delle Scienze at Turin and he proceeded to expound his views in the lengthy memoirs that he now began to submit regularly to the Accademia. Nearly all his contributions for several years to come were in some way derived from his work on affinity, although the genuine relevance of this work to certain topics is hard to accept. Perhaps most extravagant of all was his attempt, made in 1824, to relate the density of solids and liquids to the mass of their molecules and to their affinity for caloric.† Although he succeeded in incorporating all three relevant quantities in a single equation, he could not even attempt a theoretical explanation of the connection between them. Understandably, he took this topic no further and he concentrated instead on establishing the relationship between chemical affinity and affinities for light and heat. Of the existence of this relationship, we note, he was still in no doubt.

In 1824 and 1825 Avogadro's views became much better known, at least in Italy, through a series of summaries of all his earlier papers on affinity which he published in P. Configliachi and L. G. Brugnatelli's *Giornale di fisica*.‡ But it was two particularly long memoirs on the subject, read before the Turin Academy in 1823 and published in French in the Academy's *Memorie*,§ which first attracted attention abroad, being reviewed in Baron Férussac's *Bulletin des sciences* for August 1825 by a contributor who signed himself 'S'.¶ An examination of the list of editors printed in the journal suggests that this was the mathematician Jacques Frédéric Saigey, chairman of a distinguished editorial board for the physics section, which

† Avogadro, *Memorie Accad. Sci. Torino* **30** (1826), 81–154, and **31** (1827), 1–94.
‡ Avogadro, *G. Fis.* Decade II, **7** (1824), 427–37, and Decade II, **8** (1825), 1–9, 108–17, 160–4, and 313–30.
§ Avogadro, 'Mémoires sur l'affinité des corps pour le calorique . . .', *Memorie Accad. Sci. Torino* **28** (1824), 1–122, and **29** (1825), 79–162.
¶ *Bull. Sci. math.* **4** (1825), 101–8.

included such men as Dulong, Ampère, Fresnel, and Poisson. At this time the reception in France was unlikely to be favourable, for quite apart from his unconvincing reasoning, far too many basic tenets of Avogadro's treatment were unacceptable. By 1825 the physical reality of caloric was in serious doubt, as will be argued in more detail in Chapter 7, so that speculative attempts to account for its behaviour in terms of affinity, a property that was itself largely discredited in the theory of both refraction and chemical reaction, were particularly suspect. Moreover, even those who remained calorists, Laplace, Poisson, and Carnot, for example, would have found little of interest in Avogadro's views, since affinity played no major part in their versions of the theory. It is interesting to note that the doubts the French might have been expected to feel concerning the other important basis for the work, viz. Avogadro's gas hypothesis, were probably less important, since the hypothesis had by no means been rejected out of hand and it had even enjoyed some slight measure of acceptance by 1825, not least in the writings of Berzelius, Petit, and Dulong.†

Saigey's review in Férussac's *Bulletin* was predictably unenthusiastic, but he conceded that the three sorts of affinity considered by Avogadro were somehow related. After a critical description of Avogadro's work, he concluded: '. . . we believe, with M. Avogadro, that the affinity of bodies for caloric, their refracting force, and the proportions in which they combine (*leur capacité de saturation*) are related to one another', although he was careful to add that 'the connection between them remains to be discovered'.‡

At about the time of the review Dulong was himself engaged in the vain search for some law governing the refractive powers of substances which would be analogous to the Dulong and Petit law for specific heats. Although in a paper read in 1825 he retained the material theory of light to the point of interpreting his experimental work in terms of refractive power, he dismissed Avogadro's writings on the subject, which he presumably knew of through his duties on the editorial board of Férussac's *Bulletin*, as 'a long succession of purely speculative works'.§ Not surprisingly, his experiments on refraction showed

† See pp. 207, 216 n. ‡, and 252.   ‡ *Bull. Sci. math.* **4** (1825), 107.
§ Dulong, *Annls Chim. Phys.* **21** (1826), 177.

that eqn (6) was grossly in error, a fact that Avogadro promptly denied in typically vigorous fashion.† Dulong's scepticism is interesting since it represents the attitude of a leading contemporary scientist, and in particular of one who was far from unsympathetic either to Avogadro's gas hypothesis‡ or to the concept of affinity as applied in refraction. Dulong, moreover, shared Avogadro's interest in the search for some relationship between the properties of a compound and those of its constituents and he pursued the search not only in his work on refractive powers but also in his study of heat.§ Despite all this, Avogadro would seem to have been a figure of no great consequence in Dulong's opinion.

If Avogadro's treatment was suspect by virtue of his speculative reasoning, the experimental results on which it was based were hardly less so by 1825. Despite their early acclaim and their retention by the great French calorists of the 1820s, Delaroche and Bérard's values for the specific heats of gases were increasingly seen as unreliable. In particular, there was a growing belief that the volume specific heats of at least the elementary gases were equal and hence that the small differences between these specific heats, which had been so fundamental in the establishment of Avogadro's system of affinities, did not really exist. Even by 1819 this belief almost certainly had the support of Dulong and Petit, who can hardly have failed to notice that a corollary of their conviction that the equality of the atomic heats of elementary substances held for gases as well as for metals was that the volume specific heats of the elementary gases, as calculated on this basis, were equal.¶ Despite the fact that this result in turn implied, for anyone who accepted Dulong and Petit's law, that the number of atoms in unit

† Avogadro, 'Comparaison des observations de Mr Dulong . . . avec les formules de la relation entre les pouvoirs et les affinités pour le calorique déduites des chaleurs spécifiques', *Memorie Accad. Sci. Torino*, **33** (1829), 49–111. However, in this paper Avogadro did recalculate the affinities on the basis of Dulong's data.

‡ Indeed, in his paper on refraction (*Annls Chim. Phys.* **31** (1826), 174) Dulong explicitly supported the view that all gases, under similar conditions of temperature and pressure, contain equal numbers of particles. There is little doubt that he was here referring to both elementary and compound gases. See also p. 252. § See pp. 250–2, 255, and 259–61.

¶ See Fox, *Br. J. Hist. Sci.* **4** (1968–9), 19 n. There was also some support for this conclusion in the experimental results of Delaroche and Bérard.

THE CALORIC THEORY OF AMEDEO AVOGADRO 217

volume was the same in all these gases, Avogadro's confidence in Delaroche and Bérard's results was undiminished, and in 1825 he rejected not only the validity of the extension from metals to gases which Dulong and Petit proposed but also the exactness of the law even as applied to metallic elements.† Likewise he rejected the results of W. T. Haycraft, who had shown experimentally in 1823 that the volume specific heats of oxygen, nitrogen, hydrogen, and also carbon dioxide were equal; and a little later the somewhat similar results of de la Rive and Marcet were also dismissed as unreliable.‡ In this way Avogadro set himself in opposition to the weight of experimental evidence as well as to the latest theoretical developments.

Yet he was unmoved either by these difficulties or by direct criticism. He replied to Saigey by reiterating his beliefs in the *Giornale di fisica*§ and to Dulong by making minor concessions of a purely numerical nature.¶ His article in Férussac's *Bulletin* for February 1827,‖ in which for the first time he presented his views on affinity in a French journal, thus showed few changes, the treatment being that which he had described to the Turin Academy in 1823. He used the same value for the constant z that he had used then†† and gave the same table of affinities, based on values deduced both from Biot and Arago's refractive powers and from Delaroche and Bérard's specific heats.

One serious difficulty that was not removed either in the 1823 papers or in the 1827 version arose from the considerable variations between the affinities of compounds which, when tested chemically, appeared to be neutral. On the scale that he adopted at this time, both in 1823 and in 1827 (the scale was that of Table E, column 7), the affinities of such compounds

† Avogadro, *G. Fis.* Decade II, 8 (1825), 432–8. See also the note on p. 1 of this volume of the *Giornale*.
‡ Avogadro, *Memorie Accad. Sci. Torino* 24 (1830), 210 n. In this paper, read on 7 December 1828, he rejected the results announced by de la Rive and Marcet in 1827. In *Bull. Sci. math.* 12 (1829), 47–8 n., he also rejected their revised version of 1829. On the work of Haycraft and of de la Rive and Marcet see pp. 256–8.
§ Avogadro, *G. Fis.* Decade II, 8 (1825), 432–8.
¶ See p. 216 n. †.
‖ Avogadro, *Bull. Sci. math.* 7 (1827), 129–42.
†† $z = 0.5412$, slightly different from the figure used in 1817. See eqn (7).

varied from 1·300 to 2·4898, and moreover between these limits there lay the affinities of certain acids and alkalis. To answer the difficulty, Avogadro invoked a distinction between true and apparent neutrality which he had made as early as 1811.† Although all known chemical tests would indicate neutrality, he argued, the compounds in question were neutral only by virtue of the fact that any change in the proportions of their constituent elements in accordance with the law of definite proportions, interpreted on the basis of Daltonian atomic theory, would displace the compound on the electrochemical series away from the fixed point of true neutrality. This true neutrality was at 1·7035, the mean of the values for all of the apparently neutral compounds with affinities between those of the last acid, at 1·6725, and the first alkali, at 1·7281. The scale of the numerical values in the table could of course be readily modified, to give the affinity of oxygen as unity for example—the numerical values were then affinity numbers (*nombres affinitaires*)—or, more significantly, in a way that brought out the connection between affinity for caloric and what Avogadro termed neutralizing power (*pouvoir neutralisant*). This power was measured by the distance from the point of true neutrality (put at zero), with the values ranging from $-1·0000$ for oxygen (electronegative) to $+10·0222$ for hydrogen (electropositive).

So it was that Avogadro's treatment of affinities attained its most extravagant form between 1823 and 1827. What had begun as a chance observation of some ill-defined similarities between chemical affinities and a hypothetical thermal property, deduced with an obvious debt to Berthollet's chemistry, had now yielded a table of affinities for caloric that Avogadro took to be a definitive version of the electrochemical series. Such was his confidence in his system that the very dubious initial assumptions on which it was based were no longer seen by him as matters for discussion, and justification for linking together chemical affinity and affinities for heat and light was for him, if not for his contemporaries, unnecessary.

Avogadro's new approach

Despite this confidence which, as we see from his article in Férussac's *Bulletin*, Avogadro felt as late as February 1827, the

† Avogadro, *J. Phys.* **73** (1811), 74–5.

days were already numbered for his elaborate system, and a paper read in December of the following year indicated an important change of approach.† In it he described a method of determining neutralizing powers directly from the composition by weight of neutral compounds. This involved the use of an additive relationship of the form $al+bm = 0$, where a, b were the proportions by weight of the constituent elements and l, m their respective neutralizing powers, that of the neutral compound being zero. For more complex compounds additional terms would be required in the relationship, and the derivation of a table of powers on the basis of an arbitrary value for a chosen element, in this case oxygen, was in fact far from easy. But the significance lies not so much in the results obtained and the details of the calculation as in Avogadro's assertion that, for all its imperfections, such a table, based purely on weight analyses, was the standard and would hold good whether the treatment in terms of affinities for caloric was valid or not.‡ In deducing these affinities he now also abandoned the values derived from refractive power, finally acknowledging that Dulong's data had shown eqn (6) to be false, at least for certain gases.§ Moreover, he maintained that a comparison between the powers determined by purely chemical considerations and those deduced from specific heats merely suggested, but did not prove, that there was a connection between the two.¶

After the initial neglect and then the unfavourable reaction in France, the final blow to Avogadro's system of affinities came with a new and particularly effective attack on Delaroche and Bérard's data. The critic was Dulong, who in 1828 confirmed experimentally that the volume specific heats of all elementary gases were identical,‖ a result which had been expected for some years. Avogadro would perhaps have remained unmoved by the new evidence had he not recognized that the relationship expressed in eqn (5), if not the theoretical justification for it in terms of affinities, could still be salvaged as a purely empirical result. In March 1830, at all events, he finally did feel able to

† Avogadro, 'Mémoire sur les pouvoirs neutralisans des différens corps simples déduits de leurs proportions en poids dans les composés neutres qui en sont formés', *Memorie Accad. Sci. Torino* **34** (1830), 146–216.

‡ Avogadro, op. cit., p. 161. § Avogadro, op. cit., p. 201.
¶ Avogadro, op. cit., p. 216. ‖ See pp. 253–4.

take the important step of definitively rejecting Delaroche and Bérard's data and with them, by implication, his whole view of the forces that retained caloric in bodies.†

An interesting modification that he made at the same time arose from his use of constant-volume specific heats (for c, c', c'', ... in eqn (5)) in place of those measured at constant pressure. In his work on affinity for caloric he had assumed from the outset that the effect on the observed specific heat of allowing expansion to take place during heating was the same for all gases or, in other words, that c_p/c_v did not vary from gas to gas.‡ But Dulong's discovery that γ varied appreciably with the nature of the gas, being consistently lower for compound gases than for elementary gases, made this view untenable. A choice between c_p and c_v had to be made, and Avogadro came down in favour of c_v, though only, it seems, because this yielded a closer approximation to the relationship expressed in eqn (5).§

In the same paper of March 1830 Avogadro argued that a test of eqn (5) was possible for four reactions, although in the first three of these, as they are listed below, the volume relationships were imaginary and, as we now know, erroneous. Where the volume specific heat of all elementary gases, $c_v(v)$, was taken to be unity throughout, the reactions and corresponding equations were

$$1 \text{ volume of oxygen} + \tfrac{1}{2} \text{ volume of carbon vapour} \longrightarrow 1 \text{ volume of carbon dioxide}$$

$$1^2 + \tfrac{1}{2}(1)^2 = \{c_v(v) \text{ of } CO_2\}^2,$$

giving for the volume specific heat of carbon dioxide 1·225 (the corresponding experimental value was 1·243).

$$\tfrac{1}{2} \text{ volume of oxygen} + \tfrac{1}{2} \text{ volume of carbon vapour} \longrightarrow 1 \text{ volume of carbon monoxide}$$

$$\tfrac{1}{2}(1)^2 + \tfrac{1}{2}(1)^2 = \{c_v(v) \text{ of } CO\}^2,$$

giving for $c_v(v)$ of carbon monoxide 1·00 (experimental value 0·984).

---

† Avogadro, *Bull. Sci. math.* **13** (1830), 211. In the paper of December 1828 Avogadro did not know of Dulong's results, which were not published until June 1829.

‡ Avogadro, *Biblica ital.* **5** (1817), 73.

§ Avogadro, *Bull. Sci. math.* **13** (1830), 213.

1 volume of carbon vapour + 2 volumes of hydrogen ⟶ 1 volume of ethylene

$$1^2 + 2(1)^2 = \{c_v(v) \text{ of } C_2H_4\}^2,$$

giving for $c_v(v)$ of ethylene 1·732 (experimental value 1·754).

½ volume of oxygen + 1 volume of nitrogen ⟶ 1 volume of nitrous oxide

$$\tfrac{1}{2}(1)^2 + 1^2 = \{c_v(v) \text{ of } N_2O\}^2,$$

giving for $c_v(v)$ of nitrous oxide 1·225 (experimental value 1·227).

Although Avogadro had every reason to be satisfied with the way in which his equation accommodated Dulong's new data, the theoretical implications of the relationship and in particular the treatment in terms of affinities for caloric had necessarily to be abandoned. Since the specific heats of equal volumes of all elementary gases, and hence of their individual molecules, were now seen to be identical, they could no longer be interpreted as indicating a characteristic attractive power of the molecule for caloric. Avogadro accepted this conclusion, maintaining that affinity was not the factor that determined specific heat. Although he did not reject the concept of affinity for heat entirely, any correlation between specific heat and chemical affinity was now acknowledged to be out of the question, and eqn (5) assumed the status of a purely empirical relationship in which the quantities represented by the square of the volume specific heats had no theoretical significance. Needless to say, the tables of affinities based on specific heats and those derived from optical data were never revived.

The gas hypothesis could only be strengthened by the shedding of these suspect extensions and by being reconciled, as it now was, with both Dulong and Petit's law and the latest experimental data. So long as Avogadro persisted in his support of Delaroche and Bérard's results, to the necessary exclusion of the Dulong and Petit law for gases, he was alienating valuable support for his gas hypothesis, and this he could ill afford to do. Such influential chemists as Dumas and Berzelius, who, like Dulong and Petit, were led to at least a partial belief in the hypothesis,† could have found little that was acceptable in the

† On Dumas's views see pp. 283–8.

system of affinities, the establishment of which constituted most of Avogadro's scientific output from 1816 to 1827. Indeed, to his contemporaries the hypothesis must have appeared least convincing of all when it was being discussed by Avogadro himself. It is ironical, therefore, and possibly even significant for our understanding of the neglect of the hypothesis, that the paper of March 1830 in which he finally abandoned his very suspect views on affinities and brought his work into line with more acceptable principles was the last in which he dealt specifically with gases. It was thus by his subsequent attempts to extend eqn (5) to solids and liquids, referred to later in this section, that he became most widely known, while the significance of his work in relation to gases was largely ignored.

Although eqn (5) had assumed a purely empirical status, Avogadro's new view of specific heats still had some consequences of a theoretical nature, since he now felt that specific heats reflected the physical constitution of the particles of matter and not their chemical nature. The well-known implication of his gas hypothesis, that the whole or 'integrant' molecules of elementary gases (i.e. O_2, H_2, etc.) could be subdivided into two or more 'partial' molecules (or, as we should now term them, atoms), had not previously been relevant, since a characteristic affinity for caloric had been ascribed to *whole* molecules only. But the existence of partial molecules now became very central, for Avogadro argued that it was merely the number of these in a given volume, and not their nature, which determined specific heat. That the number of partial molecules rather than the number of integrant molecules was the crucial factor was easily demonstrated by the differences between the volume specific heats of elementary and compound gases. In gases of both types the number of integrant molecules in unit volume was the same, according to Avogadro, and it was only the number of partial molecules which varied and hence caused the specific heats to vary. If further evidence on this point was needed, it was found in the second of the four equations listed earlier in this section—that expressing the volume composition of carbon monoxide. Since this reaction was thought to bring about no change in the number of partial molecules per unit volume, the volume specific heat of carbon monoxide would be expected to equal that of the elementary gases, and this in fact had been shown to be the case.

Avogadro's discussion of the factors that determined specific heat could even be taken one stage further, since eqn (5) could be made to yield an explicit relationship between specific heat and the internal structure of the molecules. It clearly followed from the gas hypothesis that the number of partial molecules in unit volume of any gas, compound or elementary, at a given temperature and pressure, was proportional to the number in each integrant molecule, so that volume specific heat in turn depended solely on what we should now term the atomicity of the molecule. The realization that the physical structure of substances rather than their chemical properties determined specific heats represented a most important change in Avogadro's thinking.

It is interesting to note, incidentally, that something resembling his revised view of the problem has since been adopted in a more modern form, for the close dependence of specific heat on atomicity (among other factors) also emerges from the principle of the equipartition of energy applied to the kinetic theory of gases. However, the form of eqn (5), to say nothing of the uncertainty of some molecular compositions, did not allow Avogadro to approach even approximately to the later predictions. Instead eqn (5) led to the quite erroneous consequence that the molecular (or volume) specific heat of a compound would be proportional to the square of the number of elementary integrant molecules that combined to form one molecule of it, on a scale where the specific heat of an elementary integrant molecule was unity.† Not surprisingly Avogadro's contemporaries do not appear to have taken his bold interpretation of specific heats in atomic terms seriously. There was simply too much in his work that was suspect and, in any case, many of the doubts concerning the physical reality of atoms, which were to be expressed so forcibly during the 1830s,‡ were already being felt.

Most of Avogadro's subsequent work on specific heats is of

† For example, in the case of nitrous oxide, one integrant molecule of which is made up from 3/2 elementary integrant molecules, the specific heat on this scale is $\sqrt{(3/2)}$.

‡ On these see pp. 290–2 and the letter from Liebig to Pelouze reproduced in the Appendix, pp. 319–20.

little concern to us, since it deals almost exclusively with solids and liquids.† But his two contributions to the *Annales de chimie et de physique* in 1833 and 1834,‡ the first of his papers on specific heats to appear in this leading French journal, were nevertheless those that attracted most attention in France.§ The purpose of these papers was to extend the range of applicability of his relationship between the specific heat of a compound and those of the elements composing it, and hence also of his conclusions concerning the dependence of specific heat on molecular structure. It was therefore a question of his trying to show how the results already obtained for gases could be applied to matter in other states. The possibility that such an extension might be made had become of much greater interest to Avogadro since the original theoretical implications of his work had been abandoned. In his early paper of 1816–17 he had stated that although it was inconceivable that the attraction of a molecule for caloric could be altered by a change of state, nevertheless his square-law relating specific heat and attractive power did not necessarily hold also for solids and liquids.¶ Such a breakdown of the law would, of course, account adequately for the changes in specific heat that were known to accompany changes of state, and it was for this reason that it was postulated. After 1830, however, when specific heat had been related to the physical structure of the molecule rather than to its innate affinity, there was no clear reason why the same factors that determined the specific heat of a gas should not also determine those of solids and liquids. At first, in the paper of March 1830, Avogadro doubted whether the form of the dependence of specific heat on molecular structure was the same for matter in all three states,‖ but he quickly changed his mind and in 1832 we find him attempting to establish that the specific heat of one molecule of a solid or liquid compound was determined, in

† In any case it has been described recently in some detail in N. G. Coley, 'The physico-chemical studies of Amedeo Avogadro', *Ann. Sci.* **20** (1964), 205–10. See also Avogadro, *Opere scelte*, pp. cxv–cxix.

‡ Avogadro, 'Mémoire sur les chaleurs spécifiques des corps solides et liquides', *Annls Chim. Phys.* **55** (1833), 80–111, and 'Nouvelles recherches...', *Annls Chim. Phys.* **57** (1834), 113–48.

§ Being almost certainly the source of Regnault's knowledge of Avogadro's work, for example. See Regnault, *Annls Chim. Phys.*, 3rd ser. **1** (1841), 129–30.

¶ Avogadro, *Biblica ital.* **5** (1817), 77.

‖ Avogadro, *Bull. Sci. math.* **13** (1830), 215.

exactly the same way as that of a gas, by what he now termed the 'constituent number' (*numero costitutivo*), i.e. the number of elementary integrant molecules that composed it.† To support this extension of his principles Avogadro introduced some novel ideas. For example, it now became necessary to assume that the molecule of water vapour split up into four parts when it passed into the solid or liquid state, and the atomic weights of many elements, in particular those proposed by Berzelius, also required modification. It is hardly necessary to point out that there was no independent justification for these bold assumptions and that at this point chemists were being asked to change far too many of their established principles in order to accommodate a theory of specific heats which none of them had ever seriously entertained.

Avogadro's views had changed little when he stated them for the last time in the third volume of his textbook, *Fisica de' corpi ponderabili*, in 1840.‡ However, in this work, distinguished primarily by the extraordinary size of its four volumes but read with appreciation by Faraday,§ there was at least one point of significance, in the author's refusal to commit himself to any particular theory of heat. This new and more rigorous attitude, which reflected the general mood of the time as well as Avogadro's own disillusionment with his work of the 1816–27 period, definitively precluded any further arguments in terms of affinities for caloric. Henceforth Avogadro concerned himself solely with the study of chemical affinity, and the attempt to relate this quantity to atomic and molecular volumes absorbed all his energies.¶

The text of the *Fisica*, though it was never translated, and the papers that appeared in the *Annales de chimie et de physique* in 1833 and 1834 were the versions of his work that were most readily available in the period after 1830 when interest in

† Avogadro, *Memorie Mat. Fis. Soc. ital. Sci.* **20** (n.d. but probably 1833), fasc. 2° delle memorie di fisica, 451–586.

‡ Avogadro, *Fisica de' corpi ponderabili* (Turin, 1840), vol. 3, pp. 203–28.

§ See Faraday's letter to Avogadro, 10 August 1842, in Avogadro, *Opere scelte*, p. cxxxix. It was also viewed favourably by Théodore de Saussure (see Avogadro, op. cit., p. cxxxviii).

¶ See, for example, his elaborate attempts to relate chemical affinity with atomic and molecular volumes in *Memorie Accad. Sci. Torino*, 2nd ser. **8** (1846), 129–93 and 293–532; 2nd ser. **11** (1850), 231–355; 2nd ser. **12** (1852), 39–122.

specific heats was revived by the experiments of Franz Neumann and Victor Regnault.† Such sources can have conveyed little impression of the true origins of Avogadro's views and it is therefore not surprising that these origins have almost completely disappeared from the historical record. In any case, writers of the 1830s and 1840s would surely have dismissed the work on affinities for caloric as hopelessly wrong, even if they had known of it. For the historian, however, the work has obvious significance, since it represents the interpretation of caloric theory adopted by a scientist of considerable distinction, working largely in isolation but under increasingly tenuous French influences. These unusual conditions, as we have seen, allowed Avogadro to retain such concepts as affinity between ponderable and imponderable matter long after they had ceased to be major research interests in France, the country of their origin. Moreover, and it is this that prevents us from dismissing Avogadro lightly, it was not simply a question of his retaining them, for by 1823 he had developed them to a degree of sophistication that they had never had in France. The result was a notable technical achievement, but it is tempting to wonder how much more Avogadro might have accomplished had he been in a position to detect earlier than he did that the nature of his fundamental beliefs was dissociating him more and more from the main trends in European thought and hence ensuring his comparative, if not total, neglect.

† On these experiments see pp. 289 and 297.

7
THE CALORIC THEORY IN DECLINE

ANY history of the decline of caloric theory must constitute first and foremost a study of the growing inadequacy or, perhaps more accurately, irrelevancy of a theory. In such an account 'internalist' or 'intellectualist' factors, those concerned with the internal dialectic of science, must inevitably and very properly loom large. But, at least as far as France is concerned, a history set in this narrow context would be incomplete, for there the decline of caloric theory provides just one case-history in the decline of a whole orthodoxy in scientific thought that may broadly be described as Laplacian. In this more general movement caloric was not the only casualty and it seems impossible, therefore, to consider the gradual rejection of caloric by the French after 1815 without giving due weight to the fact that the corpuscular theory of light was abandoned by them at about the same time and that Laplacian influence on science in general diminished rapidly after the restoration of the Bourbon monarchy.

Hence it will be one of the chief purposes of this chapter to relate the decline of caloric to an overall change in the style of science, and particularly of French science, which appears to have taken place between 1815 and 1830.

The beginnings of revolt

The really serious cracks in the Laplacian orthodoxy began to appear quite suddenly after 1815. The assailants, the most important of whom were Jean-Baptiste Joseph Fourier, Dominique François Jean Arago, Pierre Louis Dulong, Augustin Jean Fresnel, and Alexis Thérèse Petit, were men who for various reasons had previously exerted only a modest influence on science. Their rise to prominence in the French scientific community after 1815 is therefore as significant as the views that they held, and in attempting to account for it I believe that we must look beyond purely intellectual factors. Above all, it is tempting to seek some correlation between the rise of these men and the momentous political events associated with the downfall of Napoleon and the Bourbon restoration. In fact, no direct correlation appears to exist and the political beliefs of the five men mentioned above were as varied as their motives for attacking the existing scientific theories. Among the most prominent supporters of the wave theory of light, for example, Fresnel was a royalist† and Arago an opponent of the Bourbons who almost lost his post at the École Polytechnique for his views in 1815.‡ Fourier had accompanied Napoleon on the expedition to Egypt in 1798–9 and had been a distinguished secretary of the Institute of Egypt in Cairo, but his attitude to Napoleon during the Hundred Days was equivocal and in 1815 he earned the displeasure of both Napoleon and the royalist party in rapid succession.§ Of the others, Dulong and Petit appear to have taken little interest in politics and they certainly did not suffer at the restoration of monarchy. Under the new regime Petit was made a professor at the École Polytechnique in 1815 and in the following year Dulong was elected to the Académie des Sciences by royal decree, although he angrily declined the honour.¶ Nevertheless, the events of 1815 are not entirely without significance, for it was they which helped to bring Fourier and Fresnel into the circle of Parisian scientists after several years of provincial obscurity|| and

† Arago, *Œuvres complètes*, vol. 1, p. 116.
‡ Arago, op. cit., vol. 3, pp. 62–3. Here, in a biographical sketch of Gay-Lussac read at a public meeting of the Académie des Sciences on 20 December 1852, Arago recounts how in the Conseil d'Instruction of the École Polytechnique in 1815 the professor of literature tried to have an unnamed professor removed from his post for his political views. A footnote added on p. 62 identifies the victim of this attack as Arago himself. E. Blanc and L. Delhoume, in *La vie émouvante et nobl ede Gay-Lussac* (1950), p. 219 n., state that the professor concerned was Dulong, but there seems to be no evidence for this conclusion. As I mention below, Dulong was not treated unfavourably by the royalists.

§ Arago, 'Éloge historique de Joseph Fourier', *Mém. Acad. Sci. Inst. Fr.* **45** (1838), pp. cxxiv–cxxxi; also the article on Fourier by J. R. Ravetz and I. Grattan-Guinness in C. C. Gillispie (ed.), *Dictionary of scientific biography* (Charles Scribner's Sons, New York, in press).

¶ From a recollection of Houtou de la Billardière, once Dulong's laboratory assistant, quoted by Girardin in J. Girardin and C. Laurens, *Dulong de Rouen. Sa vie et ses ouvrages* (Rouen, 1854), pp. 34–5.

|| See Arago, op. cit., pp. cxxix–cxxx, and Émile Verdet's account in *Œuvres complètes d'Augustin Fresnel* (1866), vol. 1, pp. xxix–xxxv. Since 1802 Fourier had been based at Grenoble as prefect for the department of Isère.

which may possibly have helped to create the opening for Petit.†

Probably the most striking characteristic of the men who led the move away from Laplacian science was their youth. Only Fourier (born in 1768) was over 30 in 1815, so that it would have been surprising had they offered any serious opposition to established scientific principles much before that date. As we should expect, there is little evidence of even individual dissent before 1815, for the members of the new generation were good Laplacians by training. All were products of the École Polytechnique, with the necessary exception of Fourier, who taught there at its inception in 1794; yet, significantly, their ties with the great (though now ageing) figures of Laplace and Berthollet were, for the most part, not those of close disciples. Either they were not members of the Arcueil circle or, as in the case of Dulong and Arago, they joined at a comparatively late stage, about 1809–10.‡

In view of the positivistic nature of much later French science, notably Fourier's, it may seem that in this anti-Laplacian movement we have the first signs of a wholesale rejection of the unobservable, hypothetical entities, such as the imponderable fluids, for example, which occupied such an important place in Laplace's thought. In fact, this was by no means the case, since Fresnel, Arago, Dulong, and Petit all had rival theories to oppose to those adopted by the Laplacians. They naturally believed their own theories to be more rational but they did not abandon interest in the nature of heat and light and, as is shown by support which Dulong gave to the atomic theory, in the structure of matter.

Since our concern is primarily with the decline of caloric theory, Petit and Dulong must naturally be singled out for more detailed consideration. Petit was possibly the most gifted member of the new generation, at least as far as

---

After graduating from the École des Ponts et Chaussées in 1809 Fresnel had been employed as an engineer in various French departments.

† See p. 231.

‡ The names of both Dulong and Arago are included in the list of members published in 1816 (*Mém. Phys. Chim. Soc. Arcueil* **3** (1817), p. vii) but not in that of 1809 (*Mém. Phys. Chim. Soc. Arcueil* **2** (1809), 449). From certain references to Arcueil in their writings it appears that they joined the circle soon after the publication of the earlier list.

traditional academic qualifications were concerned.† After completing the entrance requirements for the École Polytechnique at the remarkably early age of $10\frac{1}{2}$, he came to the notice of Hachette, who invited him to a school in Paris run by teachers from the École. Here Petit filled in the time before reaching the statutory age for entry (16). He emerged from the École Polytechnique with extraordinary distinction in 1809, being placed *hors de ligne*, with the next student in the year designated 'first'. That his training in physics would have been firmly based on Laplacian principles seems clear from Hachette's *Programmes d'un cours de physique*, published in 1809 as a handbook for students following Hassenfratz's courses in physics. In the chapter on caloric, which was contributed by Monge,‡ the customary concession of mentioning both the material and vibrational theories of heat was made at the outset, but only the caloric theory was described in detail. And a thoroughly conventional Laplacian view it was, including accounts of the 'latent-sensible' distinction and of the opinion which Laplace could so easily have held (although there is no evidence that he ever did) that the particles of matter attracted caloric to themselves with a force proportional to their own mass as well as to their natural affinity. It is not surprising to find that in his contribution on light Hachette categorically asserted that light was an elastic fluid governed by the usual laws of molecular attraction and affinity.§ The wave theory was not even mentioned. Throughout his life Petit seems to have been particularly close to Hachette, who was himself both intellectually and politically an ardent disciple of Monge (to the extent of losing his post at the École Polytechnique, with Monge, after the Bourbon restoration) and who, during Petit's student days taught several courses with Monge.¶ One of these was on machines and was no doubt the origin of Petit's particular interest in this subject in later years.

† For biographical details see Biot's obituary notice of Petit in *Annls Chim. Phys.* **16** (1821), 327–35; also my article on Petit in the forthcoming *Dictionary of scientific biography*.

‡ Hachette, *Programmes d'un cours de physique*, pp. 54–72.

§ Hachette, op. cit., p. 74. As would be expected, the fluid theories of both electricity and magnetism (the two-fluid version in each case) were adopted without question elsewhere in the *Programmes*, on pp. 133–9 and 212–17.

¶ Hachette, *Correspondance sur l'École Impériale Polytechnique* (1808), vol. 1, p. 464.

Petit's merit did not go unnoticed. In November 1811 the examiners of his doctoral thesis were amazed by the lucidity of his defence of the work, a study of capillary action treated in the Laplacian manner and with full and respectful acknowledgements to Laplace.† And even by that date he had already held posts as assistant (*répétiteur*) at the École Polytechnique, first in the course on analysis and then, from 1810, in Hassenfratz's course in physics.

It was, above all, in the teaching of physics that Petit was to make his mark at the École Polytechnique. In September 1814, with the title of assistant professor (*professeur adjoint*), he took over all of Hassenfratz's duties in the physics course and in the following October his position was confirmed and he was promoted to the rank of full professor (*professeur titulaire*). With the Bourbons now restored, memories of Hassenfratz's earlier Jacobin associations may well have served to hasten his enforced retirement at this time, but it seems clear that in any case his replacement by a younger, more able man was already long overdue. Thus the official reasons for Hassenfratz's removal from office, given as 'health and age' in a report by Durivau, the Director of Studies at the École,‡ were probably the most important ones, even if political considerations did play their part as well. The physics course under Hassenfratz had been the object of a good deal of criticism. Arago, for example, writing in an autobiographical sketch, recalled that in his days at the École Polytechnique (1803–5) Hassenfratz commanded no respect at all from the students and was 'professor only in name'.§ And a report on the examination of 1812 by Ferry, who was acting as an external examiner, included the comment: 'With the exception of optics, which was generally well understood, it must be said that the students have neglected the whole of the rest of physics, even those parts of the subject which one would expect to arouse the greatest interest.'¶ It

† Petit, *J. Éc. polytech.* **9**, cahier 16 (1813), 1–40.
‡ The references to the circumstances of Hassenfratz's removal from office appear in the minute book of the Conseil de Perfectionnement of the École Polytechnique and are dated 20 October and 3 November 1815. They are on ff. 63 and 68 of the fourth volume of the minute book, which is now in the library of the École.   § Arago, *Œuvres complètes*, vol. 1, p. 13.
¶ Ferry's report was reproduced in the fourth volume of the minute book of the Conseil de Perfectionnement cited p. 232 n. §. It appears on f. 28, for the meeting of the Conseil held on 14 November 1812.

has to be pointed out that Ferry noted an improvement in the following year, stating that Hassenfratz now had 'fewer grounds for dissatisfaction', but at least in the second division (i.e. the first-year class) there was still 'a large number of students' who had 'paid very little attention to physics'.† In these circumstances the replacement of Hassenfratz could hardly fail to effect an improvement, but the results of Petit's appointment probably surpassed expectations. Gabriel Lamé was later to refer in glowing terms to his qualities as a teacher,‡ and the importance of physics in the curriculum of the École Polytechnique certainly increased greatly under Petit. Whereas in the last year of Hassenfratz's control, 1813–14, only thirty-four lessons in physics were given in the two years of the course, no fewer than sixty-two lessons were included in the course for 1819–20, Petit's last year as professor.§

But for our present purpose it is the content of Petit's course that is of greater interest than the revolution he effected in the teaching of physics, and on this matter we are fortunate in having available a set of notes taken at his lectures in the winter of 1814–15 by none other than Auguste Comte.¶ These notes show that at the time Petit was wholly committed to Laplacian orthodoxy. For example, in his notes on heat in November and December 1814 he showed himself to be wholeheartedly a calorist. The vibrational theory was not even mentioned and instead the supposed properties of caloric were described in some detail. Heat from friction and the compression of a gas was explained by the expulsion of caloric attendant on any decrease in volume, although Petit's account of adiabatic heating seems to indicate that his views were not highly developed.

† Conseil de Perfectionnement, minute book, vo.1 4, f. 48 (meeting of 6 November 1813).
‡ See the passage from Lamé's *Cours de physique* (1836), quoted on pp. 269–70.
§ From information in the annual *Programme de l'enseignement de l'École Polytechnique*. The minutes of the Conseil de Perfectionnement reflect considerable dissatisfaction with the state of the physics teaching about the time of Petit's appointment. According to a minute of 20 October 1815 (minute book, vol. 4, f. 62) several members of the Conseil had urged an increase in the amount of physics taught. On 3 November 1815 (f. 71) it was reported that the physics course had been entirely reconstructed and on 2 December (ff. 72–4) the new syllabus was approved.
¶ A copy of these notes, which are now kept in the Maison d'Auguste Comte in Paris, was kindly supplied to me by Monsieur D. Cantemir, the resident archivist.

His comment that 'if the volume of a gas is halved, then half of the heat in it should be expelled' even suggests that he had adopted a view that Laplace had held in 1803 but quickly rejected. Accounts of other physical phenomena, including both electricity and magnetism, were also given without question in terms of imponderable, elastic fluids and, although Petit did not cover light in these lectures,† there seems no reason to doubt that here too he would have supported the material theory.

In his early commitment to Laplacian science, then, Petit was typical of the group of young men who were to lead the reaction against Laplace and his followers from 1815. Certainly his commitment did not extend to membership of the Arcueil circle, all his experimental work being apparently performed at the École Polytechnique, but the influence of Laplace's physics, and of Berthollet's chemistry too, was so great and so widely felt in France that formal membership of the circle was hardly necessary. It is interesting to note that Petit's most important scientific contacts, Dulong and Arago, were in fact members, but they only joined the circle when its greatest days were almost over, as has already been noted, and in any case they were soon to become distinguished there for their dissent from Laplacian views.

The conflict between the Laplacians and their critics began with the debate concerning the nature of light which opened in France in October 1815, when Fresnel's first paper on diffraction‡ was deposited at the Institute. In this debate both Petit and Arago were deeply involved as opponents of Laplace's doctrine and as champions of the wave theory. Their close association at this time probably owed as much to their personal friendship as it did to their scientific opinions, for they were colleagues at the École Polytechnique, where Arago had held the chair of analysis since 1809, and, what was even more important, Petit had married Arago's sister in November 1814.

It was in December 1815 that they struck their first blow in favour of Fresnel by interpreting the important experiments on refraction in gases which they had recently performed in a way that appeared irreconcilable with the corpuscular theory.§

† The notes were taken at the first-year lectures; light was covered in the second year at this time.
‡ Fresnel, 'Mémoire sur la diffraction de la lumière . . .', *Annls Chim. Phys.* **1** (1816), 239–81.
§ See p. 202.

When they did this, in a paper read to the Institute on 11 December 1815, they would have known of Fresnel's paper for less than two months, since it was only on 23 October that Arago had been nominated, with Louis Poinsot, as the Institute's official referee for the paper.† This is striking evidence of the impression that Fresnel made, and thereafter Arago's confidence in the correctness of the wave theory, and presumably Petit's too, never weakened. Early in 1816 Arago wrote in support of Fresnel in the *Annales de chimie et de physique*,‡ of which, with Gay-Lussac, he was now joint editor, and we may assume that he was also responsible for the appearance of Fresnel's paper on diffraction in the same journal in March 1816. Further support came quickly. Ampère was an immediate convert to the wave theory§ and even one so thoroughly under Laplacian influence as Gay-Lussac was impressed when he and Arago met Thomas Young during a visit that they made to England in 1816.¶ Indeed, at one point it seemed to Fresnel that the arch-Laplacians, Biot and Poisson, and even Laplace himself, might conceivably be converted.‖

But, as is well known, the victory was not easily won, and even after his success in the prize competition for a study of diffraction, set by the Académie for 1819, Fresnel still experienced the full weight of Laplacian opposition. It seems likely, for example, that it was Laplace, Biot, and Poisson who were chiefly responsible for the delay in publishing the prize-winning memoir until 1826†† and for losing some of Fresnel's papers;‡‡ and all this despite the

† *Procès-verbaux*, vol. 5, p. 562 (23 October 1815).

‡ Arago, *Annls Chim. Phys.* **1** (1816), 199–202. The note, which appeared in the February issue, was also read at the Institute on 26 February 1816.

§ See Ampère's letter to Ballance fils, 19 May 1816, in *Correspondance du grand Ampère*, ed. L. de Launay (3 vols., 1936–43), vol. 2, p. 511.

¶ Arago, 'Éloge historique de T. Young', *Mém. Acad. Sci. Inst. Fr.* **13** (1835), pp. cii–civ.

‖ See Fresnel's letter to his brother Léonor, 5 September 1818, in Fresnel, *Œuvres complètes*, vol. 2, pp. 849–50.

†† When it appeared as 'Mémoire sur la diffraction de la lumière', *Mém. Acad. Sci. Inst. Fr.* **5** (1821–2), 339–475.

‡‡ For a somewhat speculative account of these underhand activities see W. Whewell, *A history of the inductive sciences* (1837), vol. 2, pp. 408–11 and 435–9. See also the paper on 'Comte and positivism' in *Macmillan's Magazine* **13** (1866), 355–6, where Whewell recounts an interesting anecdote by Arago. Apparently Arago had told Whewell that at first he had been afraid to voice his support for Fresnel, so effective and restricting was the Laplacian domination of French scientific circles.

favourable opinion of none other than the permanent secretary of the Académie himself, Fourier, who in 1822 joined Ampère and Arago in giving a glowing report on a paper by Fresnel on double refraction.†

Although the debate over the nature of light, with Biot and Arago as the chief protagonists in the 1820s, has never been chronicled as fully as it deserves, a more detailed description is hardly called for here. However, the episode is important for our purpose in that it does reflect admirably the questioning mood of the time. Far more was at stake for French science than the immediate issue, and a recollection written some years later by the mathematician Guglielmo Libri, who had strong sympathies for Laplace, only serves to emphasize how easily the dispute turned into a reaction against Laplacian science in general. Libri wrote:

... as Laplace, who believed that one should be a geometer first and foremost, had appeared to oppose M. Arago, enemies were raised against him on every side ... all those who attacked the results contained in the *Mécanique céleste* were hailed, and the entire liberal press was mobilized against the distinguished men of our past, who, it was said, were no more than old idols that had to be destroyed.‡

Ampère, incidentally, was another who experienced the opposition of Laplace after 1820, in this case to his view that magnetism and electricity in motion were essentially identical.§

Petit's other main contact, Dulong, was a less controversial figure than Arago.¶ He did not share the precocious brilliance that distinguished, say, Petit, nor was his grounding in Laplacian science quite so thorough. He was orphaned at the age of 4 and

† Fourier, Ampère, and Arago, 'Rapport ... sur un mémoire de Fresnel', *Annls Chim. Phys.* **20** (1822), 337–44. Poisson had been appointed as one of the referees when the paper was read on 26 November 1821 (*Procès-verbaux*, vol. 7, p. 248) but he did not sign the report.

‡ [G. B. I. T. Libri], 'Lettres à un Américain sur l'état des sciences en France. I. L'Institut', *Revue des deux mondes*, 4th ser. **21** (1840), 798. The author of the article, which contained bitter criticisms of the present state of French science, was later identified on p. 88 of the general index of the *Revue des deux mondes* for the period 1831–74. This was published in 1875.

§ See Ampère's letter to Davy, 1825 (?), in *Correspondance du grand Ampère*, vol. 2, p. 680.

¶ The biographical details which follow are taken from Girardin and Laurens, *Dulong de Rouen*; also from Dulong's record card at the École Polytechnique.

his student days at the École Polytechnique were dogged by ill health. Within a year of his admission in 1801 he was granted a month's leave of absence on three occasions and he finally left without completing the course in October 1802. Thereafter he practised medicine for some years in Paris, but his excessive generosity to needy patients seems almost to have ruined him. In 1810 he was fortunate enough to be noticed by Berthollet, who invited him to Arcueil and arranged for his appointment as assistant (*répétiteur*) to Thenard. With his future further assured by teaching posts at the École Normale in Paris and, from 1813, at the Alfort veterinary school, Dulong began to make his name as a chemist, notably by his discovery of nitrogen trichloride and by some important work on metal oxalates. In all this work he proceeded in accordance with Berthollet's principles and without reference to such controversial matters as the atomic theory.

Dulong's association with Petit and his interest in physics appear to have begun early in 1815, almost certainly as a result of the prize competition in physics that was announced by the First Class of the Institute on 9 January 1815.† Competitors were asked to measure the expansion of mercury in a thermometer between 0 and 200 °C and then to determine the rate at which a body cooled both in a vacuum and in various specified gases at different temperatures and pressures. It seems clear that the subject was set to answer problems arising from the competition of 1811 in which Joseph Fourier had won the prize for his theoretical treatment of heat conduction in solids and, incidentally, fallen foul of the Laplacian party.‡ Although it was fundamental to Fourier's theory that the rate of flow of heat from a point inside a solid bar to another either inside or outside

† The subject appears on p. 2 of a pamphlet announcing the prizes to be awarded by the First Class of the Institute in 1816 and 1817. The pamphlet was printed for the public meeting of 9 January 1815. An announcement also appeared in Hachette, *Correspondance sur l'École Polytechnique*, vol. 3 (issue for May 1815), p. 250.

‡ Fourier's paper appeared only in 1824 and 1826 in *Mém. Acad. Sci. Inst. Fr.* **4** (1819–20), 185–556, and **5** (1821–2), 153–246. For comments on the delay in publication, which may well have owed something to criticisms made by the influential judges for the 1811 competition (Laplace, Lagrange, Legendre, Malus, and Haüy), see G. Darboux's introduction to *Œuvres de Fourier* (1888), vol. 1, pp. vi–viii, and Arago's *éloge* of Fourier in *Mém. Acad. Sci. Inst. Fr.* **14** (1838), pp. cxii–cxiii.

the bar was proportional to the difference of temperature between the points, he had taken good care to emphasize that modifications would be simple should Newton's law of cooling, on which he based his assumption, be proved inexact.† In fact, the approximate nature of the law for any but the smallest temperature differences had long been in little doubt,‡ so that a determination of its precise form became an obvious subject for further study now that Fourier's work had drawn new attention to the problem.

So it was that Dulong gained his interest in heat, and his friendship with Petit. As it happened, he and Petit did not submit an entry for the competition but even by May 1815, when they submitted a preliminary report on their work to the First Class of the Institute,§ they had already performed some important joint experiments on thermometry. The nature of these experiments, which were devoted largely to a careful comparison of the expansive properties of mercury, glass, air, and certain metals, strongly suggests that they were undertaken in an attempt to answer the first part of the Institute's competition, but a number of unspecified interruptions, possibly connected with the political unrest at the École Polytechnique,¶ prevented completion of the experiments, and in July 1816 the preliminary report was published in its original form.‖

Fortunately the prize for the competition was not awarded,

† Fourier, *Mém. Acad. Sci. Inst. Fr.* **4** (1819–20), 202–3. In basing his treatment on Newton's law Fourier was on very weak ground; see my comment on this in *Br. J. Hist. Sci.* **4** (1968–9), 4 n.

‡ Although much of the evidence against the law appears inconclusive to modern eyes. See G. Martine, *Essays medical and philosophical* (1740), pp. 233–47; J. C. P. Erxleben, *Novi commentarii societatis regiae scientiarum Gottingensis*, commentationes physicae et mathematicae classis, **8** (1778), 74–95; Dalton, *New System*, vol. 1, part 1, pp. 12 and 108–23. The work with the greatest influence in France, however, was probably that of François Delaroche, published in *J. Phys.* **75** (1812), 201–28. Although Delaroche's paper was read to the Institute on 3 June 1811, it was not published until September 1812 and so was in all likelihood unknown to Fourier, who was working in Grenoble, at the time he submitted his entry on 28 September 1811.

§ *Procès-verbaux*, vol. 5, p. 514 (29 May 1815), where Dulong is identified as the reader.

¶ On the effect of the unrest see G. Pinet, *Histoire de l'École Polytechnique* (1887), pp. 74–102, and Berthollet's letter to Berzelius, 27 August 1815, in Berzelius, *Bref*, vol. 1, part 1, pp. 54–5.

‖ Dulong and Petit, 'Recherches sur les lois de dilatation des solides, des liquides et des fluides élastiques, et sur la mesure exacte des températures', *Annls Chim. Phys.* **2** (1816), 240–63.

no entry of sufficient merit having been received,† and in March 1817 the same subject was set again for 1818.‡ This time Petit and Dulong produced a masterpiece,§ which won the prize and was to be acclaimed as a model of experimental method by such varied authorities as Comte, Poisson, Lamé, and Whewell.¶ Once more they began with the problem of thermometry and, as they had done in 1815, they came down in favour of the air thermometer as the ultimate standard, convinced that increments of temperature indicated on this instrument, or indeed on any gas thermometer, were increments in the true temperature. But in all this they were saying little more than in 1815 and it was with regard to the second part of the competition that they had made the greatest progress. The key to their new success lay in their work on the experimental determination of the specific heats of solids and also of mercury, a task that had to be undertaken if observed changes in the temperature of a cooling body were to be interpreted in terms of quantity of heat lost. The use that Petit and Dulong made of their results in the establishment of a new and improved law of cooling to replace the long-suspect law of Newton is certainly worthy of a detailed description in its own right as a major achievement in the empirical study of heat. But for our purpose it is sufficient that we should observe in Petit and Dulong's work on heat between 1815 and 1818 the origins of their interest in specific heats and hence also of the discovery that was soon to ensure their lasting fame.

Petit and Dulong on the nature of heat

Both in their paper of 1815 and in their prize-winning memoir of 1818 Petit and Dulong avoided any digressions on the nature of heat and related theoretical issues. It was quite possible,

† The two entries that have survived in the archives of the Académie des Sciences are very insubstantial pieces of work.
‡ *Procès-verbaux*, vol. 6, p. 164 (17 March 1817), and *Annls Chim. Phys.* **4** (1817), 302–3. In fact the subject was slightly modified; see Fox, *Br. J. Hist. Sci.* **4** (1968–9), 7 n.
§ Dulong and Petit, 'Recherches sur la mesure des températures et sur les lois de la communication de la chaleur', *Annls Chim. Phys.* **7** (1818), 113–54, 224–64, and 337–67.
¶ Comte, *Cours de philosophie positive* (1835), vol. 2, p. 534; Poisson, *Théorie mathématique de la chaleur* (1835), p. 6; Lamé, *Cours de physique* (1836), vol. 1, pp. i–ii; Whewell, *History of the inductive sciences*, vol. 2, p. 485.

after all, to study the laws of cooling and even thermometry in a purely empirical manner, and this Petit and Dulong clearly sought to do. It is true that in 1818 they did make a passing comment to the effect that in their opinion the absolute zero of temperature was probably unattainable,† and this attitude can only be interpreted as an implied rejection of the Irvinist doctrines. However, such a rejection would have been far from novel at this time in France,‡ and Petit and Dulong did not enlarge on the point or, still less, pass any opinion on the rival Laplacian theory.

So it is impossible to assess with any certainty the state of Petit and Dulong's views on the nature of heat on the eve of the discovery of their famous law in April 1819. All that can be said is that both men were educated at a time when the truth of the caloric theory was virtually axiomatic and that Petit was certainly teaching the theory as late as December 1814 at the École Polytechnique.§ However, the period 1815–20 was one of fundamental change in the style of French science, and even before 1819 both Petit and Dulong were deeply involved in the change as critics of the Laplacian principles. We have already seen how Petit appeared as a champion of the wave theory of light at a very early stage, and later in this section we shall examine the support that Dulong gave to the atomic theory from 1816, despite the opposition of Berthollet and even Laplace. This history of dissent from established opinion in the careers of Petit and Dulong clearly does not account for the decision to make the announcement of their law to the Académie des Sciences on 12 April 1819¶ the occasion for a major attack on Laplacian orthodoxy, but it does make the attack more intelligible.

I have argued at length elsewhere that Petit and Dulong's discovery that 'the atoms of all elementary substances (*corps simples*) have exactly the same capacity for heat' (the statement of the law which they used in 1819) was made not as the result of a planned programme of research but unexpectedly in the course of the experimental work on specific heats which they

† Dulong and Petit, op. cit., p. 259.   ‡ See, for example, p. 151.
§ See pp. 232–3.
¶ In Petit and Dulong, 'Sur quelques points importans de la théorie de la chaleur', *Annls Chim. Phys.* **10** (1819), 395–413. Dulong was the reader; (see *Procès-verbaux*, vol. 6, p. 437.

had continued and extended after their victory in the 1818 competition.† Hence, in so far as it could have been made, say, ten years earlier or ten years later, the discovery itself throws comparatively little light on the state of French science in the early years of the Bourbon restoration, and it is therefore far less significant historically than the criticisms of certain aspects of Laplacian science that accompanied it.

Petit and Dulong directed their attacks on two major fronts: the nature of heat and the atomic theory. As far as heat was concerned, the caloric theory was rejected in all its forms, although it was only with respect to the Irvinist version of the theory that the evidence was really convincing. Unfortunately Petit and Dulong gave no details of how their evidence had been obtained, but it appears that they had shown, by experiments of their own, that heats of reaction and changes in specific heat were not related and that 'in most cases' the release of heat in a chemical reaction was not accompanied by an overall decrease in heat capacity. If this was indeed the case, it was in itself decisive evidence against the Irvinist position, and Petit and Dulong had no hesitation in declaring that the views of Irvine and Crawford were 'opposed by too many facts for it to be possible to adopt them'.‡ Dalton's Irvinist-based theoretical work on the specific heats of gases received similarly crushing treatment.§ It should be noted that the mode of attack used by Petit and Dulong was by no means original, but the introduction of new experimental evidence made the argument still more effective than, for example, the one put by Delaroche and Bérard in their prize-winning memoir of 1812.

By comparison, the alternative, 'Laplacian' version of the caloric theory was less easily disposed of. Here the mere absence of any apparent connection between heats of chemical reaction and changes in specific heat did not invalidate the theory, and Petit and Dulong chose to level their criticisms at some elaborations of the theory that Lavoisier and Laplace had made in their joint paper dated 1793.¶ In this paper Lavoisier and Laplace had applied the earlier idea of theirs that any diminution

† Fox, 'The background to the discovery of Dulong and Petit's law', *Br. J. Hist. Sci.* **4** (1968–9), 1–22.
‡ Petit and Dulong, op. cit., p. 397.
§ Petit and Dulong, op. cit., pp. 397–8.
¶ Lavoisier, *Mémoires de chimie* (1805), pp. 121–47.

in the volume of a body necessarily brought about a decrease in the quantity of latent heat that the body could contain and hence, as a result of this decrease, the conversion of some heat from the latent to the sensible state, a conversion that was manifested, for example in the percussion of solids, as a rise in temperature. Considering the case of combustion, they had argued that the quantity of heat evolved in such a reaction would be largest when the resulting oxide was a solid, rather than a liquid or a gas, since oxygen, which was taken to be the chief source of the heat, would then have undergone the greatest possible decrease in volume. As it happened, this view of combustion was not wholly borne out by their experiments, but to account for the discrepancies it was necessary simply to assume that when it combined with the substances undergoing combustion (or oxidation), oxygen was capable of retaining quantities of heat that varied with the nature of the oxide that was formed. The solution was, of course, an arbitrary one without any independent justification and it was this arbitrariness above all else that Petit and Dulong criticized, although they had little more evidence than was available to and used by Lavoisier and Laplace themselves. The assumption that caloric existed in two states, they declared, was simply 'too improbable'.†

The conviction with which Petit and Dulong rejected both of the main versions of the caloric theory is striking in view of the small amount of evidence available to them, and there can be little doubt that it owed a great deal to the fact that a quite different and (to them) more acceptable explanation of chemical heat was now to hand. The origin of this, although the debt was never fully acknowledged, was very probably Berzelius.

Berzelius spent nearly eleven months in France from August 1818 until July 1819. At Berthollet's instigation, he and Dulong worked together on the composition of water in the laboratory at Arcueil‡ and there the two men struck up an extremely close friendship ended only by Dulong's death in 1838. By the time they first met, in September 1818, Berzelius had been known for some years as a critic of the caloric theory of chemical

† Petit and Dulong, op. cit., p. 408.
‡ One result of this collaboration was a joint paper in *Annls Chim. Phys.* **15** (1820), 386–95.

heat, although in some of his earliest writings he had adopted Lavoisier's version of the theory, presumably as this was given in the *Traité élémentaire de chimie*.† By 1811 he had certainly rejected the theory, principally on the ground that he had observed no decrease in volume during the exothermic reaction between sulphur and copper, and he had substituted the view, proposed in the 1806 Bakerian lecture by Humphry Davy, that heat of reaction was electrical in origin.‡ Of course, it was already well known by this time that an electrical discharge could give rise to heat even though no chemical reaction occurred, and it was therefore a short step from Davy's demonstration of 1808 that a piece of graphite was heated, even in the absence of oxygen, simply by being placed across the poles of a cell,§ to the view that it was an electrical discharge between the oppositely charged reactants entering into combination which caused the emission of heat. Berzelius's opinion on the matter was unchanged in 1818, when he published an extended account of his electrochemical theory in the third volume of his *Lärbok i Kemien*, a substantial part of which was soon to be translated as the famous *Essai sur les proportions chimiques*. He accompanied this account with a lengthy criticism of both the main versions of the caloric theory of heats of reaction,¶ basing his criticism on arguments that were remarkably similar to those adopted by Dulong and Petit in the following year.

It has to be emphasized that Berzelius's opinions on chemical heat could be adopted without necessarily abandoning belief in the fluid nature of heat. Thus, even after he had expressed his support for the electrochemical theory, Berzelius was still able to write (in 1812) in a manner which strongly implied his conviction that heat was a fluid,‖ although some far from compli-

---

† For information on Berzelius's early views I rely on J. R. Partington, *A history of chemistry* (1964), vol. 4, p. 168, since I have been unable to consult the relevant work, the first volume of Berzelius's *Lärbok i Kemien* (Stockholm, 1808).

‡ Berzelius, *Gilb. Ann.* **37** (1811), 278–80, and *Annls Chim.* **79** (1811), 249–51. For Davy's view see *Phil. Trans. R. Soc.* **97** (1807), 42–4.

§ Davy, *Phil. Trans. R. Soc.* **99** (1809), 71–2.

¶ Berzelius, *Essai sur les proportions chimiques* (1819), pp. 58–73. This is a translation, with only minor modifications, of pp. 49–63 of the third volume of the *Lärbok* (Stockholm, 1818).

‖ See J. S. C. Schweigger's *Neues Journal für Chemie und Physik*, **6** (1812), 139–41. The comments appeared in French in *Annls Chim.* **86** (1813), 168–71.

mentary notes on Davy's *Elements of chemical philosophy*, which he communicated to the author soon after the book appeared in 1812, suggest that he might have preferred not to commit himself on the matter. Referring to Davy's somewhat unusual theory of heat (described in Chapter 4), he wrote: 'I am utterly convinced that there is no hypothesis which adequately explains the phenomena of caloric. So rather than deceive our readers by a false show of certainty let us admit our ignorance on this point.'†

The blow dealt to caloric by the growth of the electrochemical theory was scarcely less serious for being indirect. Certainly the hypothesis of the electrical origin of heats of reaction did not, in itself, either exclude the fluid theory or imply the vibrational nature of heat, but one of caloric's greatest strengths had been precisely that it provided such a convincing explanation, indeed the *only* plausible explanation, of chemical heat. Once an alternative explanation was accepted, then the whole question of the nature of heat would become a completely open one again, with the fluid and vibrational theories competing, so to speak, on equal terms. It was in 1819 that the alternative explanation became really well known in France, initially through the paper by Petit and Dulong, and a little later, in June, by the publication of Berzelius's own views in a readily available form in the *Essai sur les proportions chimiques*.‡ Just how important the *Essai* was seen to be in this respect, as well as in the support which it gave to the atomic theory, is brought out well in Georges Cuvier's report on the work of the Académie des Sciences for 1819, which contains a lengthy account of the book and in which special reference is made to Berzelius's comments on chemical heat.§

Petit and Dulong expressed their acceptance of the electrical origin of heats of reaction in their paper in April 1819. They were careful to present their belief as no more than a highly probable conjecture, to the point of admitting that the more traditional

† Berzelius to Davy, n.d., in Berzelius, *Bref*, vol. 1, part 2, p. 41. Berzelius's attitude was unchanged in the 1820s, although he then stressed the usefulness of the language of materiality when studying problems in heat; see p. 277 n. †.

‡ Berzelius, *Essai*, pp. 68–73. On the date of publication see Berzelius's letter to Alexandre Marcet, 1 June 1819, in Berzelius, *Bref*, vol. 1, part 1, p. 195. I am indebted to Dr C. A. Russell for the latter reference.

§ Cuvier, *Mém. Acad. Sci. Inst. Fr.* 4 (1819–20), pp. lxxxi–xciii.

theories of chemical heat might still be applicable in accounting for a small part of the overall effect. Yet private correspondence, as we might expect, tells us rather more. That Dulong at least had unreservedly abandoned the caloric theory by January 1820 is evident from the following passage in a letter which he wrote in that month to Berzelius, now back in Sweden:

> We had already dealt the chemical theory of heat a fatal blow in the paper that we read to the Institute during your stay in Paris. Some new experiments lead me to consider it as indisputably true that all [thermal] phenomena which have nothing to do with radiant heat result simply from the vibrational motion of the molecules of matter themselves. According to this view radiant caloric is transmitted by vibrations in the same fluid which, when vibrating more rapidly, arouses in us the sensation of light. And in the same way the voltaic pile produces the phenomenon of fire only because, by means of the electric current, it provokes vibrations in the particles of matter. MM. Clément and Desormes have published a result which supports my opinion—an opinion which I have hitherto discussed with no one else. This is that a given weight of steam at any pressure whatsoever, though with a temperature appropriate to the particular pressure, always contains the same quantity of heat. M. Despretz has found that the same holds true for other liquids also.†
> Now, I can show that when the volume of a gas or vapour is suddenly altered, changes in temperature are produced which are incomparably greater than would be caused by the quantities of heat released or absorbed if some heat was not *created* by motion. Rumford had already used an almost identical type of argument in support of his view but he made his observations on solids, a procedure that made his reasoning far easier to attack.‡

The nature of the evidence that had now not only confirmed Dulong in his rejection of the caloric theory but also made him an unhesitating advocate of the vibrational theory of heat is of considerable interest. The word 'created' (*engendrée*), which Dulong underlined, strongly suggests that he saw the temperature changes associated with the rapid compression of a gas as being due, at least in part, to a conversion process, although there is no reason to believe that he even considered the possibility of a conservation principle (in this case relating heat and

† On Despretz's work see Fox, *Br. J. Hist. Sci.* **4** (1968–9), 16 n.
‡ Dulong to Berzelius, 15 January 1820, in Berzelius, *Bref*, vol. 2, part 1, pp. 13–14.

work). Unfortunately he nowhere enlarged on the conjectures contained in this letter, so that his views would have been known to few contemporaries. Precisely how he applied the work of Clément and Desormes was thus never made clear and we can now do no more than attempt the following reconstruction of the argument.

The crucial piece of new evidence that Dulong used was the 'law' governing the heat content of saturated vapours which Clément and Desormes had announced in their paper to the Académie des Sciences in August 1819. For anyone unfamiliar with the interconvertibility of heat and work it was an obvious consequence of this 'law' that the mere compression of saturated vapour under adiabatic conditions would cause no change in the initial state of saturation, as has already been pointed out.† Therefore, it seemed reasonable to treat the vapour as a perfect gas, so that the volume of, say, unit mass of a saturated vapour at any temperature could be determined by applying Boyle's law in conjunction with Gay-Lussac's figure for the expansion coefficient of gases and with the widely used tables relating saturated vapour pressure and temperature. From an argument conducted in this way‡ it followed, for example, that the adiabatic compression of a mass of steam initially at 60 °C (and at the corresponding saturated vapour pressure of approximately 14·5 cm of mercury) to one-fifth of its original volume would require that the temperature should rise by roughly 40 °C if the steam were to remain saturated and if no condensation were to occur. Hence, on the basis of Clément and Desormes's 'law' and on the assumption that the heat content of the vapour remained constant, it could be argued that a rise in temperature of roughly 40 °C should be observed. In the light of a rapidly growing body of experimental evidence concerning the temperature changes accompanying the rapid (if not truly adiabatic) compression of air, Dulong was almost certainly convinced that this figure was far too small and that the discrepancy was due not to any error in Clément and Desormes's 'law' but rather to the fact that the steam did not remain saturated and that the 'law' consequently became inapplicable. Since he accepted

† See p. 181.
‡ For just such an argument see Carnot, *Réflexions sur la puissance motrice du feu*, pp. 67–8 n.

Clément and Desormes's 'law', Dulong would undoubtedly have associated this departure from a state of saturation with an increase in the heat content of the steam and he may well have been seeking to account for this increase when, as in the letter of January 1820 quoted above, he concluded that heat could actually be created by the movement of a piston.

It is unfortunate that in the absence of any further comment, even from Berzelius, we can do no more than speculate on the effect that Dulong's argument and the support that it gave to the vibrational theory might have had on his contemporaries. That it would have been seen as an important contribution to the debate concerning the nature of heat can hardly be doubted, for Dulong had done more than point to the inadequacies of the caloric theory, many of which were already familiar enough. Like Rumford some twenty years earlier, he had also seen fit to tackle the far more difficult problem of establishing the vibrational theory. In 1820 Dulong's case would probably have appeared more plausible than Rumford's, but its ultimate effectiveness was limited by the nature of the evidence on which it was based. In particular, Clément and Desormes's 'law' was untrue and it was known to be so by 1827.† Indeed, the fact that Dulong himself conducted an experimental refutation of the 'law' about this time‡ may even provide a much-needed clue in understanding why he did not pursue his championing of the vibrational theory.

By the support that it gave to the chemical atomic theory, Dulong and Petit's law dealt an equally telling blow at the style that we have broadly described as Laplacian. The existence of this completely unexpected relationship between the familiar physical quantity specific heat and the individual atoms of matter was important evidence, especially in France, where even by 1819 Dalton's atomic theory had made only modest headway and was still far from being accepted.§ During the first decade and a half of the nineteenth century Berthollet's system, teaching as it did that elements combined in varying and not fixed proportions by weight, had dominated French

† On the discrediting of the 'law', by Despretz and Dulong, see Fox, op. cit., p. 16 n.   ‡ See Fox, loc. cit.
§ On the fortunes of the atomic theory in early nineteenth-century France see M. P. Crosland, 'The first reception of Dalton's atomic theory in France', in Cardwell, *Dalton & science*, pp. 274–87.

chemistry. There are many examples that illustrate this point but, to select just one of the most celebrated, we might note in particlar how in 1808 Gay-Lussac had interpreted his newly discovered law of combining volumes in terms not of the atomic theory, as we might have expected, but of Berthollet's chemistry.†

Not surprisingly, Dulong's early work at Arcueil suggests a similar acceptance of established opinion‡ and it was not until 1816 that he gave any indication of dissent. In that year, however, it suddenly became apparent that he had been deeply impressed by the atomic theory when he persuaded all the other members of the distinguished editorial board of the *Annales de chimie et de physique* to allow the publication of an extract from William Prout's papers on the structure of atoms which had recently appeared in England.§ Dulong himself prepared the extract for the April issue of the journal.¶ Also it was soon after this episode, in July 1816, that he described a classic example of what we should now know as the law of multiple proportions for the various oxides of phosphorus.‖ In the paper he discussed atomic weights freely and throughout he seems to have had no hesitation in interpreting his experiments in terms of the atomic theory, a fact that Berthollet and Thenard, the Institute's referees, duly noted in a generally favourable report.††
It is unfortunate indeed that the text of a paper by Berthollet describing his reservations towards Dulong's conclusions, with special reference to the atomic theory, was never printed.‡‡ Clearly the theory was still the object of much scepticism§§ and

† Gay-Lussac, *Mém. Phys. Chim. Soc. Arcueil* **2** (1809), 231–3. In fact Berthollet had already shown how combinations between gases in fixed proportions could be accommodated in his system; see Berthollet, *Essai de statique chimique* (1803), vol. 1, pp. 366–7.

‡ See, for example, his paper 'Recherches sur la décomposition mutuelle des sels insolubles et des sels solubles', *Annls Chim.* **82** (1812), 273–308.

§ See Dulong's letter to Berzelius, 8 January 1822, in Berzelius, *Bref*, vol. 2, part 1, pp. 36–7. Among the other members of the board were Berthollet, Chaptal, Gay-Lussac, Thenard, Chevreul, Biot, and Arago. Prout's paper had appeared in *Ann. Phil.* **6** (1815), 321–30, and **7** (1816), 111–13.

¶ *Annls Chim. Phys.* **1** (1816), 411–16.

‖ Dulong, *Annls Chim. Phys.* **2** (1816), 141–50.

†† *Procès-verbaux*, vol. 6, pp. 101–3 (21 October 1816).

‡‡ According to the *Procès-verbaux* Berthollet's paper was read immediately after the report on Dulong's work. No copy of Berthollet's paper has survived.

§§ Although there is some evidence that opinions in France were beginning to change by about 1816; see Fox, op. cit., p. 17 n.

the precise circumstances that led Dulong to appear as an atomist are therefore all the more difficult to ascertain. It is possible that he had now been finally persuaded by his work on the composition of the oxides of phosphorus, and it may be also that he had been impressed by the fact that almost simultaneously Berzelius cited his own work on the subject as strong evidence in favour of the physical reality of atoms.† As an editor of the *Annales de chimie et de physique* Dulong certainly knew of Berzelius's work and his esteem for the Swedish chemist was such that he would not have dismissed it lightly.‡

But, whatever the motives that first caused Dulong to dissent from the attitudes towards the atomic theory that were so established at Arcueil, we may be certain that he would have required little further persuasion from Berzelius in 1818 and 1819, although his adoption of the electrochemical theory may well date from this time. Any doubts he might still have had were to be finally dispelled in 1819, of course, though not only by the Dulong and Petit law. In November 1819 Berzelius, now back in Sweden, sent word to Dulong of Eilhard Mitscherlich's discovery of isomorphism,§ which Dulong immediately saw as yet another proof of the existence of atoms. After welcoming the news, he replied to Berzelius:

I am convinced, despite the objections of M. de Laplace and of some others, that this theory [i.e. the atomic theory] is the most important idea of the century and that in twenty years' time it will have wrought incalculable benefit in all branches of the physical sciences.¶

The prediction, as is well known, was over optimistic.

The study of the thermal properties of gases, 1820–40

Petit's tragically early death from tuberculosis in 1820 was a serious blow to French science in general and to Dulong in

† Berzelius, *Annls Chim. Phys.* **2** (1816), 320–39.

‡ This is evident also in Dulong's letter to Ampère, 5 August 1816, quoted in P. Lemay and R. E. Oesper, 'Pierre Dulong, his life and work', *Chymia* **1** (1948), 175. On Dulong's general esteem for Berzelius see his comment in *Annls Chim. Phys.* **2** (1816), 174 n.

§ See Berzelius's letter to Dulong, 5 November 1819, in Berzelius, *Bref*, vol. 2, part 1, pp. 10–11.

¶ Dulong to Berzelius, 15 January 1820, in Berzelius, *Bref*, vol. 2, part 1, p. 12.

particular. Yet despite the loss of his gifted colleague, Dulong still had fruitful years ahead of him. Henceforth his research interests had a certain unity in that they were concerned almost exclusively with gases, for reasons that are not hard to suggest. Above all, he appears to have been guided throughout by the conviction, widely shared in the early nineteenth century, that the familiar, simple laws that were thought to govern the physical properties of gases were in some sense true while the corresponding ones for liquids and solids were only approximate, being vitiated by unknown factors. It was natural to suppose, therefore, that all the 'true' laws of nature, including those as yet undiscovered, would be observed most easily in gases.

But this was not the only consideration that brought Dulong increasingly into contact with matter in the gaseous state. For example, the prize competition on respiration and animal heat set by the Académie des Sciences for 1823 seems to have provided the stimulus for several months of experiments, although the paper which resulted was never submitted and was published only after his death.† The paper is an interesting one, for in it Dulong took the important step of rejecting the theories of animal heat of both Lavoisier and Crawford, concluding that the fixation of oxygen in the body would not account for all the heat produced and that the source of the additional heat might never be known.

A far more absorbing undertaking than these studies of animal heat was his work on the vapour pressure of steam. A commission set up by the Académie early in 1823 at the Government's request had recommended, among several safety measures for steam-engines working at more than 2 atmospheres, the insertion of a fusible metal plug in the boiler walls.‡ Since the plug was intended to melt and so release the steam at a temperature slightly higher than that at which the engine normally operated, it was essential to have reliable data relating vapour pressure and temperature, and once again the Government asked the Académie for assistance. The resulting commission was headed by Dulong, who with Arago seems to

† Dulong, 'Mémoire sur la chaleur animale', *Annls Chim. Phys.*, 3rd ser. **1** (1841), 440–55.

‡ See Charles Dupin's report, on behalf of a commission consisting of himself, Laplace, Prony, Ampère, and Girard, in *Procès-verbaux*, vol. 7, pp. 470–9 (14 April 1823).

have borne the main burden of the long and often dangerous experimental work.† It was small consolation after some five years of the most wearisome experiments that the vapour-pressure measurements extending up to 24 atmospheres which Dulong eventually set before the Académie in November 1829‡ remained standard until Regnault's still more comprehensive study in the 1840s. These results need not detain us here but the accompanying experimental proof of the exactness of Boyle's law for air up to 27 atmospheres is of some interest. Unfortunately bureaucratic opposition, notably to the erection of apparatus on the tower of the Lycée Henri IV,§ prevented a more detailed investigation, which would have included an extension of measurements to other gases, but the result as it stood gave further authority to the view that the physical behaviour of gases was governed by simple laws. Above all, it did nothing to weaken Dulong's own conviction to this effect.

Despite the demanding nature of his work for the commission Dulong did not neglect more theoretical problems entirely. Indeed, his constant concern until his death in 1838 seems to have been to answer the many questions arising from the law which he and Petit had discovered so suddenly in 1819. New determinations of specific heat were obviously called for, but the experimental difficulties of such work, especially where gases were concerned, were only too well known, and it may well have been this consideration that caused Dulong to seek his answers instead in terms of refractive power. This he did in a paper read before the Académie des Sciences in October 1825.¶ That one who had shown no reluctance to depart from Laplacian

---

† The original members of the commission were Dulong, Prony, Ampère, and Girard, Arago being added at some time after July 1824.

‡ Dulong et al., Annls Chim. Phys. **43** (1830), 74–111. A preliminary report had been submitted in July 1824; see Annls Chim. Phys. **27** (1824), 95–101, and Procès-verbaux, vol. 8, pp. 114–17 (19 July 1824). The wearisomeness of the experiments is brought out well in, for example, Dulong et al., Annls Chim. Phys. **43** (1830), 111; Dulong's letter to Berzelius, 10 November 1825, and Berzelius's letter to Dulong, 2 February 1825, in Berzelius, Bref, vol. 2, part 1, pp. 63 and 93; also in an unpublished letter of 4 November 1828 to an unnamed correspondent, where Dulong speaks of his hopes of giving up responsibility for the work (Library of the Institut de France, Paris, file 2220). See also the passage from a letter of 10 August 1828 quoted on p. 255.

§ Dulong et al., op. cit., p. 92.

¶ Dulong, 'Recherches sur les pouvoirs réfringens des fluides élastiques', Annls Chim. Phys. **31** (1826), 154–81.

principles on the question of the nature of heat, and who in the letter to Berzelius quoted on p. 244 had already given evidence of support for the wave theory of light, should have conducted such a study in the language of the corpuscular theory, using measurements of refractive index to calculate refractive power, seems strange at first sight; but it becomes rather more understandable once we appreciate Dulong's aims. These were twofold, as can be seen from two separate statements which Dulong made in reference to his work. From both of these statements, which are quoted below, it is clear that Dulong was seeking above all a way of investigating the properties of individual atoms or molecules from observations on the refraction of light, and for this purpose the wave theory offered no quantity that had a significance comparable with that of refractive power. It was this consideration above all others, we may assume, that led him to use at least the language of the corpuscular theory.

Of the two statements concerning his aims, one appeared in a letter of November 1825 to Berzelius, in which Dulong wrote:

> When I undertook this work it was my intention to investigate whether the molecules of [all] elementary substances exert the same action on light. It was possible to see this as a likely result in view of the law governing capacities for heat.†

The other is to be found in the 1825 paper itself, where we read:

> It was even probable that in this way a deeper knowledge would be obtained of the modifications which can be brought about in the molecules of matter by a chemical reaction.‡

The first of Dulong's aims as stated here was simply the establishment of some extension of the law of atomic heats to refraction. But the second of his two aims, by contrast, was one that had so far eluded him even in the case of heat. In 1819 he and Petit claimed to have proved, by experiments that they had already conducted, that there was a simple relationship between the thermal capacity of a molecule of any compound and that of all the elementary atoms.§ They had admitted, however, that doubts concerning the true composition of compounds were a major difficulty, and, not surprisingly, the

† Dulong to Berzelius, 10 November 1825, in Berzelius, *Bref*, vol. 2, part 1, p. 63.   ‡ Dulong, op. cit., p. 155.
§ Petit and Dulong, *Annls Chim. Phys.* **10** (1819), 407-8.

relationship had not been, and never was to be, divulged. By 1825 any hope of extending the law to molecules must have appeared remote and it was presumably with some feeling of frustration that Dulong now turned to refractive power. Here at least was a quantity that was more easily measurable than specific heat. Moreover, by determining it for different gases under similar conditions of temperature and pressure Dulong was convinced that he was obtaining a precise measure of the 'action' of the individual gas particles on light, for he believed that under such conditions the particles of all gases had the important property of being equally spaced.† In saying this, of course, he was expressing his acceptance of Avogadro's hypothesis, in which he found yet another good reason for believing that the laws he was seeking would emerge most simply if he studied matter in the gaseous state.

Just how great Dulong's hopes of success in his work on refraction really were is not clear, but the results were undoubtedly disappointing. For example, a complete analogy with the Dulong and Petit law for elements would have demanded proportionality between the refracting force and the density or atomic weight of the gases examined, but no relation of any sort emerged. And, needless to say, Dulong had to admit failure with his second aim also. Refraction, it seemed, was a phenomenon quite unrelated to specific heat and one that probably depended on the electrical state of the molecule rather than on its mass. Such a conclusion, as Dulong pointed out, accorded well with the wave theory, the truth of which was now in little doubt in his mind.‡

Dulong was distressed by his failure. In October 1825, shortly after completing the work on refraction, he complained to Auguste de la Rive: 'No general law . . . only approximate laws. There it is! I have no good fortune, whereas Gay-Lussac has only to touch on a subject in order to discover a law.'§ It is fortunate perhaps that he did not live to witness Regnault's systematic and definitive refutation of the gas laws in the 1840s. For Dulong, as we see, 'true' laws were essentially simple.

† Dulong, op. cit., p. 174.
‡ Dulong, op. cit., pp. 180–1. See also his letter to Berzelius, 10 November 1825, in Berzelius, *Bref*, vol. 2, part 1, p. 64.
§ From a recollection by de la Rive in *Œuvres de É. Verdet* (1872), vol. 1, pp. xxv–xxvi.

When, in 1828, he turned again to the experimental study of specific heats,† his problems were unchanged and the answers he obtained were hardly more satisfactory. The method, however, represented a distinct advance in so far as, although only indirectly, it did allow the determination of c_v, the quantity that he considered the more important for the theory of heat.‡ The principle of his experiment was to measure the pitch of the fundamental notes sounded in a tube 60 cm in length when the tube was filled with various gases, a procedure made practicable by Dulong's considerable talent for music.§ A simple calculation based on Laplace's expression for the velocity of sound then allowed γ for any gas to be determined from the expression

$$\frac{\nu}{\nu'} = \frac{\sqrt{(1+0\cdot00375\,t)}\sqrt{(\gamma_{\text{air}}/\rho_{\text{air}})}}{\sqrt{(1+0\cdot00375\,t')}\sqrt{(\gamma_{\text{gas}}/\rho_{\text{gas}})}},$$

where t' was the temperature of the gas and ν' the frequency of the fundamental note obtained with the gas in the tube, t and ν being the corresponding quantities for air, and where ρ_{air} and ρ_{gas} were the densities of air and the gas measured at the same temperature and pressure. The values of γ for oxygen (1·415) and hydrogen (1·407) which Dulong obtained were so similar to that for air (1·421), a figure already known from observations on the velocity of sound, that it was assumed that γ was the same for all elementary gases. This conclusion, taken in conjunction with the supposed identity of the volume specific heats of such gases (now stated by Dulong as a necessary consequence of Dulong and Petit's law), suggested a corollary that was to assume very great significance in the eyes of the pioneers of energy conservation, notably James Joule. This significance, it must be added, appears to have been quite lost on Dulong himself, but his paper was to attract so much attention in later years that some further discussion seems necessary here.

Dulong observed that if air was heated at a constant pressure from 0 to 1 °C, i.e. so as to absorb 1·421 units of heat on the scale where $c_v(v)$ for that volume of air was unity, it would expand by 1/267 of its initial volume. If the air was now compressed to its original volume without loss of heat, it could be

† Dulong, 'Recherches sur la chaleur spécifique des fluides élastiques', *Annls Chim. Phys.* **41** (1829), 113–59.
‡ Dulong, op. cit., p. 130.
§ On which see Girardin and Laurens, *Dulong de Rouen*, p. 10.

argued (wrongly, though for a calorist quite naturally) that the 0·421 units of heat whose function had been solely to effect expansion would be liberated and would raise the temperature of the air by 0·421 °C. It is hardly necessary to point out that, whatever gas was involved, this rise in temperature was expected to be always equal to $(\gamma-1)$ for a compression of $1/267$, so that for elementary gases (γ constant) the temperature changes predicted would always be the same. For compound gases, however, γ and hence also the increase in temperature produced by the same fractional decrease in volume varied considerably. In fact, it was Dulong's observation that γ was smallest for those gases with the greatest volume specific heat which led him to investigate whether the magnitude of the temperature changes, i.e. $(\gamma-1)$, might be determined solely by capacity and whether the amounts of heat liberated in equal volumes of gases for any given compression might therefore be identical. Once this assumption was made, of course, it was possible to calculate $c_p(v)$ for any gas relative to air simply from a knowledge of γ. For example, in the case of carbon dioxide ($\gamma = 1\cdot337$) a fractional decrease in volume of $1/267$ would be expected to liberate $0\cdot337 c_v(v)_{CO_2}$ units of heat, where $c_v(v)$ for air on the same scale was unity. Hence, on equating $0\cdot337 c_v(v)_{CO_2}$ to 0·421 (the heat liberated by a similar compression in air), 1·249 and 1·670 would emerge as the volume specific heats of carbon dioxide at constant volume and constant pressure respectively. Similar calculations for other gases allowed several comparisons to be made between the specific heats predicted in this way and the experimental values of Delaroche and Bérard. The agreement between the two sets of figures was only moderate (see Table A) but in the opinion of Dulong, who in any case had good reason to be sceptical towards Delaroche and Bérard's work,† it was sufficiently convincing for him to conclude that his postulate was confirmed and that a given decrease in volume did in fact release the same quantity of heat from equal volumes of all gases taken at the same temperature and pressure. Privately he referred to this conclusion as 'a law noteworthy for its simplicity and its numerous applications'‡ but, as has

† See p. 216.
‡ Dulong to Berzelius, 15 June 1828, in Berzelius, *Bref*, vol. 2, part 1, p. 84. See also Dulong, *Annls Chim. Phys.* **41** (1829), 156.

already been mentioned, it was only with the advent of energy conservation that the full implications of his discovery were appreciated. The all-important next step of quantifying the compression process in terms of work done and hence of concluding that the work of compression was always the same when equal quantities of heat were produced was not one that Dulong, or apparently anyone in the 1820s, was capable of taking. It is interesting, if somewhat idle, to speculate whether a greater familiarity with, and interest in, the world of power engineering might conceivably have led Dulong to at least a limited version of the principle of the conservation of energy at this stage, but a reading of his paper gives no indication that he even contemplated such an extension of his results.

Despite its enormous significance to modern eyes the paper was in many ways yet another failure. Above all, the recurring problem of relating the specific heat of a compound to its composition remained unanswered. Dulong merely stated that the results he had just obtained confirmed the relationship that he and Petit had tentatively established in 1819† but that further progress depended on his obtaining new information concerning the contractions that occurred in chemical reactions between gases.

Dulong never fulfilled his promise that this topic and a study of the variation of specific heat with pressure would be the subject of a second memoir. His scientific career was in fact almost at an end, for overwork and ill health were already seriously hindering his research. In 1827 he was vice-president and in the following year president of the Académie des Sciences, and in addition there was always the burden of teaching and the work of the commission on vapour pressures. In 1828 he complained bitterly to Berzelius: 'My health has been in a wretched state, especially in the last two years.... Four years ago I embarked on a huge piece of work on heat. I cannot manage to finish it.... It is the lessons that are killing me.'‡ His teaching in chemistry at the Faculty of Science had now been taken over by Gay-Lussac to leave him free for the less

---

† Dulong, op. cit., p. 158. See also letter of 15 June 1828 cited in the previous note. For the comments of Petit and Dulong in their 1819 paper see *Annls Chim. Phys.* **10** (1819), 407–8.

‡ Dulong to Berzelius, 10 August 1828, in Berzelius, *Bref*, vol. 2, part 1, p. 74.

taxing physics course, but by 1830 he had declined so seriously that his anxious colleagues persuaded him to give up the chair of physics at the École Polytechnique and become the Director of Studies there.† The freedom from teaching duties effected an improvement in his health, but there would now be even less time for scientific work. To Berzelius he confided in June 1831: '...great as this advantage would be for me, to give up scientific work would be too high a price to pay. I am burdened with a mass of administrative details which leave me no spare time at all.'† Yet he held the post for the rest of his life and never carried out the threat, which he made elsewhere in this letter, to resume the physics chair should it become vacant, even though the resignation of his successor in the chair, Claude Pouillet, gave him the opportunity of doing so in 1832. His health failed him yet again in July 1833 when he was obliged to relinquish the coveted position of permanent secretary of the Académie des Sciences after holding office for only one year. The chemist Friedrich Wöhler, who met Dulong about this time, commented how 'His perpetual state of ill health seems to affect his whole personality.'‡

Dulong was not alone in his interest in the specific heats of gases during the 1820s nor in his belief that Delaroche and Bérard had not said the last word on the matter. There were others who were also apparently influenced by the Dulong and Petit law. In 1823 W. T. Haycraft announced to the Royal Society of Edinburgh his conclusion that the volume specific heats of all gases were identical.§ His apparatus consisted of two identical calorimeters filled with water, in each of which was immersed a spiral tube. Dry air heated to about 180 °F was passed through the tube in one calorimeter, while the gas under test, similarly heated and carefully dried, flowed through the other. By comparing the rises in temperature in the two calorimeters the specific heat at constant pressure of any gas relative to air could thus be determined. The heating effects of equal volumes of carbon dioxide and ethylene, the only compound

† Dulong to Berzelius, 20 June 1831, in Berzelius, *Bref*, vol. 2, part 1, pp. 100–1.

‡ Wöhler to Berzelius, 27 October 1833, in *Briefwechsel zwischen J. Berzelius und F. Wöhler*, ed. O. Wallach (Leipzig, 1901), vol. 1, p. 533.

§ Haycraft, 'On the specific heat of the gases', *Trans. R. Soc. Edinb.* **10** (1824), 195–216.

gases used, were found, quite erroneously, to be the same as those of all the common elementary gases.

The apparatus of the two young Swiss physicists Auguste de la Rive and François Marcet was equally insensitive. In 1827 they rejected existing determinations made at constant pressure and described their own simple method designed to yield the specific heat at constant volume.† The gas was enclosed in a small glass vessel that could be exposed to a constant source of heat, either by direct immersion in hot water or by being suspended at the centre of an almost completely evacuated copper sphere that was itself immersed in the water. The specific heats of fourteen gases were deduced by comparing the increases in temperature indicated by a manometer fitted to the vessel, when it was exposed to the source of heat for equal intervals of time. Like Haycraft, de la Rive and Marcet concluded that the volume specific heats of all gases, elementary and compound, were identical. The apparatus lent itself readily to the determination of specific heats at pressures below atmospheric, and predictably enough there emerged further evidence to support the well-established myth that the specific heat of a given mass of gas increased with rarefaction. Observations on air, for example, showed that the decrease in volume specific heat accompanying a reduction in pressure to one-third of atmospheric was only about 17 per cent, and similar results were apparently obtained for hydrogen, ethylene, and carbon dioxide. As good calorists, de la Rive and Marcet naturally had no reason to doubt the accuracy of their experiments. Indeed, even the smallness of the change in volume specific heat which they observed was quite predictable, as they explained in their paper:

. . . although on the one hand the number of molecules to be heated [in a given volume] is decreased by rarefaction, yet on the other hand the spaces which exist between the molecules and which are thought to be full of caloric increase in the same proportion. Indeed, we shall point out that the changes in temperature that occur when the volume of a gas is varied become very easy to explain [by our argument], whereas if the specific heat decreased or increased in

† De la Rive and Marcet, 'Recherches sur la chaleur spécifique des gaz', *Annls Chim. Phys.* **35** (1827), 5–34. On the reasons for their rejection of the determinations at constant pressure see p. 275.

the same proportion as the pressure, we should have to fall back on a quite different explanation.†

It is hard to avoid the conviction that we are here face to face with yet another illustration of the effect that prior commitment to a theory can have on experimental work, if only in making the observer less critical of his apparatus. We are reminded all too clearly how in 1812 Delaroche and Bérard had drawn a similar conclusion to the above on the basis of two sets of readings only.

Further experiments, in which a larger glass vessel and hence also larger quantities of gas were used, were announced in April 1829,‡ before the publication of Dulong's paper of 1828. Although the observed decrease in volume specific heat for a reduction in pressure to one-third of atmospheric had now risen to about 30 per cent, observations on air, carbon dioxide, nitrous oxide, and hydrogen at various pressures gave no reason to modify the earlier general conclusions. In particular, the volume specific heats of all gases still appeared to be equal.

James Apjohn, professor of chemistry at the Royal College of Surgeons in Ireland, was presumably following a suggestion by Gay-Lussac in devising his ingenious method for the determination of the specific heats of gases, although he made no acknowledgement. In a paper that he read in March 1815 Gay-Lussac had pointed out that when a current of air passed over a thermometer, the bulb of which was wrapped with wet muslin, the temperature observed was a minimum when the latent heat absorbed by the liquid on the muslin in evaporating equalled the heat given up by the air in cooling to the temperature of the bulb.§ Hence, after corrections for such factors as radiant heat reaching the bulb, it was possible for the specific heat of the air at constant pressure and the latent heat of the liquid to be simply related. Between 1835 and 1837 Apjohn announced three sets of results obtained by various methods but all based on this same principle.¶ Despite an obvious lack of consistency

† De la Rive and Marcet, op. cit., p. 30.
‡ De la Rive and Marcet, *Annls Chim. Phys.* **41** (1829), 78–92.
§ Gay-Lussac, *Annls Chim. Phys.* **21** (1822), 82–92.
¶ Apjohn, 'On the specific heats of the permanently elastic fluids', *Notices of the communications to the British Association . . . 1835* (1836), pp. 30–2; also *Trans. R. Ir. Acad.* **18** (1838), 1–16.

between the results, he felt justified in breaking with contemporary opinion to the point of rejecting the view that all gases had identical volume specific heats, even when this conclusion was restricted to elementary gases, and he declared furthermore that the specific heats and atomic weights of gases were unrelated.

A similar principle was used quite independently of Apjohn by the Utrecht physician Alexander Karel Willem Suerman. In 1829 he had outlined the theory of his method in an essay on vapours that he submitted successfully for a prize competition in physics at the University of Leiden,† but it was not until June 1836 that he presented his work on the specific heats of gases for an honorary doctorate at Utrecht. His results, which included a study of the variation of specific heats with pressure,‡ were far more favourable to Dulong's views than were Apjohn's. Slight deviations from equality and even an unexpectedly high value for hydrogen, which Suerman believed he could explain away, did not rule out the possibility that elementary gases had identical volume specific heats, while the values for the compound gases nitrous oxide and carbon dioxide, though not that for carbon monoxide, were appreciably higher.

As a result of new experiments de la Rive and Marcet were also in agreement with Dulong by 1835.§ The principle of their method was to allow a small copper vessel containing heated terebene to cool at the centre of a large evacuated copper sphere. Gas could be passed through a spiral tube immersed in the terebene and, from a comparison of the rates of cooling observed both with and without the gas flowing, the amount of heat carried off by the gas could be calculated and interpreted to yield its specific heat at constant pressure. Their experiments with this apparatus left de la Rive and Marcet in no doubt that compound gases had volume specific heats that were generally greater than those of elementary gases, and they looked forward with interest to the task of discovering the factors that determined specific heat, a task that was now, of course, as important

---

† Published in *Annales academiae Lugduno-Batavae, 1829–1830* (Leiden, 1831).

‡ Suerman, 'Expériences sur la chaleur spécifique des gaz et de l'air à pressions différentes', *Annls Chim. Phys.* **63** (1836), 315–32. For my biographical information on Suerman I am indebted to Dr Th. J. Meijer of the Academisch Historisch Museum of the University of Leiden.

§ De la Rive and Marcet, *Annls Chim. Phys.* **75** (1840), 113–44.

for them as it was for Dulong. In 1835 they could as yet offer no solution but they promised to give the matter the closest attention.

Dulong never lost interest in this problem despite his extreme disability. In a letter of 1832 addressed to de la Rive and Marcet† he stated that he had already adopted an experimental procedure similar to the one that they later described in their 1835 paper and he mentioned that he needed only one week free from other duties in order to complete his experiments on specific heat. But unfortunately there is no evidence that the measurements were ever made. Even so, a letter to the chemist Alexandre Baudrimont, written towards the end of 1834, gives a further indication of Dulong's continued activity and also of his undiminished confidence in the efficacy of the study of gases in elucidating the structure of matter. He wrote:

For a long time I have thought, like you, that the elements which make up compounds, instead of uniting with one another directly, as chemists suppose they do, might well split up (*se diviser*) when they combine and that elementary substances might be molecular, like compounds, rather than simply atomic. This at least is the conclusion to be drawn from experiments that I performed earlier with Petit. Even now I am busy with the problem and I shall find a solution in a more reliable way than you since I am working on substances for which I have one factor less to consider, this factor being the cohesion which intervenes to upset the results. I am working on gases, in which the cohesion is almost non-existent.‡

The nature of the experiments Dulong was performing at this time is not stated, but the mention of Petit suggests that he was probably referring to determinations of the specific heats of gases. Aided by the important concept of a polyatomic elementary molecule which, as we see from this letter, he was definitely using at this time, he may well have been attempting to predict specific heats in terms of atomicity. However, by the time of his death in July 1838 he had made little progress, although he still maintained that he needed only a few more

† Quoted in de la Rive and Marcet, op. cit., pp. 123–6.
‡ Girardin and Laurens, *Dulong de Rouen*, p. 73. Baudrimont's views on the structure of matter (with an acknowledgement to Gaudin and Ampère) were given most fully in his *Introduction à l'étude de la chimie par la théorie atomique* (1833), especially pp. 73–5; but see also his *Traité de chimie générale et expérimentale* (1844), vol. 1, pp. 8–10.

days in order to complete the work.† In September of that year Arago, who had been entrusted with the task of inspecting Dulong's papers, announced to the Académie des Sciences that he had managed to deduce two 'laws' concerning specific heats from a scrap of paper found in Dulong's study. These 'laws', which Arago arrived at with the aid of Cabart, Dulong's assistant, and a Monsieur Savary (presumably Félix Savary, professor of geodesy at the École Polytechnique), who had been partially in Dulong's confidence over the work, were as follows:

(a) Compound gases formed from elementary gases which undergo no decrease in volume when they combine have the same specific heat as the elementary gases.

(b) Compound gases in the formation of which the constituent gases undergo any given decrease in volume all have the same specific heat, although this is very different from the specific heat of the elementary gases.‡

The very statement of these conclusions reflects the measure of Dulong's failure, for the fact that one who had always appeared as such an enthusiastic believer in the literal truth of the atomic theory (not only in 1819 but also in his letter of 1834 to Baudrimont) should relate specific heat not to the composition of an atom or molecule but to purely observational volume relationships hardly represents success. It seems clear that by 1838 Dulong would have become only too aware of the complexity of the task that he had set himself. And it is also possible that he had been influenced by the mood of scepticism with regard to physical theory which had been clearly expressed by several French men of science in the 1830s. As will be pointed out later in this chapter, the prevailing attitude to the nature of heat had now become one of deep agnosticism, and serious problems in the determination of atomic weights had called in doubt the very existence of atoms. Jean-Baptiste Dumas was only one of the most prominent of the earlier supporters of the atomic theory who could see no way out in the 1830s, and Dulong would have been very insensitive indeed had he not himself felt the mood of the time.

† According to a personal recollection by Girardin in Girardin and Laurens, op. cit., p. 27.
‡ *C.r. hebd. Séanc. Acad. Sci., Paris* **7** (1838), 604 (17 September 1848).

Unfortunately the idea that Dulong was influenced by the new scepticism that was abroad in French science cannot be proved definitively, since he published nothing in the 1830s. Yet one event, his sudden flirtation with the positivists about 1836, does suggest that at that time he was experiencing a growing disillusionment. The pessimistic philosophy of positivism appears to have attracted him at a period of great frustration and so to have overcome his undoubted independence of mind, which would not have allowed him to fall in readily with any philosophical movement and which had already been apparent in his relations with the Arcueil circle. It is primarily in the hope of being able to throw some much-needed light both on his philosophy of science in the mid 1830s, but also as a contribution to our understanding of the role of the positivists in the decline of the caloric theory, that Dulong's association with Comte is now examined in some detail.

Dulong and the positivists

Positivism, in the form given it by Auguste Comte, originated in France in the 1820s. While its principles can hardly be said to have dictated the course of French science in the second quarter of the nineteenth century, they nevertheless contained elements that appealed to some of the most prominent French scientists of the period. Of these the best known was undoubtedly Joseph Fourier, whose mathematical studies of heat conduction in solids, notably the *Théorie analytique de la chaleur*, published in 1822, was consistent with most of Comte's ideals. In the *Théorie* Fourier dispensed with any discussion concerning the nature of heat and he even began the book with the now well-known declaration (which would have done credit to any Comtian positivist): 'First causes (*les causes primordiales*) are unknown to us; but they are subject to simple unvarying laws which can be discovered by observation and the study of which is the object of natural philosophy.'†

---

† Fourier, *Théorie analytique de la chaleur* (1822), p. 1. Cf. the similar attitude to the unknown causes of electrical phenomena which was adopted by Ampère in his *Théorie des phénomènes électrodynamiques, uniquement déduite de l'expérience* (1826), pp. 4–8. Throughout this work Ampère deliberately adopted a positivistic approach, in sharp contrast to the style of most of his other writings.

THE CALORIC THEORY IN DECLINE

There seems no reason to believe that Fourier wrote this under any influence from Comte, but he was showing at least some interest in Comte's philosophy by 1829† and in later years he did establish something resembling a small positivist school in science, with men like Gabriel Lamé and Jean Marie Constant Duhamel pursuing the study of heat in mathematical terms and without reference to its nature.‡ In the light of this evidence it is obviously impossible to deny that the positivist style of science did have a certain following in France from the early 1820s onwards and, in view of Comte's well-known scepticism towards Laplacian science in general and towards the concept of the imponderable fluid in particular,§ it is therefore natural to inquire whether positivism may have been an important factor in the overthrow of the caloric theory.

A general assessment of the extent of positivist influence on nineteenth-century science would clearly demand a far more detailed study of the problem than I can possibly give here, and I must therefore state somewhat categorically my belief that for the most part positivism reflected, rather than caused, the changes in the style of French science which can be detected between 1820 and 1850. For example, in this book I have tried to show that the rejection of Laplacian science came about as the result of a movement that was well under way before Comtian positivism emerged and that the leaders of this movement, Fourier perhaps excepted, did not have noticeably positivistic leanings. Hence if positivism has a place in the history of the decline of Laplacian science, I would suggest that it is rather as a symptom of the difficulties encountered by the men who were trying to establish new principles to replace those that they had so recently abandoned. A clear instance of these difficulties can be found in the work on the atomic theory that is discussed in the next chapter. Here, by the 1830s, the problem of verifying the physical reality of atoms was proving far more difficult than Dulong and the other early critics of Berthollet's chemistry had ever envisaged, and this gave rise to despair and to

† See p. 266.
‡ For an account of this later work see G. Bachelard, *Étude sur l'évolution d'un problème de physique; la propagation thermique dans les solides* (1927), pp. 89–132.
§ See, for example, Comte, *Cours de philosophie positive* (6 vols., 1830–42), vol. 2, pp. 438–54.

expressions of this despair, notably by Dumas, which had a distinctly positivist flavour.† Similarly work on the wave theory of light was presenting certain very complex problems by the 1820s and 1830s, and despite some brilliant mathematical treatments by Navier, Poisson, Augustin Louis Cauchy, James MacCullagh, George Green, and others, it was not until the mid 1840s that G. G. Stokes finally succeeded in establishing an acceptable explanation of the (supposed) properties of the luminiferous ether.‡ In this case the difficulties do not seem to have raised serious doubts concerning the truth of the wave theory, but with regard to the theory of heat the problems encountered after the rejection of caloric were considerably more daunting, so daunting in fact they were rarely even discussed. Thus Dulong's hopes of being able to demonstrate the truth of the vibrational theory, which he expressed in 1820, were still unrealized when he died in 1838. And the agnostic character of comments on the nature of heat in the textbooks of the 1820s and, more particularly, of the 1830s and 1840s§ also reflects the fact that the caloric theory was no longer taken seriously but that it had proved impossible to establish the truth of an alternative theory.

As one of the earliest and most convinced supporters of both the atomic theory and the vibrational theory of heat, Dulong was only too conscious of the difficulties that beset chemists and physicists alike after the collapse of the Laplacian orthodoxy. It is for this reason that his reactions to the doctrine of positivism, which appeared to offer a refuge of sorts, are unusually instructive.

Since mathematics headed Comte's hierarchy of the sciences, as fulfilling most nearly the positivist ideals,¶ it is not surprising to find that his models of scientific method were generally taken from mathematics rather than from the comparatively 'backward' science of chemistry, to which much of Dulong's early work belonged. Despite this and despite the fact that Fourier's analytical treatment of heat conduction in solids became a natural archetype for positivist method from

† Although it is clear that Dumas himself was no positivist. See p. 287.
‡ For a brief account of this work see E. T. Whittaker, *A history of the theories of aether and electricity* (2nd edn, 1951), pp. 128–69.
§ On these see pp. 275–7.
¶ Comte, *Cours de philosophie positive*, vol. 1, pp. 88–115.

THE CALORIC THEORY IN DECLINE

the beginning,† Comte did find much to admire in Dulong. He was generous in his praise of Dulong's purely physical work: for example, his determination of a law of vapour pressures, the expansion experiments conducted with Petit in 1818, and the accompanying examination of Newton's law of cooling.‡ Here Dulong had done all that the positive philosophy could have demanded, which was, as Comte put it, 'to look on all phenomena as being subject to unvarying natural laws, the discovery of which laws and their reduction to the smallest possible number stand as the goal for all our endeavours.' The correct procedure was to 'regard as quite unattainable and meaningless for us the search for what are called causes, whether these be first causes or final causes', and this Dulong appeared to have done.§

Unfortunately the most famous discovery that Dulong had made, the Dulong and Petit law, was hardly in the same category. There can be little doubt that it suffered in Comte's eyes from forming part of the study of chemistry, a branch of science in which the theological and metaphysical phases of reasoning, which preceded the positive, had exerted a great influence, more so than on mathematics, astronomy, and physics, for example, though less than on physiology.¶ The law also implied belief in the truth of the atomic theory, although this was less damning than might be concluded from a reading of certain later positivists. Comte thought highly of Dalton and although Dalton appeared on the Positive Calendar not with the chemists but with such engineers as Watt and Carnot, i.e. by virtue of his experimental work on vapour pressures, his theory was referred to by Comte as 'the famous atomic theory, which to this day has governed all subsequent developments in the doctrine of chemical proportions and still serves as an indispensable basis for the everyday application of the doctrine'.

---

† There is much evidence of Comte's admiration for Fourier. The *Cours de philosophie positive* was dedicated to him (jointly with Henri Marie Ducrotay de Blainville) and it contained some very favourable references to his work (see, for example, pp. 16–17 of vol. 1 and pp. 549–93 of vol. 6). See also Comte's letter to Blainville, 31 March 1826, in *Correspondance inédite d'Auguste Comte* (4 vols., 1903–4), vol. 1, p. 28, where Comte expresses the (vain) hope that Fourier may be persuaded to attend his lectures on positivism.

‡ Comte, op. cit., vol. 2, pp. 534 and 544. The law of vapour pressures referred to by Comte had been established empirically by Dulong and his colleagues in the commission on vapour pressures set up by the Académie; see pp. 249–50.

§ Comte, op. cit., vol. 1, p. 14.   ¶ See Comte, op. cit., vol. 3, pp. 7–11.

Still speaking of the atomic theory, he added: 'This principle accords so well with our ideas in every branch of science that it has almost assumed the status of a felicitous generalization founded directly on ideas which are immediately familiar to any mind concerned with the study of the various parts of natural philosophy. Hence the universal acceptance of the principle has come about without impediment.'† The substitution of a theory of chemical equivalents for the atomic theory would naturally be welcomed as more 'positive' but, as Comte pointed out, Wollaston's attempt to do just this had effected a change merely in terminology and not in basic conception.‡

Comte's attitude to Dulong was almost certainly not determined by such philosophical convictions alone, however. The most cursory examination of his writings, and in particular of his correspondence, indicates how greatly his estimates of scientific worth were coloured by the state of his personal relationships, which were themselves in turn influenced above all by the working of his extraordinary ambition, notably for the principal mathematical chair at the École Polytechnique.§ It is therefore not surprising that his most laudatory references to Dulong date from after 1836, when Dulong appears first to have taken a genuine interest in Comte's principles. Dulong had not been present at the first public exposition of the positivist philosophy in April 1826, which was terminated after only a few lectures by Comte's temporary insanity, nor at the repetition of the course in 1829. Among scientists prominent at the Académie des Sciences and at the École Polytechnique, Louis Poinsot and the biologist Henri Marie Ducrotay de Blainville attended both courses, and Navier, who sponsored Comte's entry to the École Polytechnique as his assistant in 1832 and whom Comte later described as 'my one really devoted friend within the Council of the École',¶ was present in 1829, together with Fourier.‖ So Dulong could hardly have been

† Comte, op. cit., vol. 3, pp. 142–3.
‡ Comte, op. cit., vol. 3, p. 149.
§ Comte was an unsuccessful candidate for one of the chairs of analysis and mechanics on three occasions, in 1831, 1836, and 1840.
¶ Comte to Dulong, 31 August 1836, in *Correspondance inédite d'Auguste Comte*, vol. 4, p. 163.
‖ References to some of the more important members of Comte's audiences appear in Comte, *Cours*, vol. 1, p. 5.

THE CALORIC THEORY IN DECLINE

unaware of the new philosophy and, as is seen from the first of the two passages quoted below, he even seems to have felt a positive antipathy towards Comte until the latter months of 1836. The turning-point came when, in November and December of that year, he attended the lectures which Comte gave at the École Polytechnique while filling temporarily the chair of analysis and mechanics vacated by the death of Navier. That the distinguished Director of Studies should follow the lectures of a mere assistant was a notable event. Comte wrote in 1842:

> The famous Dulong . . . had been present in person at these memorable lectures, which earned me his sincere esteem despite his previous inclination to share involuntarily in the general prejudice against me that abounded in our scientific circles.†

And even more eloquently in July 1840, in a letter to the president of the Académie des Sciences (Jean Victor Poncelet):

> . . . I shall never forget the warmth and zeal with which this rare and scrupulously loyal man allowed his usual modest reserve to be overcome and made it his duty to strive to make known the favourable impressions he had formed of me as a result of that decisive ordeal. By the irresistible power of his mind and his personality he thus sought to repel the unfair and malicious insinuations which had been directed against me shortly before in the very heart of this Academy.‡

In fact, even before he attended Comte's lectures Dulong had already given evidence of some degree of sympathy when, on 19 September 1836, he specially requested that Comte's letter of candidature for Navier's chair at the École Polytechnique should be read out before the Académie.§ Comte had earlier solicited Dulong's support¶ and this mark of esteem greatly impressed him, as his numerous subsequent references to the incident attest.|| But it proved of no avail and Comte was

† Comte, op. cit., vol. 6, pp. xxiii–xxiv.
‡ *Correspondance inédite d'Auguste Comte*, vol. 4, p. 203. On Dulong's attendance at these lectures see pp. 248–9 of this volume and pp. xii–xiv of the sixth volume of the *Cours de philosophie positive*.
§ *C.r. hebd. Séanc. Acad. Sci., Paris* **3** (1836), 387–8.
¶ See Comte's letter to Dulong, 31 August 1836, in *Correspondance inédite d'Auguste Comte*, vol. 4, pp. 162–3.
|| See, for example, his letters of 24 and 25 September 1836, to Blainville and Dubois respectively, in *Correspondance inédite d'Auguste Comte*, vol. 4, pp. 42 and 191; also the document dated 4 August 1840 in op. cit., vol. 4,

unsuccessful, as he had been some five years earlier when the chair had been vacated by the death of Fourier.

So there can be little doubt that Comte's tributes to Dulong were inspired at least in part by gratitude for the furthering of his ambition. We can only speculate how much more prominent Dulong would have been in that part of the *Cours de philosophie positive* which dealt with chemistry had it been written after 1836 and not in the last months of 1835,† when, as we recall, Dulong was less well disposed to Comte. In particular, the Dulong and Petit law might well have figured prominently instead of being ignored.

Since evidence from Comte's mouth was so rarely disinterested, it is fortunate that we can document the high esteem in which Dulong was held by those subscribing to positivist ideals from another source, the *Cours de physique* of Gabriel Lamé, a textbook written for the use of students at the École Polytechnique. Lamé was a contemporary of Comte, and also of Duhamel, at the École Polytechnique, all three men having entered in the same promotion of 1814. Both Lamé and Duhamel were ardent disciples of Fourier and so subscribed to many of the principles which Comte advocated. Comte's dismissal of Lamé as his inferior in 1832‡ can all too easily be attributed to their rivalry for the position of permanent secretary at the Académie des Sciences. In fact, he appears to have admired the work of both of his contemporaries,§ although again we must note the element of jealousy that clouded his attitude towards Duhamel in 1833.¶ By 1843 his attitude to Lamé at least had

p. 208. Comte frequently claimed that Dulong held him in high esteem; see, apart from the references just cited, his letters of 27 July 1840 and 25 January 1844, to Blainville and the Minister of War, in op. cit., vol. 1, p. 49, and vol. 4, p. 223.

† On the dating see Comte, *Cours*, vol. 3, p. 5.

‡ See his letter to Blainville, 21 June 1832, in *Correspondance inédite d'Auguste Comte*, vol. 1, pp. 35–6.

§ For evidence of his high opinion of Duhamel see Comte, *Cours*, vol. 2, pp. 586–8. On Lamé see the comment from a letter of 27 January 1843 to Blainville quoted below, p. 269.

¶ See Comte to Blainville, 17 March 1833, in *Correspondance inédite d'Auguste Comte*, vol. 1, p. 37. In this letter Comte supported Duhamel for membership of the geometry section of the Académie, though rather half-heartedly. He described Duhamel as 'indisputably superior' to the other candidates but added, 'this is not saying much'. His more favourable comments on Duhamel, referred to in the previous note, date from 1835.

changed considerably and in a letter to Blainville he strongly advocated his election to the Académie des Sciences. Since Lamé had then for several years defended Comte against his detractors on the Council of the École Polytechnique, there is once again the inevitable personal element, but his tribute on the purely scientific or, rather, philosophical level was now warm:

... among our present-day geometers, [he wrote], it is M. Lamé whose mind is most open to true philosophical ideas and most capable of appreciating the scientific or logical value of studies relating to the most prominent branches of natural philosophy.†

And who did Lamé himself acknowledge as his model in science? Not Fourier, as Comte's remarks might lead us to expect, but Petit and Dulong. In fact the entire preface to his *Cours de physique* in 1836 was a tribute to the work of these two predecessors in the chair of physics at the École Polytechnique, and it was they whom he strove to emulate. In a passage in which the use of the words 'metaphysical' and 'positive' emphasize the similarity of his ideals to those of Comte, he declared:

MM. Petit and Dulong constantly sought to free teaching from those doubtful and metaphysical theories, those vague and thenceforth sterile hypotheses which used to make up almost the whole of science before the art of experimenting was perfected to the point where it could serve as a reliable guide. Their splendid work on thermometry and the laws of cooling completely reconstructed the physical theory of heat by providing us with accurate and consistent instruments and by providing an indisputable basis for mathematical reasoning. Also in a branch of research that is incomparably more difficult than any other they arrived at general laws which for reliability and fruitfulness are in no way inferior to the finest laws of acoustics and light, laws whose discovery is largely due to the perfection of our organs and also to the geometrical precision of the instruments employed as an aid. From then on it could be imagined that at some time in the future it would be possible to make the teaching of physics consist simply of the exposition of the experiments and observations which lead to the laws governing natural phenomena, without it being necessary to state any hypothesis

† Comte to Blainville, 27 January 1843, in *Correspondance inédite d'Auguste Comte*, vol. 1, pp. 54–5.

concerning the first cause (*cause primitive*) of these phenomena that would be premature and often harmful. It is important that science should be brought to this positive and rational state.†

The positive principles were those which Petit and Dulong followed in their teaching and which, according to Lamé, they had fulfilled so far as was possible in the contemporary state of science. Although the experiments on expansion and cooling represented the ideal procedure for Lamé, as they did for Comte, he showed interest in their law of specific heats, which he accepted as evidence, though not conclusive evidence, in favour of the physical reality of atoms, despite the fact that its exactness was in serious doubt.‡

The relationships that Dulong had with Comte and the praise from one so obviously familiar with orthodox positivist principles as Lamé can leave little doubt that his methods won the approval of many positivists. By his work on cooling, thermometry, and the study of vapours he had demonstrated, probably more convincingly than anyone, how effectively the experimental study of heat could proceed in complete ignorance of its true nature. Yet in private and in certain of his writings Dulong had also shown himself to be a daring speculator, and the fact that during the last years of his life he seems to have weakened in his confidence that the causes of phenomena were discoverable is therefore a change of heart of more than ordinary significance. It tells us a great deal not only about Dulong but also about the *malaise* that appears to have afflicted the physical sciences in France in the 1830s.

The decline of Laplacian influence

'Intellectualist' factors were not alone in bringing about the decline of Laplacian science. Equally important were the changes in the scientific 'establishment' in which critics of Laplace and his followers gradually gained high and influential office, especially after 1815. To illustrate the effect of these changes, we might note how easy it became for Arago, as the newly appointed editor responsible for the physical sciences, to publicize Fresnel's views in the *Annales de chimie et de physique* from 1816 onwards. New elections to the Académie

† Lamé, *Cours de physique*, vol. 1, pp. ii–iii.
‡ Lamé, op. cit., vol. 1, pp. 412–14.

des Sciences also told against Laplacian interests. Arago, as it happened, had been elected (as a perfectly conventional Laplacian) as early as 1809, when he overwhelmingly defeated Laplace's own candidate Poisson.† Ampère too had been elected in 1814. But it was after the Bourbon restoration that the changes became most marked. In 1816 an election for the chemistry section brought in Joseph Louis Proust, the well-known exponent of the doctrine of fixed proportions in chemistry, who was graciously proposed as the best candidate by Berthollet.‡ In 1823 Dulong was finally elected after two unsuccessful candidatures, gaining thirty-six votes to Fresnel's twenty,§ and in the same year Fresnel had the rare distinction of being elected unanimously.¶ They were followed in 1824 by Navier, who was soon to clash bitterly with Poisson over the theory of vibrations in elastic solids.‖

But the most influential figure to emerge was Fourier, whose treatment since his return from Egypt was hardly worthy of a former instructor at the École Polytechnique and a permanent secretary of the Institute of Egypt.†† Almost immediately, in 1802, he had been appointed Prefect of the Department of Isère and it was at Grenoble that he produced his finest work

† For two seemingly inconsistent accounts of this election see Arago's own autobiographical sketch in Arago, *Œuvres complètes*, vol. 1, pp. 89–90, and J. L. F. Bertrand, *Éloges académiques* (1890), p. 47. According to Arago, Laplace campaigned vigorously on behalf of Poisson, believing that Arago, who was still only 23, was not yet sufficiently distinguished. But Bertrand states that Laplace eventually voted for Arago.

‡ *Procès-verbaux*, vol. 6, pp. 19 and 24 (5 and 12 February 1816). Berthollet's generosity to those who sought to break with his style of chemistry about this time was remarkable and it won general admiration. See, for example, Dulong's expression of grief at Berthollet's death in a letter of 20 November 1823 to Berzelius, in Berzelius, *Bref*, vol. 2, part 1, p. 45; also Cuvier's comment in his *éloge* of Berthollet, *Mém. Acad. Sci. Inst. Fr.* **8** (1829), p. ccx.

§ *Procès-verbaux*, vol. 7, p. 420 (27 January 1823).

¶ *Procès-verbaux*, vol. 7, p. 500 (12 May 1823). Although all fifty-two votes cast in this election were for Fresnel, fifty-eight members were present. Hence Laplace and Poisson, who did attend, may well have abstained. Biot had been present at the meeting on 5 May, when the list of candidates was drawn up (*Procès-verbaux*, vol. 7, p. 499), but he was absent for the election and he attended no more than three meetings of the Académie between this date and August 1825.

‖ See the series of papers by these two adversaries in *Annls Chim. Phys.* **37** (1828), 337–55; **38** (1828), 304–14 and 435–40; **39** (1828), 145–51 and 204–11; **40** (1829), 99–110.

†† For biographical details see Arago's *éloge* of Fourier in *Mém. Acad. Sci. Inst. Fr.* **14** (1838), pp. lxix–cxxxviii.

despite heavy administrative commitments. But even his prize-winning memoir of 1811 on heat conduction was treated unworthily† and Fourier languished in the provinces until the second Bourbon restoration. Although Louis XVIII's ministers refused to approve Fourier's nomination to the Académie des Sciences in the new category of *académiciens libres* in 1816,‡ they raised no objection in the following year when he was elected in the normal way to a long-overdue place in the physics section. Further honours quickly came his way. In 1822 he defeated Biot and Arago in the election for the post of permanent secretary of the Académie§ and in 1827 he joined Laplace and Cuvier as one of only three scientific members of the Académie Française. Even his funeral was on the grand scale, possibly grander in fact than Laplace's two years earlier.¶ The anti-Laplacian trend was continued by the election of Arago to succeed Fourier as permanent secretary for the mathematical sciences in 1830 and by Dulong's election in 1832 to the corresponding office for the physical sciences to replace Cuvier. Arago's control over the Académie, which lasted until his death in 1853, appears to have been considerable and not entirely beneficial, if we are to believe Libri's virulent criticism written in 1840.‖

Changes were apparent in less influential spheres also, for example in the Société Philomathique of Paris.†† Since its foundation in 1788 the society had included among its members all the leading Parisian scientists of the day, as well as many of a much lower calibre. It had never been a serious rival to the

† See p. 236 n. ‡.
‡ See, apart from Arago's *éloge*, the comments by Libri in *Revue des deux mondes*, 4th ser. **21** (1840), 794.
§ According to Arago, Laplace maintained a show of impartiality between Fourier and Biot but in fact voted for Fourier (who seems also to have had Arago's support). See Arago, *Œuvres complètes*, vol. 1, pp. 100–1.
¶ Compare the description in Arago's *éloge* of Fourier, p. cxxxvii, with that in Fourier's *éloge* of Laplace in *Mém. Acad. Sci. Inst. Fr.* **10** (1831), p. ci. Fourier was indisposed on the day of Laplace's funeral and since he arranged for no official representative of the Académie des Sciences to be present, it was left for an obviously distressed Biot to compose a brief tribute just before the funeral procession set out; see Biot, *Mélanges scientifiques et littéraires*, vol. 1, pp. 11–12.
‖ [Libri], op. cit., pp. 797–807. For a more favourable view of Arago's period of office see Bertrand, op. cit., p. 61.
†† I am grateful to Dr J. R. Ravetz for drawing my attention to these changes in the Société Philomathique.

First Class of the Institute but it had had at least one short period of distinction during the suppression of the Académie des Sciences (1793–5)† and from 1792 it had published a useful bulletin containing both original contributions and reports on current work in all the sciences. By 1816 the membership included not only Berthollet and Laplace, as 'emeritus members', as well as Biot, Poisson, Thenard, and Gay-Lussac, but also several more recent recruits whose future roles in the criticism of Laplacian science were not yet apparent.‡ Among these we note Arago (elected in 1808), Blainville, and Dulong (both in 1812). But it was after 1816 that the most significant changes occurred. Between February 1818 and May 1819 Fourier, Petit, Fresnel, and Navier were all elected, and in 1818 Blainville became secretary of the society, a post that he held for four years. The orthodox Laplacian tone of the society's bulletin changed only slowly, with Poisson, who had been on the board of editors since at least 1807 and a regular contributor since 1803, and Biot, who joined the board in 1816, continuing as the leading influences until 1820, at least in physical and mathematical science. In 1821, however, Fourier took Poisson's place on the board and in the following year Biot was replaced by none other than Fresnel. Notice of Fresnel's work appeared in the bulletin for the first time in 1821 and was understandably even more prominent after 1822, by which time Biot's virtual withdrawal from the scientific world, in particular from the Académie des Sciences,§ had greatly reduced the opposition to the wave theory. No notice of Laplace's work appeared after 1821, and although Poisson made intermittent contributions until the bulletin ceased publication for some years in 1826, such names as Fresnel, Arago, and Poisson's future rival Navier became increasingly prominent.

How caloric theory fared amid this general decline in Laplacian influence is clearly our most important problem here, and on this point the conclusion must be that generally it fared badly. The scarcity of contemporary comment suggests that the work of even the great calorists like Laplace and

† See M. P. Crosland, *The Society of Arcueil* (1967), pp. 173–6.
‡ For details of the membership see the lists which appeared annually in the society's bulletin.
§ On which see p. 271 n. ¶; also [Libri], op. cit., p. 799.

Poisson was being virtually ignored in the 1820s, and this was certainly the fate of Poisson's *Théorie mathématique de la chaleur*, a treatment of heat transfer in terms of caloric and a complete anachronism by the time it appeared in 1835. Even explicit criticism of Poisson's book came only from abroad. In Switzerland an otherwise sympathetic review deplored Poisson's retention of caloric in such problems as heat conduction in solids,† while in Britain J. D. Forbes commented: '... the whole work is to be regarded rather as a mathematical exercitation than as a serious step in physics.'‡ When Poisson died five years later he received few honours. Even his *éloge*, which Arago was to have delivered, was never given and instead extracts from a biographical memoir by Arago were read to the Académie des Sciences ten years after his death.§

To contemporary observers, however, especially those outside France, it would have been hardly obvious that caloric theory was a spent force after 1820. Laplace, after all, was still the most eminent mathematical physicist in Europe and he did not die until 1827. Moreover, to a man of such distinction the publication of his views presented no difficulty, so that anyone who was unaware of what was really happening in France would probably not have detected that Laplace's authority in these last years of his life was in any way diminished. So it was that caloric remained prominent in the literature, and in particular in textbook literature, although by the mid 1820s few leading French scientists, other than the dedicated Laplacians, could have been convinced that the fluid really existed. Even by 1823, for example, when Laplace published the definitive version of his caloric theory of gases, there can be little doubt that he was already addressing a generation of which many of the most influential members found such discussions increasingly irrelevant.

Naturally enough it was chiefly among minor figures, such as Mollet and Carnot, who were always on the fringe of the Parisian

---

† *Bibliothèque universelle* 59 (1835), 144–66, especially p. 154.

‡ Forbes, *A review of the progress of mathematical and physical science* (Edinburgh, 1858), p. 154.

§ The complete biographical memoir appears in Arago, *Œuvres complètes*, vol. 2, pp. 593–689. The extracts were read at a public meeting on 16 December 1850, but in Arago's absence; see *C.r. hebd. Séanc. Acad. Sci., Paris* 31 (1850), 840. A brief and formal grave-side tribute by Arago appears in Arago, op. cit., pp. 690–8.

scientific community, and among those working outside France that caloric theory retained its hold longest. If we look at the experimental work on specific heats that was examined earlier in this chapter, we find Haycraft, for example, being noticeably influenced in 1823 by Crawford's views, which he was 'still disposed to credit to a limited degree'.† He explained the cold at high altitudes by the increased capacity of air at low density and thought it important that he had shown the heat capacity of exhaled air at body temperature to be less than that of atmospheric air, a fact that was of interest only to a reader who accepted Crawford's theory of animal heat. His explanations of the intense heat of blast furnaces and of that produced by the combustion of gunpowder were similarly valid only in the context of Crawford's theory. De la Rive and Marcet, by contrast, were calorists of the Laplacian type. In 1827 they referred to the heat released or absorbed in adiabatic volume changes as 'the latent or constituent heat of the gas' and carefully distinguished it from the 'specific heat', which was the heat required solely to produce changes in temperature, not in volume.‡ It was on these grounds that in 1827 and 1829 they chose to measure specific heat at constant volume rather than at constant pressure, although it is noticeable that by 1832 they no longer held this view.§ Suerman too would appear to have been a Laplacian. At least he showed interest in the caloric theory and tried, unsuccessfully, to reconcile his results on the variation of specific heat with pressure with the formula that Poisson derived in 1823.¶ The agreement with Clapeyron's formula, as it happened, was remarkably good.

By the end of Laplace's life the most widely held opinion concerning the nature of heat, in Britain as well as in France, almost certainly lay somewhere between outright rejection of caloric theory and acceptance of its literal truth. This emerges most clearly in the views of the textbook writers, especially from 1830 onwards. The day of the textbook as a vehicle for the dissemination of original ideas was by now very nearly over, but such books, although lacking the historical significance of the

† Haycraft, *Trans. R. Soc. Edinb.* **10** (1824), 195.
‡ De la Rive and Marcet, *Annls Chim. Phys.* **35** (1827), 6.
§ See pp. 257 and 259
¶ Suerman, *Annls Chim. Phys.* **63** (1836), 330–2. For Poisson's formula see p. 176.

treatises by Crawford, Lavoisier, and Dalton, for example, still reflected the opinions of an influential body of working scientists. In most books the well-established custom of mentioning both the vibrational and material theories continued to be followed, but only a few authors now felt able to commit themselves categorically to one theory or the other. Thomas Thomson, who in the early years of the century had been a convinced calorist, expressed the prevailing attitude well when he wrote in 1830: 'I think, therefore, that it will be safest for us, in the present state of our knowledge to acknowledge our inability to solve this difficult problem, and to confess that we are incompetent to decide whether it be a substance or a quality.'† Similar sentiments were expressed about this time by Lamé (despite a sneaking preference for the vibrational theory), Dumas, William Brande, Neil Arnott, and John Herschel, all of whom expressed the view that the study of thermal behaviour could proceed quite satisfactorily without reference to the nature of heat.‡ And J. F. Daniell and Robert Dixon went so far as to omit discussion of the matter altogether.§ Other writers, among them César Manuel Despretz, Eugène Péclet, Edward Turner, Philip Kelland, Michael Donovan, Thomas Graham, Francis Lunn, and E. G. Fischer, while equally convinced that the nature of heat was unknown, found the language of materiality at least useful in the explanation of certain important phenomena.¶ This last point of view was, of course,

† Thomson, *An outline of the sciences of heat and electricity* (London and Edinburgh, 1830), p. 335. On Thomson's earlier views see pp. 105 and 107–8.

‡ Lamé, *Cours de physique*, vol. 1, pp. 298–9; Dumas, *Traité de chimie appliquée aux arts* (Paris and Brussels, 1828), vol. 1, p. vi; Brande, *A manual of chemistry* (3rd edn, 1830), vol. 1, p. 28; Arnott, *Elements of physics or natural philosophy* (4th edn, 1829), vol. 1, pp. 23–4; Herschel, 'Heat', in *Encyclopaedia metropolitana* (1830), mixed sciences, vol. 2, p. 326. Brande had long been non-committal on the nature of heat and so had continued the tradition of scepticism towards caloric which was so strong at the Royal Institution in the early nineteenth century; see, for example, his comments on pp. 37–8 of the first edition of the *Manual* (1819), and cf. the very similar attitude taken by Thomas Garnett, the first professor of natural philosophy and chemistry, in his *Outlines of a course of lectures on chemistry delivered at the Royal Institution of Great Britain, 1801* (1801), p. 16. In his *Outlines of a course of lectures on chemistry* (Liverpool, 1797), pp. 14–15, Garnett was also agnostic, although he slightly favoured the caloric theory.

§ Daniell, *Introduction to the study of chemical philosophy* (1st and 2nd edns, 1839 and 1843), and Dixon, *A treatise on heat* (Dublin, 1849), part 1.

¶ Despretz, *Traité élémentaire de physique* (1825), pp. 53–4; Péclet, *Traité*

THE CALORIC THEORY IN DECLINE 277

by no means a novel one;† but it is the extraordinary frequency with which it was adopted in the late 1820s, 1830s, and 1840s that is striking and, I believe, significant. At the same time both theories did have their more confirmed adherents, although dogmatic statements on the nature of heat were now the exception rather than the rule. By the mid 1840s, of course, there was an increasing number of men, among them Joule, Herapath, Liebig, and Helmholtz, who adopted the opinion that heat is motion in one form or another. Yet before that date, even in the 1830s, evidence of such support is far more difficult to find. Sadi Carnot and Friedrich Mohr are now perhaps the most familiar exceptions from this period,‡ but in March 1832 support for the vibrational theory also came from Ampère, whose work at the time was better, if not well, known.§ In a paper published in that month in the Swiss *Bibliothèque universelle* and three years later in the *Annales de chimie et de physique* (and hence presumably widely read) Ampère described a version of the theory that was specially adapted to the views on the molecular structure of matter that he had first expounded in 1814. His statement that radiant heat, like light, was a vibration in the all-pervading ether was standard enough vibrationalist doctrine, but his belief that the sensation of heat and the vibrations in the ether were caused by the motion of the atoms composing the molecules of a substance, rather than by the motion of the molecules themselves, was definitely an innovation. However, it was an innovation that does not

élémentaire de physique (3rd edn, 1838), vol. 1, p. 336; Turner, *Elements of chemistry* (6th edn, 1837), vol. 1, introduction, p. 6; Kelland, *Theory of heat* (Cambridge, 1837), p. 1; Donovan, *A treatise on chemistry* (4th edn, 1839), pp. 40–1; Graham, *Elements of chemistry* (1842), pp. 85–7; Lunn, 'Heat', in *Encyclopaedia metropolitana* (2nd edn, 1845), mixed sciences, vol. 2, p. 225; Fischer, *Physique mécanique*, trans. J. B. Biot (4th edn, Paris and Brussels, 1830), pp. 73–4. It is interesting to note that Turner had been a calorist at the time of the first edition of his book in 1827 (see p. 7 of this edition).

† Cf., for example, J. Imison, *Elements of science and art* (1808), vol. 2, pp. 30–1; J. Gorham, *The elements of chemical science* (Boston, Mass., 1819), vol. 1, p. 65; and, most important of all, Berzelius, *Lehrbuch der Chemie*, trans. F. Wöhler (Dresden, 1825), vol. 1, pp. 68–9.

‡ See pp. 189 and 308 n. ¶.

§ 'Idées de Mr Ampère sur la chaleur et sur la lumière', *Bibliothèque universelle* **49** (1832), 225–35. Ampère's paper was reproduced with minor modifications and a brief additional introduction reaffirming his belief in the wave theories of light and radiant heat, in *Annls Chim. Phys.* **58** (1835), 432–44.

appear to have provoked immediate comment or to have won any general support,† despite the fact that Ampère took good care to reconcile his theory of heat with Fourier's work on heat conduction.

The caloric theory still had a certain following also. Dalton, for example, appeared to be a no-less-convinced calorist in 1842, when the second edition of his *New system of chemical philosophy* appeared, than he had been in 1808 at the time of the first edition.‡ However, since his views on the matter were simply reproduced unchanged in 1842, the extent to which Dalton was really convinced of their truth by that time must remain doubtful. But there were others whose support for caloric has to be taken rather must seriously, notably Claude Pouillet, Thomas Traill, Leopold Gmelin, and Pierre Prévost, who in 1828 applied his own highly individual version of the caloric theory in a study of cooling and the variation in the specific heat of a gas with volume.§ Also, in 1843, the American chemist and physician Samuel Metcalfe published a two-volume work supporting in considerable detail the materiality of heat and the view that electricity and heat were interconvertible modifications of the same 'essence'.¶ But the normal textbook treatment of caloric in the 1820s, 1830s, and 1840s was rudimentary, amounting to no more than the briefest summary of the supposed properties of caloric and with none of the elaborate details that we associate with earlier accounts. Such refinements as the distinction between the various forms which caloric could assume gradually disappeared, so that the detailed exposition given by

---

† On the favourable opinion of Philip Kelland, however, see p. 3 n. ‡.

‡ Dalton, *New system* (2nd edn, 1842), pp. 1–2.

§ Pouillet, *Éléments de physique expérimentale* (Paris and Brussels, 1829), vol. 1, pp. 236–7; Traill, 'Heat', in *Encyclopaedia Britannica* (7th edn, Edinburgh, 1842), vol. 11, pp. 180–1; Gmelin, *Handbuch der Chemie* (Heidelberg, 1843), vol. 1, p. 201; Prévost, 'Discussion de quelques expériences relatives à l'influence de la densité sur la chaleur spécifique des gaz', *Mém. Soc. Phys. Hist. nat. Genève* **4** (1828), 255–64. Prévost's theory of heat, which is consulted most easily in his 'Mémoire sur la constitution mécanique des fluides élastiques', *Annls Chim. Phys.* **38** (1828), 41–9, deserves a far more detailed study than it has received hitherto. Based firmly on the work of Georges Louis Le Sage, it laid stress on the motion of the particles of caloric. William Prout was another who favoured a material theory of heat (though not necessarily the traditional caloric theory) about this time; see his contribution to the Bridgewater Treatises, *Chemistry meteorology and the function of digestion* (1834), pp. 58–67.

¶ Metcalfe, *Caloric* (1843), vol. 1, pp. xi–xii and 1–14.

Avogadro in his *Fisica de' corpi ponderabili* in 1840† was by then almost a solitary reminder of the intricacies of the theories of Laplace and Poisson.

Another significant change that took place about the same time and that can also be traced clearly in the textbooks was in the degree of importance that was attached by chemists to the study of heat. The point is made clearly enough by a comparison between, on the one hand, the 87 pages (out of 1090) which were devoted to all aspects of the study of heat in Thomas Graham's important *Elements of chemistry* of 1842 and, on the other, the lengthy and involved chapter or chapters on heat with which virtually every textbook of chemistry began in the early years of the century. This breakdown of the close association between chemistry and the study of heat was, of course, an event of great moment in the history of the physical sciences,‡ but it has scarcely received the attention it deserves. All the causes of the breakdown have not yet been clearly identified, although there seems little doubt that the decline in the acceptability of caloric as the basis for an explanation of the phenomena of heat was at least one crucial factor. Once heat was no longer accepted as a substance (or possibly even as an element, as Lavoisier believed), it ceased almost inevitably and immediately to be the concern of the chemist. Thus the emergence of the principle of the conservation of energy and the new science of thermodynamics in the late 1840s and 1850s, which put the subject of heat very firmly in the hands of the physicists, was merely completing a change in attitude that had already been detectable for two or three decades.

There can be little doubt that the attitude of writers like Despretz, Turner, Graham, and the others who found the materiality of heat useful as a teaching concept did a great deal to prolong the life of the caloric theory long after it had ceased to serve a useful function as a source of new ideas and research

† Avogadro, *Fisica*, vol. 3, pp. 739–48 and 773–824.

‡ There is interesting evidence of this breakdown in one of the 'faciendi books' kept by Thomas Charles Hope, professor of chemistry at Edinburgh (Edinburgh University Library, Hope papers, Gen. 268). Referring to the chemistry course of 1842–3 (the last before his retirement), Hope noted his wish to 'Specially abridge on Heat, which occupies too much of the course'. Of roughly 150 lectures that he gave each year about this time Hope devoted no more than fifteen to heat and certain related topics. I am indebted for this information to Mr J. B. Morrell.

achievements. Hence it is important in any study of the transition to the modern view of the nature of heat to recognize that the new theory lent itself less readily to the explanation and exposition of such phenomena as latent heat, chemical heat, and thermal expansion. And perhaps even more obscure on the vibrational theory was the cause of pressure in gases. Certainly a few men, such as Herapath and J. J. Waterston, did offer an alternative on this last question, but none of them wielded any appreciable influence and the static gas hypothesis therefore had no serious rival until after the discovery of energy conservation. In these circumstances some explanation of the repulsion between gas particles was indispensable and caloric undoubtedly provided the more convincing answer. In short, even in the 1840s the time was not yet ripe for the outright rejection of the caloric theory. However apparent its inadequacies, it still had a certain usefulness, especially for pedagogic purposes, and, what was still more important, there was no feasible alternative. The situation was a confusing one and the decline of interest in the problem of the nature of heat which appears to have occurred during the second quarter of the nineteenth century can be seen as a natural consequence of it. Not surprisingly it was only when the way forward once again became clear, with the discovery and acceptance of the principle of the conservation of energy in the late 1840s and 1850s, that interest in the problem revived.

8

THE AGE OF VICTOR REGNAULT

IN the previous chapter I tried to show how discussions concerning the nature and properties of heat attracted far less attention in the late 1820s, 1830s, and 1840s than they had done in the early years of the century. It was now generally accepted that the question whether heat was a fluid or a vibration could not yet be solved, and the debate between Irvinists and Laplacians, which had once aroused so much interest, had died away completely, leaving no comparable theoretical issue in the study of heat to take its place. The problems that really gripped the imagination were now to be found in quite different fields: in organic chemistry, for example, or in the startling new phenomena of electromagnetism brought to light by the work of Oersted, Faraday, and others. In these circumstances it is somewhat difficult to account for the appearance, in the 1840s, of one of the towering figures in the history of the study of heat: Henri Victor Regnault. It is by no means immediately obvious why this brilliant product of the École Polytechnique should have chosen to devote the greater part of his life to the detailed and laborious experimental researches for which he is now remembered rather than to one of the more obviously exciting branches of science. (Organic chemistry, to which Regnault made his earliest scientific contributions, springs to mind as the most attractive alternative.)

Why then did he turn to the study of heat? A glance at his writings will suffice to show quite conclusively that he has no place in the speculative, *Naturphilosophie* tradition which, in the 1840s, seems to have been at least one factor that aroused in such men as Mayer and Helmholtz an interest in heat and, more particularly, in the possibility of its convertibility into other forms of energy.† Nor does he fit any more happily in the

† On this role of *Naturphilosophie* see T. S. Kuhn, 'Energy conservation as an example of simultaneous discovery', in M. Clagett (ed.), *Critical problems in the history of science* (Madison, 1959), pp. 338–9.

power-engineering tradition that helped to lead Joule, Liebig, Seguin, Holtzmann, and Hirn to their statements of the interconvertibility of heat and work.† Certainly one of the aims of Regnault's experiments was to provide data that would be useful to engineers and these experiments were financed by the French government with this end in view. But it was definitely not these practical motives that first aroused Regnault's interest in heat.

The real incentive was of a much more theoretical nature. It arose directly from a research tradition involving the study of specific heats and dating back to the discovery of Dulong and Petit's law in 1819. In so far as this research tradition touches on several issues, notably the reception of the atomic theory in the first half of the nineteenth century, that are not directly related to the study of heat, it must inevitably lead us away from the main theme of this book. Even so, a brief examination of the tradition does seem indispensable if we are to understand the original motives for Regnault's work and, more generally, to identify the sort of problems that continued to arouse a certain interest in heat during the second quarter of the nineteenth century, when the really pressing, exciting problems of earlier years had ceased to stimulate research.

Specific heats and the atomic theory

In 1819 Dulong and Petit's law appeared primarily as convincing evidence for the physical reality of atoms, but it also had a possible practical application as an alternative method of determining atomic weights. The existing determinations, such as those of Dalton and Berzelius, were laborious, and the inevitably arbitrary nature of the assumptions concerning the composition of compound molecules meant that there was often a choice between two or three quite different atomic weights, depending on the assumptions made. Dulong and Petit were confident not only that their law would allow this choice to be made correctly but also that, given accurate measurements of specific heat, it would yield values that were wholly trustworthy in their own right without reference to the better-established chemical methods.‡ Unfortunately, the exactness of the law was never sufficiently established for their hopes to be

† See Kuhn, op. cit., pp. 331–6.
‡ Petit and Dulong, *Annls Chim. Phys.* **10** (1819), 403–6.

completely fulfilled, but there could be no doubt that atomic weight and specific heat were closely related and the continuing prominence of attempts to discover true atomic weights after 1820 therefore ensured that the study of specific heats remained a live research interest.

By a curious coincidence, almost immediately after the discovery of Dulong and Petit's law concern with the problem of atomic weight determinations had been stimulated still further by the emergence of yet another new method. In 1819 Berzelius's German pupil Eilhard Mitscherlich showed that the atoms of certain elements in isomorphous crystalline compounds could be replaced by those of other elements without altering the crystalline form and that the atomic weights of the 'replacement' elements could then be easily deduced by comparing the relative weights of them in the isomorphous compounds so formed.† Berzelius was so impressed by his pupil's discovery, as well as by the Dulong and Petit law, that he proceeded to revise his own values for atomic weight in the light of the new work. By 1826 he had produced a modified table in which many of his original weights were halved,‡ thus bringing them closer to those derived from the Dulong and Petit law and, incidentally, to present-day values. But the apparent reconciliation between the atomic weights determined by the various methods was short-lived, for it was in 1826 too that Jean-Baptiste Dumas, then a brilliant young chemist working in Thenard's laboratory, announced his celebrated method for the measurement of the vapour densities of a wide range of substances not normally in the gaseous state.§

As Dumas recalled some years later,¶ his original intention in undertaking the vapour-density work was merely to confirm existing atomic weights. He had no hesitation in putting forward that part of Avogadro's hypothesis dealing with the equal spacing of the particles of all gases at any given temperature and pressure, since all physicists, he declared, were in agreement on this.|| But more justification was required in

† Mitscherlich, *Abhandlungen der physikalischen Klasse der Königlich-Preussischen Akademie der Wissenschaften* (1818–19), pp. 427–37.
‡ *Pogg. Ann.* **7** (1826), 414.
§ Dumas, 'Mémoire sur quelques points de la théorie atomistique', *Annls Chim. Phys.* **33** (1826), 337–91. ¶ Dumas, *Annls Chim. Phys.* **50** (1832), 172.
|| Dumas, *Annls Chim. Phys.* **33** (1826), 337–8. Dumas's justification for this statement, which is surely an exaggeration, is not at all clear. He probably

connection with the divisibility of elementary molecules which, although it necessarily followed if Avogadro's hypothesis was to be extended to compound gases, only Gay-Lussac had accepted in practice, according to Dumas.† Unfortunately there is no evidence of Gay-Lussac's support except in so far as it is implied by the very favourable report which he made, with Dulong, on Dumas's 1826 paper.‡ However, there was little doubt in Dumas's mind that the vapour densities of elements at least were proportional to their atomic weights, and he found good agreement with the accepted results for iodine vapour, one of only two elements examined. Mercury vapour, on the other hand, with a density that indicated an atomic weight one-quarter of that originally adopted by Berzelius, gave the first hint of the serious discrepancies that were soon to emerge. Dumas's confidence in the atomic theory at this point was such that he declined to comment further without a more detailed examination of mercury, and two years later he reaffirmed his faith in the theory, elaborating his distinction between the 'physical atom' (*atome physique*) and the 'chemical molecules' (*molécules chimiques*), a very small number of which were supposed to compose the 'atom'.§ While the physical atom could never be split by such physical agents as heat, a chemical reaction, he maintained, could take place between individual chemical molecules. In de la Rive and Marcet's recent demonstration that all gases had identical volume specific heats Dumas found conclusive evidence that the particles that had identical specific heats were the physical atoms, for it was manifestly the number of these, rather than the number of chemical molecules, that remained the same in equal volumes both of a compound gas and of its constituent elementary gases.¶ Hence in solids also the atomic weights deduced from measurements of specific heat would be the weights of the physical atoms and not of the chemical molecules, about which, it seemed, we could know little. It is interesting to note that the atomic weights

had Ampère and even Avogadro in mind. Also Petit and Dulong had almost certainly accepted the hypothesis by 1819, though possibly, like Berzelius, with the restriction that it could only be applied to elementary gases.

† Dumas, op. cit., p. 338.
‡ Gay-Lussac and Dulong, *Annls Chim. Phys.* **34** (1827), 326–31.
§ Dumas, *Traité de chimie* (1828), vol. 1, pp. iv and xxxiii–li.
¶ Dumas, op. cit., vol. 1, pp. xli–xlii.

that Dumas adopted for metals in 1828 were those deduced from the Dulong and Petit law.†

It was as a result of a request from Mitscherlich, who visited Paris in 1832, that Dumas was persuaded to resume the vapour-density measurements on phosphorus and sulphur that he had begun some years earlier.‡ Dumas now confessed that the results he had already obtained for these two elements had appeared so anomalous that he had not dared to publish them before, but there could now be no doubt as to their correctness and the difficulties had to be faced. The vapour density of sulphur, whose vapour is now thought to contain S_8 and S_2 molecules in equilibrium and which has a vapour density corresponding to a molecular formula of about S_7, yielded an atomic weight three times that deduced from measurements of specific heat and chemical considerations. The value deduced from the vapour density of phosphorus, which has a molecular formula in the vapour state of P_4, was similarly 'too large' by a factor of 2. We can see now that the numerical scale adopted for atomic weights was such that agreement between the weights derived from vapour densities and from chemical analysis taken in conjunction with specific heats could occur only for gases or vapours which, like oxygen (to which all other weights were referred), had diatomic elementary molecules. Dumas came very close to realizing this in the solution to the anomalies that he adopted. It was a solution that conflicted with existing views of the atomic theory but at least it avoided further modification of the table of atomic weights. He suggested that atoms of sulphur in the liquid state might at a certain temperature form groups of three in order to resist vaporization. If, therefore, vapour density could be measured at a temperature just above melting-point, before the grouping had occurred, it might be that no anomaly would occur. Certainly the well-known changes in the viscosity of liquid sulphur with temperature were not inconsistent with the possible occurrence of such a molecular rearrangement. So the size of the atom in the solid state, the specific heat of which had been measured by Dulong and Petit,

† Dumas, op. cit., vol. 1, pp. l–li.
‡ Dumas, *Annls Chim. Phys.* **49** (1832), 210–14, and **50** (1832), 170–8. The earliest experiments on phosphorus and sulphur had apparently been performed at the time of Dumas's first measurements of vapour density.

would be just one-third of that existing in the vapour state. Although the possibility of such unpredictable behaviour put a drastic restriction on the usefulness of Dumas's method, he appeared satisfied, not to say pleased, with the explanation he had given.

But the satisfaction was short-lived, and by 1836 his attitude had changed and given way to a deep distrust of the whole concept of atoms. It was in that year that he delivered the course of lectures at the Collège de France in Paris on the history and present state of chemistry which have become celebrated as the *Leçons sur la philosophie chimique*.† His attitude to the existing atomic theory had now become one of scepticism. Certainly there was little more evidence to hand than there had been in 1832, except for Mitscherlich's discovery of yet another anomaly, for arsenic vapour,‡ but Dumas could no longer view the molecular rearrangement hypothesis as a wholly satisfactory way out of his difficulties. He still believed that the particles of sulphur vapour, for example, might well contain three times as many 'chemical atoms' (*atomes chimiques*) as the particles of the permanent gases, much as he said in 1832, although he now considered that the particles of even the permanent gases probably consisted of groups (*groupes moléculaires* or *groupes atomiques*) of chemical atoms. Furthermore, the similarity of certain physical properties of gases might lead to the conclusion that these groups, which were also, of course, the 'physical atoms' (*atomes physiques*), were equally spaced in all gases. But all this, he concluded, in a passage that summarizes well his attitude to scientific inquiry at this date, 'will in any case be no more than a hypothesis and on this subject too many hypotheses have already been made'. The way ahead was clear: 'instead of investigating these hypotheses more thoroughly, it would be far better to seek some reliable foundations on which to base more substantial theories'.§

† Two editions of the *Leçons* were published. One, which I have used, was published in Paris by Bechet jeune and dated 1837; the other, published by Ébrard, also in Paris, was undated but probably appeared at the same time. The texts and pagination are identical. Later editions, with a few very minor modifications in the text though with quite different pagination, were published in 1878 and 1937.

‡ Mitscherlich, *Annls Chim. Phys.* **55** (1833), 5–41.

§ Dumas, *Leçons*, p. 270.

Dumas, as is clear from the *Leçons*, was no positivist.† Even now, in 1836, he could declare:

> The aim of chemical philosophy ... is to arrive at the general principles of science ... to give the most general explanation of chemical phenomena, to establish the connection that exists between the observations and their causes.‡

Moreover, such explanations of phenomena, if and when they were obtained, would still be in terms of atoms. He added:

> From the point of view of present-day chemistry, [chemical philosophy] consists in the general study of the particles of matter which chemists call atoms and of the forces to which these particles are subjected. Hence it embraces investigations into all the properties of atoms. ....‡

So it was simply in not basing speculation solidly on facts and observation that the atomic theory had gone astray. There was nothing wrong with hypotheses in themselves, but those concerning the atomic theory had been made too quickly, when present chemical knowledge did not justify them. It was thus with a show of extreme reluctance that Dumas embarked on his description of what seemed to him the probable structure of gases, in the light of the now notorious anomalies.§ The agreement between the atomic weights of the common elementary gases deduced from density measurements and those deduced from combining weights, as well as the equality of their volume specific heats, certainly provided strong evidence that equal volumes of at least these gases contained equal numbers of chemical atoms. But what of sulphur vapour, for example? Here the atomic weight deduced from the specific heat in the solid state agreed with that determined by chemical methods but, as Dumas knew only too well, the number of chemical

---

† For further evidence of this see Dumas's enthusiastic references to the science of thermodynamics in his *éloge* of Regnault in *Mém. Acad. Sci. Inst. Fr.* **42** (1883), p. lviii. Since this chapter was written I have been delighted to learn that in his Cambridge Ph.D. thesis, 'The rôle of analogical argument in the development of organic chemistry' (1969), Dr J. H. Brooke also has come to this conclusion with regard to Dumas and his supposed positivism. On p. 65 n. of the thesis Dr Brooke ascribes part of the responsibility for Dumas's reputation as a positivist to Georges Urbain's introduction to the 1937 edition of the *Leçons*, where both Dumas and positivism are praised and thereby associated.

‡ Dumas, *Leçons*, p. 2.   § Dumas, op. cit., pp. 278–80.

atoms in a unit volume of vapour could not be the same as for the elementary permanent gases. One further difficulty, which in fact had been merely ignored for some years, also had to be faced. This was that there were certain elements, tellurium for example, for which Berzelius had been quite unable to reconcile the evidence of chemical analysis and measurements of specific heat.† Berzelius's atomic weight for tellurium was twice that calculated from the Dulong and Petit law. The solution to this problem that Dumas offered, albeit with extreme reserve, was based on a threefold distinction. Specific heat, he supposed, was associated with the ultimate particles of matter (*dernières particules des corps*), the 'true atoms', a certain number of which, say 1000, would constitute a single chemical atom of sulphur, copper, zinc, or indeed of any element for which the Dulong and Petit law and chemical analysis gave the same atomic weight, while twice this number of true atoms, i.e. 2000, would make up one chemical atom of tellurium. It was combinations of whole numbers of these chemical atoms that gave rise to the laws of definite and multiple proportions and it was therefore the weights of these atoms that emerged from any chemical analysis. The transition to the vapour state, of course, brought about yet another grouping between the chemical atoms. This Dumas had believed for some time, as we have already seen, but his view that the chemical atoms of mercury might even split in two in the vapour state was a new idea and one that owed its justification to his distinction between true and chemical atoms.‡

In this way, despite his erroneous views on atomic structure, Dumas came to the correct conclusion that whatever the three methods of determining atomic weights were measuring, they

† The solution recently proposed by Berzelius had been to suppose that Avogadro's hypothesis concerning the spacing of gas particles did not apply to elements that were not gaseous at room temperature; see *Jber. Fortschr. phys. Wiss.* **13** (1834), 59–63.

‡ The view that chemical atoms, or chemical molecules as he had called them at an earlier stage, could be subdivided was one that Dumas had not previously held. It will be noted that now, in 1836, Dumas postulated that the chemical atoms of mercury might split in *two* in the vapour state. If he had entertained this view at the time of his 1826 paper, however, he would have envisaged a splitting in *four*, since the standard atomic weight for mercury, as given by Berzelius until 1826, was twice as great. It is for this reason that in 1826 Dumas had found the vapour density of mercury to be one-quarter and not one-half of the figure he expected.

were not measuring the same quantity. Once this was realized, the way ahead became clearer. It lay not in further speculation but in careful experiment. For example, a detailed examination of the phenomenon of isomorphism would be expected to yield a great deal of information concerning chemical atoms. But it was the study of specific heats that was likely to be most fruitful and interesting, since, as Dumas was now tempted to believe, it would act as a probe into the ultimate building blocks of matter. It is interesting to note how in one important respect the situation had been completely reversed since 1828, when de la Rive and Marcet's results (since discredited) had persuaded Dumas that chemical forces were more effective than heat in breaking down matter into its simplest form. Now, in 1836, he urged the determination of large numbers of specific heats, especially of compound bodies, and showed particular interest in Franz Neumann's recent extension of the Dulong and Petit law to certain compounds.† Neumann's experiments, announced in 1831, had shown that the product of the molecular weight and specific heat (i.e. molecular heat) was very nearly constant for a number of metal carbonates and that a similar relationship, though with a different constant product, held for sulphates. The law for elements, as Dumas pointed out, thus became a particular case of a more general law in which 'atoms' of similar structure, whether elementary or compound, would have identical thermal capacities. It still remained, however, to extend measurements to new types of compound and to discover the exact nature of the dependence on structure, and later in the *Leçons* Dumas put the case for the study of specific heats, and also of Faraday's laws of electrolysis, in a typically forthright fashion:

... chemistry will quickly rise to new heights once it has reliable observations on which to base the discussion and comparison of the specific heats of large numbers of substances and with which to confirm Dulong and Petit's law as it applies to all compounds. Once chemistry also has at its disposal a large number of exact experiments with which to compare the specific electricities (*électricités spécifiques*)‡ of different elementary or compound substances, there is no telling what results may be obtained by a detailed study of the

† Neumann, *Pogg. Ann.* **23** (1831), 1–39.
‡ *Électricité spécifique* was Dumas's term for the electrochemical equivalent.

roles of heat and electricity in the constitution of the molecules of matter.†

But, for all the promise of the investigation of forces so closely linked to the individual particles of matter, Dumas was still adamant that much theorizing on the atomic theory was premature. The unjustified assumptions that had to be made now proved too much for him and in a well-known passage, which expresses rather strongly his despair with the present state of the theory, he declared: 'If I had my way, I should erase the word "atom" from science, in the firm belief that it goes beyond the realm of experiment; and never in chemistry must we go beyond the realm of experiment.'‡

Dumas was not alone in his sceptical attitude to atoms, for the atomic theory had long had influential critics quite apart from Berthollet and his disciples in the early years of the nineteenth century. William Hyde Wollaston, for example, who did so much to spread knowledge of the atomic theory, had made it clear that he was determining equivalents and not atomic weights.§ And Humphry Davy, even when presenting the Royal Medal to John Dalton on behalf of the Royal Society in 1826, had been careful to distinguish Wollaston's work, 'the practical part of the doctrine', from Dalton's, 'the atomic or hypothetical part'.¶ Further instances of this very cautious attitude to the atomic theory are by no means rare. Indeed, they are to be found in every decade of the nineteenth century, although the quasi-positivistic stands adopted by such men as Berthelot, Ostwald, and Brodie in the latter half of the century are probably the best-known.‖ However, the really important point for our purpose is that a certain scepticism with regard to atoms was already being felt by the time Dumas put forward his opinions on the matter in 1836. Hence when he displayed the gross anomalies of the theory as it was then understood, there

---

† Dumas, *Leçons*, pp. 422–3.   ‡ Dumas, op. cit., p. 290.
§ Wollaston, 'A synoptic scale of chemical equivalents', *Phil. Trans. R. Soc.* **104** (1814), 1–22, especially p. 7.
¶ Davy, *Six discourses delivered before the Royal Society* (1826), pp. 128–9.
‖ For two recent studies of nineteenth-century scepticism towards the atomic theory see W. H. Brock and D. M. Knight, 'The atomic debates: "memorable and interesting evenings in the life of the Chemical Society"', *Isis* **56** (1965), 5–25, and G. Buchdahl, 'Sources of scepticism in atomic theory', *Br. J. Phil. Sci.* **10** (1959), 120–34.

was no question of his creating a radically new situation; the scepticism existed and Dumas's evidence could serve only to intensify it. It is difficult to assess fully the influence that Dumas's views, as expounded in the *Leçons*, exerted on the course of the debate concerning the physical reality of atoms, but, if he did nothing else, Dumas certainly brought forward hard evidence to support the general and rather vaguer mistrust of hypothetical, unobservable entities that seems to have motivated, say, Wollaston, and it is probably no coincidence, therefore, that the attack on atoms seems to have intensified in the late 1830s. In 1837, for example, the Irish chemist Robert Kane, speaking at the Liverpool meeting of the British Association for the Advancement of Science, claimed that the use of the word atom as if atoms had real physical existence 'could not, with truth, be charged upon any well-informed chemist'.[†] The accompanying discussion is interesting not only for William Whewell's objection to the assumption of the physical reality of atoms (an objection that he repeated in his *History of the inductive sciences* in the same year[‡]) but also for Sir William Rowan Hamilton's championing of what resembles a Boscovichian view of the atom. Faraday, who had already expressed grave doubts concerning the theory,[§] was present and 'emphatically stated, that he was not an atomic chemist'. Liebig too would seem to have attended, for he was introduced to the audience after the above discussion. Whether the comments influenced him is not clear, but by October 1838 he and some other prominent German chemists had expressed their opposition to the atom as a physical reality, as we learn from a hitherto unpublished letter from Liebig to his friend and closest contact in France, the chemist Théophile Jules Pelouze.[¶] Friedrich Wöhler, Gustav Magnus, Heinrich Rose, Leopold

[†] *The Athenaeum*, 7 October 1837, p. 747.
[‡] Whewell, *A history of the inductive sciences* (1837), vol. 3, pp. 149 and 150. See also his *Philosophy of the inductive sciences* (1840), vol. 1, pp. 405–20.
[§] Faraday, 'Experimental researches in electricity—seventh series', *Phil. Trans. R. Soc.* **124** (1834), 121.
[¶] Liebig to Pelouze, 12 October 1838, in the archives of the Académie des Sciences, Paris (dossier 'Pelouze'). A complete transcript of the letter is given in the Appendix, pp. 319–20. For a later statement of this opposition see Liebig, *Chemische Briefe* (2nd edn, Heidelberg, 1845), pp. 55–65, where Liebig was, however, somewhat less severe in his attitude to atoms.

Gmelin, and Liebig, it appears, had taken a three-week holiday in Alsace during which they had discussed the atomic theory and decided to work henceforth in terms of purely observational equivalent weights. Referring to this decision, Liebig wrote: '...why drag our ideas concerning the molecular constitution of bodies into the way we state their composition and write our formulae? It is impossible to see a single reason which can justify such a usage.' He went on to ask whether French chemists too might be persuaded to adopt his view of atoms, although it was Berzelius that he wanted to see at the head of the movement. The outcome of Liebig's request is unclear, but his hopes of forming an organized movement were certainly not realized (partly, perhaps, as a result of the diversion of his interests towards physiology†).

As one possible way out of the difficulties concerning the atomic theory, Dumas advocated the intensive study of specific heats, as we have seen, and in this, as in his attitude to atoms, he was not alone. In 1836 Alexandre Baudrimont wrote to the Académie des Sciences in Paris claiming to have found a way of removing the discrepancies between the different methods of determining atomic weight by establishing that the specific heats of elementary or compound substances were proportional to the number of 'molecules' they contained.‡ Baudrimont's *Introduction à l'étude de la chimie par la théorie atomique* (1833)§ and the letter of 1834 from Dulong which has already been quoted on p. 260 suggest that in arriving at this conclusion Baudrimont was using the concept of the polyatomic elementary molecule, and that his 'molecules' were what we should now be careful to term atoms. But since no details were given in the Académie's *Comptes rendus* and since the memoir was never read, we can only offer conjecture on this point.

There is also some evidence of official interest in the problem of specific heats in the announcement of two major prize

† A diversion which is generally recognized to have taken place about 1840. See, for example, F. L. Holmes's introduction to the Johnson Reprint Corporation edition of Liebig's *Animal chemistry* (New York and London, 1964), pp. xv–xxxv.

‡ Baudrimont, 'Sur la détermination des poids atomiques et en particulier sur une loi relative aux chaleurs spécifiques', *C.r. hebd. Séanc. Acad. Sci., Paris* **2** (1836), 530.

§ Baudrimont, *Introduction*, pp. 73–5.

competitions in 1839. The subject of the first, proposed by the Accademia Reale delle Scienze of Turin on 21 April 1839,† was as follows:

> Determine experimentally the specific heats of as many as possible of the permanent gases, both elementary and compound.
> It is requested that in the case of at least one of the gaseous substances the specific heat at constant pressure and that at constant volume should be determined separately in order to confirm the relationship between these two types of specific heat for gases, a relationship which was established by Dulong and which allows either specific heat to be calculated for any gas provided that the other specific heat is known.

According to the official announcement one of the main purposes of the competition was to stimulate a further investigation of the atomic theory based on the Dulong and Petit law. It was stated that gases showed themselves particularly suitable for this purpose since the absence of attractive forces between their particles suggested that for them the law, now generally considered to be no more than approximate, might well hold exactly. Certain features of the terminology and the assumption (made without reservation) that the 'integrant molecules' (*molecoli integranti*) in both elementary and compound gases were similarly spaced almost certainly indicate Avogadro's influence in the choice of subject. His problem of relating volume, or molecular, specific heat to the composition of compound molecules (the important factor in his view being the number of 'partial' molecules in each integrant molecule‡) had now been given new prominence by the suggestion contained in Dumas's *Leçons*, according to which the volume specific heats of even elementary gases might be determined by their molecular structure or, in this case, by the number of 'true atoms' in each 'physical atom'. Avogadro's problem was mentioned specifically in the announcement with the comment that a solution would be found only by determining a large number of volume specific heats at identical temperature and pressure. In a review of earlier work the announcement criticized first Delaroche and Bérard's experiments (for yielding c_p only) and then Dulong's

† *Memorie Accad. Sci. Torino*, 2nd ser. **1** (1839), scienze fisiche e matematiche, pp. lxi–lxvi; also *C.r. hebd. Séanc. Acad. Sci., Paris* **9** (1839), 331.
‡ See pp. 222–3.

(for their indirectness), but it fully appreciated the difficulty of a direct measurement of c_v. Hence it would be sufficient if, after a direct determination of c_p for various gases, the rises in temperature produced in them by a given compression were measured and examined to see whether Dulong was correct in concluding that they were inversely proportional to the volume specific heats at constant volume, specific heats which could be deduced approximately from the experimental values of $c_p(v)$ taken in conjunction with Dulong's values for γ. Once Dulong's conclusion was thus confirmed, careful measurements of the temperature changes in adiabatic compression might in themselves yield reliable values not only for c_v but also, since γ would emerge from such experiments, for c_p also.

It is a measure of the extreme difficulty of such work that by the closing date in December 1841 no entries had been submitted† and that even when the time limit was extended to December 1843, at Avogadro's instigation,‡ there was similarly no response and the question was withdrawn.§

The second of the two competitions announced in 1839 was probably set in the hope that studies of specific heat might be enlisted in the examination and classification of the rapidly increasing number of organic compounds. In November 1839 Biot suggested an investigation to discover whether specific heat was identical for all the members of a group of isomers,¶ and a few weeks later the Académie des Sciences announced that a prize would be awarded for the determination of the specific heats both of elements and of a large number of mineral and organic compounds.‖ Special emphasis was laid on the possibility of comparing the specific heats of various groups of isomorphous compounds and of isomers. The style of the announcement is interesting in that it emphasizes the observational aspects of the investigation, as Dulong had done in his

---

† *Memorie Accad. Sci. Torino*, 2nd ser. **4** (1842), scienze fisiche e matematiche, pp. lvii–lx.

‡ *Memorie Accad. Sci. Torino*, 2nd ser. **5** (1843), scienze fisiche e matematiche, p. lix.

§ *Accademia delle Scienze di Torino. Registro dei verbali della Classe di scienze fisiche, matematiche e naturali*, vol. 9, p. 137. I am indebted to Dr Gianni Rondolino, secretary of the Accademia, for consulting this work on my behalf.

¶ *C.r. hebd. Séanc. Acad. Sci., Paris* **9** (1839), 655–6 (21 November 1839).

‖ *C.r. hebd. Séanc. Acad. Sci., Paris* **9** (1839), 847–8 (30 December 1839).

later work in the 1830s, and it thus forms a notable contrast with the Turin Academy's announcement, in which the language of the atomic theory was used so confidently. But the French competition was no more successful. In June 1842 Victor Regnault announced, on behalf of a committee consisting of Gay-Lussac, Arago, Becquerel, and himself, that no entry had been received and he recommended that the subject should be replaced by one on the heat of chemical reaction.† Strangely enough, it was Regnault himself who had already made the competition unnecessary, for his own early experiments on specific heat, inspired by the problems of his mentor Dumas and announced in 1840 and 1841, had already set standards of care and precision that were unlikely to be matched by any of his contemporaries.‡

The physics of Victor Regnault

One brief section can hardly do justice to Regnault's monumental achievements in the study of the thermal properties of matter in all three states. The results of his researches, contained in three forbidding volumes of the *Mémoires* of the Académie des Sciences§ and numerous papers, stand as a fitting culmination of the research interest whose history has been traced in this book, for when Regnault ceased experimental work about 1870, he had provided authoritative answers to nearly all the most important problems relating to the study of heat which had been tackled, with such limited success, since the middle of the eighteenth century. That his answers have undergone only minor modifications to this day bears witness to the excellence of his work.

However, it is not my intention here to examine Regnault's experiments in detail. His general philosophy of science, never stated explicitly but implied in the vast compilations of data that he drew up, is of far greater interest, not least for the light

† *C.r. hebd. Séanc. Acad. Sci., Paris* **15** (1842), 1147 (19 December 1842), and **14** (1842), 1051 (27 June 1842).
‡ See p. 297.
§ *Mém. Acad. Sci. Inst. Fr.* **21** (1847), 1–767; **26** (1862), pp. iii–x and 3–928; **37**, part 1 (1868), 3–575; **37**, part 2 (1870), 579–968. These were reprinted with identical pagination in three volumes as *Relation des expériences entreprises . . . pour déterminer les principales lois et les données numériques qui entrent dans le calcul des machines à vapeur* (1847–70), and they are therefore referred to hereafter as Regnault, *Relation des expériences*, vols. 1, 2, and 3.

that it throws on the state of the physical sciences in France about the middle of the nineteenth century. Naturally any conclusions of this more general nature, based as they are on the work of one man, must be viewed with extreme caution, but Regnault, both by the influences that he experienced as a young man in the 1830s and by the central position that he occupied in the French scientific community of his day, is a figure of more than ordinary significance.

Unlike Sadi Carnot, Regnault was by no means a peripheral figure in French science. His education was conventional, with a brilliant career at the École Polytechnique being followed by studies first at the École des Mines and then under Liebig at Giessen and Boussingault at Lyons.† He travelled widely but in 1836 he was back in Paris, as assistant to Gay-Lussac, and in 1840, when he was still not quite 30, his distinction as an organic chemist earned him a place in the Académie des Sciences and, shortly afterwards, Gay-Lussac's chair of chemistry at the École Polytechnique. But even then his days as a chemist were numbered. By November 1839 he had already performed experiments on the specific heats of isomers which, although unpublished, had attracted the attention of such influential figures as Arago, Thenard, Élie de Beaumont, Gay-Lussac, and Dumas.‡ Of these names the most significant is that of Dumas, for it seems that it was largely through him that this early interest in specific heats grew into the task that was to occupy the rest of Regnault's working life. Recalling over forty years later how he helped to effect the transition of the gifted young organic chemist into one of the most celebrated of all experimental physicists, Dumas wrote, with obvious pride:

> Perhaps I may be permitted to recall the persistent efforts that I had to make in order to persuade Regnault to take up the problem of specific heats. He hesitated for a long time but at last, embarking resolutely on a course that was to bring honour to his life, he showed . . . the qualities of a scholar of the first order.§

† For biographical details see Dumas's *éloge* of Regnault in *Mém. Acad. Sci. Inst. Fr.* **42** (1883), pp. xxxvii–lxxv; also the obituary notice by 'T. H. N.' (presumably the widely travelled American chemist Thomas Herbert Norton) in *Nature, Lond.* **17** (1878), 263–4, and the tributes in *C.r. hebd. Séanc. Acad. Sci., Paris* **86** (1878), 131–43 (21 January 1878).

‡ *C.r. hebd. Séanc. Acad. Sci., Paris* **9** (1839), 656 (21 November 1839).

§ Dumas, op. cit., p. xlix.

Regnault's initial problems were naturally enough Dumas's. In papers read to the Académie in April 1840 and January 1841 he demonstrated first the approximate nature of Dulong and Petit's law when applied to solid elements and then the validity, within these limitations, of Neumann's attempt to extend the law to compounds.† By the date of the second paper he probably already believed that only in the case of gases could the law be observed exactly‡ and he had already performed experiments to determine their specific heats, though no details were given.§ Whether direct or indirect means were used, it was presumably intended that these determinations should be at constant volume, since in Regnault's opinion the heat added to a substance in order to effect expansion was a principal cause of the deviations from constancy of the various atomic and molecular heats.¶ Unhappily Regnault's hopes of finding a gas for which Dulong and Petit's law was obeyed exactly were doomed to disappointment, and in 1853, when he eventually described his work, he was in no doubt that the atomic heats of the common permanent gases hydrogen, oxygen, and nitrogen were by no means equal and were, in any case, quite different from those of chlorine and bromine.‖ His attempt of 1853 to classify the volume specific heats of compound gases in terms of the volume changes that occurred during their formation (along the lines of Dulong's earlier attempt††) was partially successful, but here again the relationships were at best approximate and certainly gave no support to the belief, expressed by Avogadro, that the specific heat of a compound gas might be predicted exactly in terms of those of its constituent elements.‡‡

With Dumas's problems answered, it might have been expected that Regnault would return to organic chemistry, but he had now become committed to the study of heat in other ways. As early as 1840 he had shown interest in heat-engine operation§§ and soon afterwards he took what we can now see as

† Regnault, 'Recherches sur la chaleur spécifique des corps simples et composés', *Annls Chim. Phys.* **73** (1840), 5–72, and 3rd ser. **1** (1841), 129–207.
‡ For a somewhat later statement of this belief see Regnault, *Annls Chim. Phys.* 3rd ser. **9** (1843), 341.
§ Regnault, *Annls Chim. Phys.*, 3rd ser. **1** (1841), 132.
¶ Regnault, *Annls Chim. Phys.* **73** (1840), 66, and 3rd ser. **1** (1841), 200.
‖ Regnault, *Relation des expériences*, vol. 2, pp. 302–27.
†† See pp. 260–1.   ‡‡ Regnault, op. cit., vol. 2, pp. 310–27.
§§ Regnault, op. cit., vol. 2, p. iii.

the fatal step of accepting financial support for his experiments from the Minister of Public Works.† Encouraged by the generous official backing, he embarked immediately on the systematic redetermination of all the experimental data that could conceivably be required in the theory and practice of steam-engines and other heat engines. The result, as it emerged over the next twenty-five years, was the most precise and yet the most unimaginative of compilations, a warning perhaps of the dangers of dependence on government sponsorship in the pursuit of science. By the time Regnault finally recognized the enormity of his task‡ it was too late, and in the 1850s and 1860s, as I shall try to show, resignation rather than excitement seems to characterize his research.

There was no place in Regnault's papers for the discussion of theory. They contained only experimental results interspersed with digressions on the validity of various empirical laws. Inevitably in an investigation of such unprecedented thoroughness the inadequacy of the simple gas laws was ruthlessly exposed. Between December 1841 and May 1842, for example, Regnault showed not only that the expansion coefficient for gases determined by Gay-Lussac was too high but also that the coefficient varied both with the nature of the gas and with its initial pressure.§ During the next decade the laws of Clément and Desormes¶ and of Boyle‖ were similarly shown to be no more than approximations, and the belief, which had led so many calorists astray, that the specific heat of a gas decreased as it was compressed was finally and authoritatively dispelled.†† How differently the caloric theory of gases would have developed had these results been available and accepted half a century earlier is an interesting matter for speculation. There would

† Regnault, op. cit., vol. 1, pp. 4–5.

‡ He had certainly recognized this by 1853. See Regnault, op. cit., vol. 2, p. 3; also op. cit., vol. 1, p. 12.

§ Regnault, 'Recherches sur la dilatation des gaz', *Annls Chim. Phys.* 3rd ser. **4** (1842), 5–63, and 3rd ser. **6** (1842), 52–83; also *C.r. hebd. Séanc. Acad. Sci., Paris* **13** (1841), 1077–9 (13 December 1841). In his conclusions concerning the dependence of the coefficient on the initial pressure or density of a gas Regnault was refuting results obtained by Humphry Davy some twenty years earlier; see Davy, *Phil. Trans. R. Soc.* **113** (1823), 204–5.

¶ Regnault, *Relation des expériences*, vol. 1, pp. 635–728.

‖ Regnault, op. cit., vol. 1, pp. 329–428, and *C.r. hebd. Séanc. Acad. Sci., Paris* **23** (1846), 787–98.

†† Regnault, *Relation des expériences*, vol. 2, pp. 205–28.

then have been no reason to believe that matter in the gaseous state was in any sense unique, and gases would presumably have attracted correspondingly less attention. Also the work of Laplace, Poisson, and Carnot, who had built so much on the ideal gas laws and on Delaroche and Bérard's results, would have taken a very different form, and the Irvinist theory would have become quite untenable. By the 1840s, however, caloric theory in the refined sense in which it had been understood during the first quarter of the century was a thing of the past, and such theoretical issues as Regnault's work did raise were not of current interest. Indeed, by this time many of the experimental results that he obtained were original only in their extraordinary degree of accuracy and obvious reliability. For example, quite reputable experiments were already known which cast doubt on the truth of Boyle's law,† Gay-Lussac's expansion coefficient‡ and Clément and Desormes's law,§ while the rigorous truth of Dulong and Petit's law had never been adequately demonstrated.

Doubts, however, were one thing; Regnault's thorough and systematic investigation quite another. And the fact was duly recognized, by the stream of students, among them the young William Thomson, who came from all over Europe to learn the new techniques,¶ by the Royal Society's award of the Rumford and Copley medals, in 1848 and 1869 respectively,|| and by Regnault's admission in 1850 as an officer of the Legion of Honour. The official recognition accorded him could hardly have been more generous, but it would seem that the sterility of

† C. M. Despretz, *Annls Chim. Phys.* **34** (1827), 335 and 443.

‡ F. Rudberg, *Pogg. Ann.* **41** (1837), 271–93, and **44** (1838), 119–223. For the work of Gustav Magnus, which was contemporary with but independent of Regnault's, see his papers in *Pogg. Ann.* **55** (1842), 1–27, and **57** (1842), 117–99. William Prout had also obtained a figure for the coefficient that disagreed with Gay-Lussac's, but Regnault almost certainly knew nothing of Prout's experiments, which were reported only briefly in *Report of the first and second meetings of the British Association for the Advancement of Science; 1831–2* (1833), pp. 569–70. Prout's little-known work is discussed in W. H. Brock, 'William Prout and barometry', *Notes Rec. R. Soc. Lond.* **24** (1969), 281–94.

§ This 'law' had been disproved by 1827 by César Manuel Despretz and also, about the same time, by Dulong. On this work see my comments in notes 67 and 68 of Fox, *Br. J. Hist. Sci.* **4** (1968–9), 16.

¶ See Regnault, op. cit., vol. 2, p. lx; also S. P. Thompson, *Life of William Thomson* (1910), vol. 1, pp. 121–32.

|| *The record of the Royal Society* (4th edn, 1940), pp. 347 and 348.

the work that had occupied the best years of his life did not escape even Regnault himself. The complaints that he made about the unexpected magnitude of the undertaking† would surely not have come from a man satisfied with his achievements, and I feel that we must see in Regnault, as in Dulong some twenty years earlier,‡ a man who had fallen victim to the demands of a government commission. In Regnault's case, of course, the preoccupation with the tedious accumulation of results was especially unfortunate in view of the momentous developments in physics that were taking place outside France during the 1840s. In these developments, which within a very few years led to the principle of the conservation of energy and to the establishment of a new theory of heat, the role of Regnault was simply to provide the data, important data admittedly, which the really creative physicists of the day, men like William Thomson, Clausius, and Rankine, used in their work. Admittedly the possibility that Regnault had some original thoughts on heat and energy cannot be ruled out, and he did stake a claim to be considered among the independent discoverers of the new principle when speaking before the Académie in April 1853. After praising the work of Joule, Thomson, Rankine, Mayer, and Clausius, Regnault went on:

> For a long time I myself have been expounding similar ideas in my lectures, ideas to which I was led by my experimental researches on the elastic fluids. In the course of these researches, in fact, I was continually lighting upon anomalies which seemed inexplicable by the theories that were previously accepted.§

Unfortunately there is no independent evidence to support Regnault's claim,¶ and even if he had been teaching energy conservation in his lectures before 1853, this would not necessarily have implied originality on his part. As one of three referees appointed by the Académie for a short paper by Joule, Regnault must have known of the famous paddle-wheel experiments by August 1847 and he was evidently sufficiently

---

† See Regnault, op. cit., vol. 2, p. 3; also op. cit., vol. 1, p. 12, where it is clear that Regnault had underestimated the task.

‡ See pp. 249–50 and 255.   § Regnault, op. cit., vol. 2, p. vi.

¶ Although Charles Brunold was certainly wrong to follow Henri Bouasse in stating that Regnault was a lifelong believer in caloric. See Brunold, *L'entropie. Son rôle dans le développement historique de la thermodynamique* (1930), p. 48.

impressed to allow the paper to appear in the Académie's *Compte rendu*.† In the same publication in the following month he would presumably have seen Marc Seguin's extension of the principle of the equivalence of heat and work to account for the cooling of gases and steam on expansion,‡ and there was of course much else that he might have read on the subject by both British and German authors, even though these works were admittedly not well known in France.§ Yet, despite all this, there is good reason to believe that the new discoveries had little immediate effect on him even when he came to learn of them. For example, in 1849, when his experiments on the specific heats of gases chanced to yield a striking instance of the new principle (in the cooling of a gas undergoing expansion), he offered no comment of any sort on the broader implications of the observation,¶ and in the following year during some experiments on gas expansion he quite misinterpreted a small cooling effect that appears to have been precisely the same effect as that to which Joule and Thomson were to give their name when they discovered it independently some two years later.‖

Even as late as 1850, therefore, Regnault was showing very little interest in the principle of the conservation of energy. He *may* have accepted it by that date, of course; but he had certainly not recognized its importance or true significance, and it was only in 1853, when the struggle to gain acceptance for the principle was all but won, that he finally gave it his open support.†† In an all too characteristic manner he had

† Joule, 'Expériences sur l'identité entre le calorique et la force mécanique ...', *C.r. hebd. Séanc. Acad. Sci., Paris* **25** (1847), 309–11 (23 August 1847).

‡ Seguin, 'Note à l'appui de l'opinion émise par M. Joule, sur l'identité du mouvement et du calorique', *C.r. hebd. Séanc. Acad. Sci., Paris* **25** (1847), 420–2 (20 September 1847).

§ It is an indication of the extent of French ignorance of foreign writings on energy conservation and related matters that Émile Verdet felt it necessary to publish a series of translations of relevant papers in *Annls Chim. Phys.*, 3rd ser. **34** (1852), **35** (1852), **36** (1852), and **37** (1853). Papers by Mayer, Joule, Clausius, and William and James Thomson were included.

¶ Regnault, op. cit., vol. 2, p. vii, and op. cit., vol. 3, pp. 579–83.

‖ The cooling is referred to briefly and dismissed as unimportant in Regnault, op. cit., vol. 3, p. 593. The Joule–Thomson (or Joule–Kelvin) effect was first described in a paper read to the British Association for the Advancement of Science in September 1852 and published later in that year; see Joule and Thomson, *Phil. Mag.*, 4th ser. **4** (1852), 481–92.

†† Regnault, op. cit., vol. 2, pp. iii–viii. There is some evidence that Regnault's ideas were beginning to change by 1851; see. F. Reech, *C.r. hebd. Séanc. Acad.*

entered the field when the really creative phase in the early history of thermodynamics was over. He now had only his experimental dexterity to offer but, although he duly embarked on a careful determination of the mechanical equivalent of heat, he was never satisfied with his experiments† and so failed even in the one respect in which he might still have been pre-eminent.

The kinetic theory of gases

It remains at this point to say something of the view of gas structure which, in the space of little more than a decade beginning in the mid 1850s, succeeded in taking the place of the static theory and in finally condemning that theory to an obscurity from which it has never since emerged. According to the new theory, which we now know simply as the kinetic theory of gases,‡ the familiar properties associated with the gaseous state were to be explained in terms of small particles—the gas molecules—moving about rapidly in straight lines and colliding with the walls of the containing vessel and, less frequently, with one another.

As is well known, the theory did not originate in the mid-nineteenth century. It had been proposed by the Swiss mathematician Daniel Bernoulli in his *Hydrodynamica* of 1738§ and had been used by him in a theoretical derivation of Boyle's law as well as in a demonstration that the pressure of a gas being heated at constant volume was proportional to the square of the translational velocity of its particles.¶ For more than a century after the publication of the *Hydrodynamica* Bernoulli's theory, which was in any case just one of a number of competing kinetic

Sci., Paris **33** (1851), 570 (24 November 1851). However, even in 1854 Reech could still write that Regnault headed those who held reservations concerning the principles of Carnot and Clapeyron; see Reech, *Machine à air d'un nouveau système* (1854), p. i.

† See Dumas's *éloge* of Regnault in *Mém. Acad. Sci. Inst. Fr.* **42** (1883), p. lviii. In fact Regnault did give one value for the equivalent; see his *Relation des expériences*, vol. 3, p. 574.

‡ Although, as I point out later, more than one 'kinetic' theory of gases was available in the 1850s.

§ Bernoulli, *Hydrodynamica* (Strasbourg, 1738), pp. 200–4.

¶ Bernoulli, op. cit., pp. 202–3. On these proofs see also G. R. Talbot and A. J. Pacey, 'Some early kinetic theories of gases: Herapath and his predecessors', *Br. J. Hist. Sci.* **3** (1965–6), 139–40.

theories at this time,† never appeared as a serious rival to the static theory. The latter, quite apart from its intrinsic merit as a generally satisfactory and even fruitful basis for the explanation of the properties of gases, had the obvious advantage of being associated with the great name of Newton and, since the inter-particle repulsions were generally attributed to caloric, with that of Lavoisier also.‡ In fact, throughout the eighteenth century the theory adopted by Bernoulli was largely though not completely ignored,§ and by the first half of the nineteenth century it seems to have assumed the status of an excessively speculative and obscure belief. This is certainly implied by the very cool response of the Royal Society and in particular of the Society's president Sir Humphry Davy, who was himself no calorist,¶ when John Herapath submitted a paper describing his Bernoulli-style kinetic theory (and much else) in 1820.|| And the point is illustrated even more clearly perhaps by the Society's harsh treatment of a perfectly lucid exposition of the same theory which was sent to it in December 1845 by John James Waterston, a Scot who at the time was employed as a naval instructor for the East India Company in Bombay. Admittedly in this case the paper was read and an abstract was inserted in the Proceedings of the Society,†† but it was decided that the paper should not be published—a decision that is hardly surprising in view of the response of the official referees,

† See Talbot and Pacey, op. cit., pp. 133–49.
‡ There were, of course, other reasons for the neglect of Bernoulli's theory; see the discussion in Talbot and Pacey, op. cit., p. 141.
§ As Talbot and Pacey point out (op. cit., p. 140), interest in the theory during the eighteenth century seems to have been largely restricted to the Bernoulli family and their immediate associates.
¶ See pp. 118–20.
|| For an account of this episode, in which Herapath was opposed by P. M. Roget and (to a lesser extent) the Society's vice-president, Davies Gilbert, as well as by Davy, see S. G. Brush, 'The Royal Society's first rejection of the kinetic theory of gases (1821), John Herapath versus Humphry Davy', *Notes Rec. R. Soc. Lond.* **18** (1963), 161–80. Herapath's paper eventually appeared, as 'A mathematical inquiry into the causes, laws, and principal phenomena of heat, gases, gravitation, &c.', in *Ann. Phil.* 2nd ser. **1** (1821), 279–93, 340–51, and 401–16, and various extensions and comments on his work continued to appear in the same journal for some two years after this. It is interesting to note that criticism of Herapath's views came also from readers of the journal, including the engineer Thomas Tredgold and two anonymous writers; see Talbot and Pacey, op. cit., p. 144.
†† *Phil. Trans. R. Soc.* **183** A (1893), 78.

one of whom (the astronomer Sir John William Lubbock) described Waterston's work as 'nonsense, unfit even for reading before the Society'.†

By the late 1850s attitudes to the kinetic theory were changing rapidly. How and why this change took place has long been, and still is, a live issue among historians, yet I propose to add little to the already extensive literature on the subject. My main reason for treating the early history of the kinetic theory so cursorily is that I see the emergence, or rather re-emergence, of the theory in the late 1850s and its subsequent acceptance in the body of standard doctrine as being essentially unrelated, or at least only tenuously related, to the theory whose history has formed the subject of this book. After all, the new theory did not gain its victory as the result of a direct confrontation between it and the old view of gas structure adopted earlier in the century by Dalton, Laplace, Avogadro, and so many others. Indeed, such a direct confrontation could not conceivably have taken place in the 1850s, for the simple reason that by then the caloric theory of gases had long ceased to provide a really credible explanation for the behaviour of matter in the gaseous state. As I have pointed out in earlier chapters, there are clear signs that the caloric theory generally was in serious disarray as early as the 1820s, and through the 1830s and 1840s writers on the subject of heat were increasingly reluctant to acknowledge caloric as anything more than a useful aid to thought or a pedagogic tool. Now it has to be admitted that the traditional theory of gases was only rarely the object of explicit criticism in this period, but this was not because its truth was beyond question; rather there were simply no problems that necessitated a knowledge of gas structure, even though the particles, molecules, or atoms that constituted gases were themselves very much under discussion. So in the 1830s and 1840s it was of no great consequence whether gas particles were arranged in a

† Quoted by Lord Rayleigh in the introduction to Waterston's paper which he wrote when the paper was eventually published, at Rayleigh's own instigation, in *Phil. Trans. R. Soc.* **158** A (1893), 5–77; see p. 2. On Waterston's theory and its reception see S. G. Brush, 'The development of the kinetic theory of gases, II. Waterston', *Ann. Sci.* **13** (1957), 273–82, and 'John James Waterston and the kinetic theory of gases', *Am. Scient.* **49** (1961), 202–14. Lubbock is identified in Dr Brush's *Kinetic theory* (Oxford, 1965), vol. 1, p. 17. The other referee, who was also hostile, was Baden Powell, Savilian professor of geometry at Oxford.

lattice formation or whether, as in the modern kinetic theory, they moved bodily; and in these circumstances the matter could be quietly ignored and allowed to remain unsolved.† despite the fact that the rejection of caloric as a means of accounting for the properties of heat had deprived the static theory of one of its greatest strengths—its capacity to explain inter-particle repulsion. So we should not be misled by the scarcity of explicit comment into believing that all was well with the caloric theory of gases in the second quarter of the nineteenth century, for in fact the case was quite otherwise. Certainly the theory survived after a fashion until the 1850s,‡ but by this time there were few indeed who would have cared to defend it against the new ideas.§

If, as I believe, it was not just a sudden or even a growing awareness of the deficiencies of the static theory that caused its rejection and replacement; and if, moreover, change was not made imperative simply by some 'crisis' situation in the study of gas structure itself,¶ what were the causes? Clearly the decline in the plausibility of the caloric theory of gases was a necessary preliminary, but the really decisive stimulus for change seems to have been the general acceptance, in the early 1850s, of the principle of the conservation of energy.

In seeing the energy-conservation principle as the most important single catalyst in the emergence of the new theory of gases, I am not suggesting, of course, that all the men who can

† Thus the work of Herapath, Waterston, and Joule should not be taken as evidence of any widespread interest in the matter between 1820 and 1850.

‡ The last use of the caloric theory of gases in a research problem seems to have been made by the Swiss mathematician Élie Ritter in June 1845. His 'Note sur la constitution physique des fluides élastiques', *Mém. Soc. Phys. Hist. nat. Genève* **11** (1846–8), 99–114, was a theoretical study of the observed deviations from the ideal gas laws. The deviations were accounted for in terms of small attractive forces between static gas particles.

§ The most striking testimony is the absence of resistance from supporters of the theory when the new ideas were put forward by Clausius and Maxwell in the late 1850s.

¶ I am using the word 'crisis' here in the sense that it has in T. S. Kuhn's *The structure of scientific revolutions* (Chicago, 1962). In this book (see especially pp. 52–90) Professor Kuhn argues that acute dissatisfaction with an existing theory, amounting to 'crisis', is one necessary condition for the rejection of that theory. Another necessary condition on this view is the availability of a satisfactory alternative with which to replace it. Professor Kuhn identifies crisis situations in, for example, Ptolemaic astronomy before Copernicus announced his new theory and in the phlogistic chemistry of the 1770s.

lay claim to a place among the numerous independent discoverers of the principle,† or even those who developed it in the early years, were supporters of the kinetic theory. In fact, although many of them did make use of the mass of experimental data relating to the thermal properties of gases that had been acquired in the preceding half-century,‡ comparatively few showed any interest in gas structure. In this respect James Joule was a notable exception, for as early as June 1844, less than a year after he had appeared before the meeting of the British Association for the Advancement of Science in Cork to make the first public statement of his own somewhat restricted version of the principle,§ he revived the 'beautiful idea' concerning the rotational motion of gas particles that Davy had put forward in 1812;¶ he even went so far as to suggest that the rotational motion to which Davy had referred might be motion in the atmospheres of the fluid electricity that he (Joule) supposed to surround the atoms of ordinary matter.‖ In 1847 he was still speculating on rotational motion, as we see most clearly in the famous lecture on 'Matter, living force, and heat' that he delivered in Manchester on 28 April of that year.†† At

† In his classic paper 'Energy conservation as an example of simultaneous discovery', in Clagett, *Critical problems*, pp. 321–56, Professor Kuhn cites no fewer than twelve men who seem to deserve recognition as discoverers of the principle (albeit to varying degrees and in different senses of the word 'discoverer').

‡ The data concerning adiabatic volume changes were particularly important. See *Isis* **49** (1958), 132, where Professor Kuhn writes: 'All seven of the men who between 1824 and 1845 independently enunciated the law of the conservation of energy made immediate use of adiabatic phenomena to document, elaborate, or quantify their new hypothesis.' The seven men were Carnot, Mohr, Seguin, Mayer, Colding, Joule, and Holtzmann. Regnault's recently acquired data were also much used from the late 1840s, especially by Thomson, Rankine, and Clausius.

§ Joule, 'On the calorific effects of magneto-electricity, and on the mechanical value of heat', *Phil. Mag.*, 3rd ser. **23** (1843), 263–76, 347–55, and 435–43. The paper was read on 21 August 1843.

¶ Joule, 'On the heat evolved during the electrolysis of water', *Mem. Proc. Manch. lit. phil. Soc.*, 2nd ser. **7** (1846), 111, in an appendix read on 20 February 1844. The idea was repeated with the acknowledgement to Davy, in Joule's important paper 'On the changes of temperature produced by the rarefaction and condensation of air'; see *Phil. Mag.*, 3rd ser. **26** (1845), 381–2.

‖ These atmospheres had already been depicted by Joule in connection with a paper in William Sturgeon's *Annals of electricity, magnetism, and chemistry* **8** (1842), plate 4.

†† *The scientific papers of James Prescott Joule* (1887), vol. 1, pp. 265–76, especially pp. 274–6. This type of motion was also discussed in the draft of an

this time he believed that rotation might occur in the particles of matter in all three states but by October 1848 he considered that Herapath's theory of gases, which had recently been restated in some detail by its author in his *Mathematical physics*,† was 'somewhat simpler', although he added that 'the hypothesis of a rotary motion accords equally well with the phenomena'.‡

As far as the process of the acceptance of the modern kinetic theory was concerned, Joule's advocacy in 1848 had remarkably little immediate effect. One reason for this may have been that the paper in which he put his views was published somewhat obscurely in the *Memoirs* of the Manchester Literary and Philosophical Society.§ But Joule's paper was not entirely unknown and its comparative neglect for a period of nearly ten years almost certainly owed far more to the fact that the need to reform the prevailing theory of gases was still not felt to be a pressing one; other problems, such as those soon to be

unpublished lecture now preserved in the Department of the History of Science and Technology, University of Manchester Institute of Science and Technology (Joule MSS., notebook 2). Joule's sketch of atoms rotating in a solid, which appeared in the draft, has been reproduced by Professor E. Mendoza in *Physics today* **14** (1961), 39, and in *Mem. Proc. Manch. lit. phil. Soc.* **105** (1962–3), 10. Unlike Professor Mendoza, I do not believe that the draft is an early one for the 'Matter, living force, and heat' lecture since it bears no resemblance to the completed text and is followed, some pages further on in the same notebook, by what is unmistakably a draft of the latter lecture. At all events the sketch in question had certainly been drawn by April 1847. Moreover, Joule's use of the figure 817 (ft lbf per Btu in modern units) for his equivalent indicates that the draft dates from *after* the earliest paddle-wheel experiments, which were first described to the British Association meeting at Cambridge in June 1845. By the time he wrote the draft Joule had abandoned his earlier idea that the rotational motion of heat was motion in the atmospheres of electricity; he now believed that heat was a rotation of the atoms of matter themselves.

† Herapath, *Mathematical physics* (1847), vol. 1, pp. 15–17 and 236–79.

‡ Joule, 'Some remarks on heat and the constitution of elastic fluids', *Mem. Proc. Manch. lit. phil. Soc.*, 2nd ser. **9** (1851), 107–14, especially p. 111. In this paper Joule used Herapath's theory in a simple calculation of the velocity of the particles of hydrogen at room temperature. The paper marks a fundamental change in Joule's thinking on the question of gas structure, since in his earlier speculations on rotational motion he had retained the static theory and had been concerned simply to find an explanation for the repulsive force between gas particles. For evidence of this see the 'Matter, living force, and heat' lecture of 1847, in Joule, *Scientific papers*, vol. 1, pp. 274–5.

§ Later, in 1857, Clausius complained that he had been unable to obtain a copy of Joule's paper; see *Pogg. Ann.* **100** (1857), 354 n. This complaint led Joule to have his paper republished in *Phil. Mag.*, 4th ser. **14** (1856), 211–16.

tackled by William Thomson and Rudolf Clausius,† appeared more urgent. Indeed, Joule's paper of 1848 was far less influential in bringing about the transition to the new theory than was the brilliant and widely praised series of experiments to demonstrate the interconvertibility of heat and work that he had conducted since 1843 and that had already done much to change attitudes towards the nature of heat in general. Joule repeatedly maintained that belief in the conservation of caloric was incompatible with the principle of energy conservation‡ and, even though no less a pioneer of the principle than Robert Mayer refused to abandon his belief in the materiality of heat,§ the view that heat was motion or, as it was more usually put, *vis viva* gained ground rapidly. In fact, largely as a result of Joule's work, it came to be regarded, by the early 1850s, as an obvious, even trivial corollary of the equivalence of heat and work.¶

Of course, even when this corollary was accepted, it still remained to determine the *type* of motion that was thought to constitute heat. As far as solids and liquids were concerned, we might think that the easiest and most natural solution would

† Here, of course, I am referring to the work of Thomson and Clausius on the reconciliation of the views of Joule and Carnot, work that led them to the discovery of the second law of thermodynamics and the concept of entropy.

‡ See, for example, his comment in 'Matter, living force, and heat', in Joule, *Scientific papers*, vol. 1, p. 273.

§ Mayer, 'Bemerkungen über die Kräfte der unbelebten Natur', *Annalen der Chemie und Pharmacie*, (eds. F. Wöhler and J. Liebig) **42** (1842), 236–40, and J. J. Weyrauch (ed.), *Kleinere Schriften und Briefe von Robert Mayer* (Stuttgart, 1893), p. 320 n. Marc Seguin was another who did not commit himself to the view that heat is motion. He frequently used the term *calorique* in his book *De l'influence des chemins de fer* (1839), although on p. 383 he stated that the nature of heat was unknown.

¶ This was certainly so for Clausius. In his paper 'Ueber die bewegende Kraft der Wärme...', *Pogg. Ann.* **79** (1850), 369–70, he assumes, without offering any justification, that the heat in a body is measured by the *vis viva* of its particles. This belief that heat was a motion of the particles of ordinary ponderable matter was axiomatic for most of the discoverers and pioneers of energy conservation. See, for example, C. F. Mohr, 'Ansichten über die Natur der Wärme', in Liebig's *Annalen der Pharmacie* **24** (1837), 141–7; Liebig, *Animal chemistry*, trans. W. Gregory (1842), p. 32; H. von Helmholtz, *Ueber die Erhaltung der Kraft* (Berlin, 1847), pp. 27–31. Carnot's conversion has already been mentioned on p. 189. It is a measure of the ease with which the new view of the nature of heat was accepted that the issue was rarely discussed in detail by the early workers on energy conservation. Despite this agreement, however, there was great uncertainty concerning the *type* of motion that constitutes heat, as I point out in the next paragraph.

have been to suppose that the motion was a vibratory motion of the particles composing them. Such a theory of heat had a long and distinguished history behind it, and the fact that it had continued to be described in textbooks throughout the first half of the nineteenth century had ensured that it had not been forgotten even when the caloric theory was at the height of its popularity. As it happened, however, great (and, of course, fully justified) uncertainty was felt concerning the nature of the motion of heat in the 1850s, and the process by which the matter was resolved was slower and far more complex than is suggested in most histories of the period. That there were perfectly feasible alternatives to what we now consider to be the correct solution (according to which the heat energy of solids and liquids consists in the vibrations, and for liquids also in the rotational motion, of the molecules) is demonstrated clearly by Joule's discussions of the problem in the 1840s. And the point is made no less clearly by William Rankine's attempt, of 1850 and 1851, to interpret the new thermodynamic principles in terms of an elaborate vortex model of the atom, the motion in this case being associated with an 'elastic atmosphere' that revolved or oscillated about the 'nucleus or central point' of any 'atom' of matter.† According to Rankine quantity of heat was nothing more than the *vis viva* of the revolutions or oscillations performed by these atmospheres. Some years later, in 1867, another vortex theory was proposed by William Thomson, who suggested that the atoms of matter might be nothing more than centres of vortex motion ('vortex rings') in an all-pervading fluid ether.‡ This view had obvious difficulties. It was not easy to see, for example, how the whirling etherial matter acquired weight. Yet for a number of years Thomson continued to develop and refine his theory,§ and it was only at the very end of the nineteenth century that it finally became untenable, chiefly as a result of J. J. Thomson's particulate interpretation of cathode

† Rankine, 'On the centrifugal theory of elasticity, as applied to gases and vapours', *Phil. Mag.*, 4th ser. **2** (1851), 509–42. On Rankine's theory see E. E. Daub, 'Atoms and thermodynamics', *Isis* **58** (1968), 293–303.

‡ Thomson, 'On vortex atoms', *Proc. R. Soc. Edinb.* **6** (1866–9), 94–105 (18 February 1867).

§ For an account of this development and of the earlier work on vortex motion by Helmholtz, to which Thomson owed a great debt, see R. H. Silliman, 'William Thomson: smoke rings and nineteenth-century atomism', *Isis* **54** (1963), 461–74.

rays. Now it is not suggested that the vortex theories had anything more than a very modest success,† nor must it be thought that they were intended exclusively, or even principally, as an explanation of thermal phenomena—they did far more than that. But they and Joule's speculations of the 1840s are mentioned here to emphasize the point that no one kinetic theory had the field to itself in the mid nineteenth century. In view of this it is not at all surprising that even such a prominent member of the 'heat is motion' school as John Tyndall felt it necessary to conclude his discussion of the problem by commenting: 'The ideas of the most well-informed philosophers are as yet uncertain regarding the exact nature of the motion of heat.'‡

In the present state of our knowledge it would be rash to try to identify precisely the point at which the modern theory of heat gained the ascendancy over the other kinetic theories for solids and liquids. Certainly it seems that an important turning-point was marked by the widely read paper of 1857 in which Clausius gave a most lucid, convincing, and essentially correct account of the various motions—rotational, vibrational, and translational—of which he believed the molecules of solids and liquids to be capable.§ But the exact date of the acceptance of Clausius's views, even if it can be determined, scarcely concerns us here; it is important only to recognize that the theories of heat proposed in the wake of the discovery of the principle of energy conservation were all kinetic in nature. One thing was clear by the 1850s: there was no going back to caloric. Heat was recognized to be motion and the task that remained was simply to elucidate the particular type of motion involved.

† Although among those who found them at least interesting and even fruitful were P. G. Tait, J. J. Thomson, and A. A. Michelson; see Silliman, op. cit., pp. 471–3. William Thomson himself finally abandoned the theory in 1898, but J. T. Merz, in 1903, could still refer to it as 'the most advanced chapter in the kinetic theory of matter, the most exalted glimpse into the mechanical view of nature'; see Merz, *History of European thought*, vol. 2, p. 66.

‡ Tyndall, *Heat considered as a mode of motion* (1863), p. 62; see also the similar statement in Thomas Graham's influential *Elements of chemistry* (2nd edn, 1858), vol. 2, pp. 449–57. Tyndall was rather less guarded in some of his other writings. See, for example, the Rede lecture of 16 May 1865, in his *Fragments of science* (6th edn, 1879), vol. 1, pp. 32–3, where he states clearly that the heat of solids is a molecular vibration.

§ Clausius, *Pogg. Ann.* **100** (1857), 360. In 1850 Clausius, like many of his contemporaries (see p. 308 n. ¶), had been content with the rather vague statement that heat was *vis viva*.

However, in the case of solids and liquids only rather inconclusive speculation was possible, and the same was true of gases—at least until 1857. Before that date the kinetic theory adopted by Herapath, Joule, and, in 1856, by the German *Realschule* professor August Karl Krönig† was moderately well known but it had no characteristics that marked it as being obviously superior to the theories of, say, Davy or Rankine. For an unprejudiced observer there was simply no way of making a really rational choice. Between 1857 and 1860 this unhappy state of affairs was changed radically by a series of brilliant papers by Clausius and Maxwell, who demonstrated, to the satisfaction of most of their contemporaries, that the Bernoulli-style kinetic theory was capable of a far more sophisticated and fruitful development than any of its rivals.

Clausius made his most important contributions in two papers published in 1857 and 1858.‡ In the first of these papers he argued that the observed specific heats of gases could only be predicted theoretically by taking account not only of the translational *vis viva* of the gas molecules but also of their rotational *vis viva* and the *vis viva* that they possessed in respect of the vibrational motion of the constituent atoms within the molecule. In the second paper Clausius introduced the concept of the mean free path of gas molecules and he used it to explain the slowness with which gases were known to diffuse through one another. In doing this he provided a most effective answer to an objection that had been raised by a critic of Clausius's theory, the Dutch mathematician, geologist, and meteorologist, C. H. D. Buijs-Ballot.§

It was left for Maxwell, in 1859, to add the ingredient of our

† Krönig, 'Grundzüge einer Theorie der Gase', *Pogg. Ann.* **99** (1856), 315–22. Krönig appears to have arrived at his theory in complete ignorance of the writings on the subject by Herapath and Joule.

‡ Clausius, 'Ueber die Art der Bewegung, welche wir Wärme nennen', *Pogg. Ann.* **100** (1857), 353–80, and 'Ueber die mittlere Länge der Wege . . .', *Pogg. Ann.* **105** (1858), 239–58.

§ Buijs-Ballot (or Buys-Ballot), *Pogg. Ann.* **103** (1858), 240–59. Clausius refuted Buijs-Ballot's objection by showing that at normal pressures the mean free path of a gas particle was very small, so that its progress in any given direction was slow. Buijs-Ballot had maintained that, according to the kinetic theory, a particle would traverse a room many hundreds of times in a second. Maxwell soon supported Clausius's case by showing that for air at 60 °F the mean free path was 1/447 000 inch and that a particle underwent 8 077 200 000 collisions per second; see Maxwell, *Phil. Mag.*, 4th ser. **19** (1860), 32.

modern kinetic theory that was most obviously lacking in Clausius's treatment, viz. the idea of a statistical distribution of velocities among the particles of a gas.† Previously the assumption that the velocities of all the particles in a gas were identical, or very nearly so, had been adopted almost without question,‡ but Maxwell's very different view was quickly accepted as more reasonable. Within a few years, moreover, his case had been greatly strengthened by one of the most striking successes in the whole history of the kinetic theory. This came about in 1866 when, in the Bakerian lecture for that year, Maxwell described experiments of his own which confirmed that, at any given temperature and at least for pressures between $\frac{1}{2}$ inch and 30 inches of mercury, the viscosity of air was independent of its density.§ This most improbable result had been predicted by Maxwell in his 1859 paper without the support of any experimental evidence.¶

After this startling experimental verification the success of the modern kinetic theory of gases was virtually assured, although the theory was still not exempt from criticism. Vortex theories, such as were proposed by Rankine and Thomson, continued to have a certain following, and Dr S. G. Brush has noted yet another theory, based on the motion of ether, which was proposed by the German Jacob Ferdinand Redtenbacher in 1869.‖ Moreover, throughout the last three decades of the nineteenth century and even into this century vigorous opposition to the kinetic theory was expressed by a number of positivistically inclined thinkers, among them Ernst Mach, Georg Helm, Wilhelm Ostwald, and Pierre Duhem, who objected principally to the theory's dependence on the atomic hypothesis concern-

† Maxwell, 'Illustrations of the dynamical theory of gases', *Phil. Mag.*, 4th ser. **19** (1860), 19–32, and 4th ser. **20** (1860), 21–37. The paper was read at the British Association for the Advancement of Science meeting at Aberdeen in September 1859. On Maxwell's work see S. G. Brush, 'The development of the kinetic theory of gases. IV. Maxwell', *Ann. Sci.* **14** (1958), 243–55.

‡ Although Clausius appears to have been uneasy about this assumption; see Clausius, *Pogg. Ann.* **100** (1857), 377.

§ Maxwell, *Phil Trans. R. Soc.* **166** (1866), 249–68, especially p. 256.

¶ Maxwell, *Phil. Mag.*, 4th ser. **19** (1860), 32. In fact, by 1866 experiments that served to confirm Maxwell's prediction had already been made by the young German physicist Oskar Emil Meyer and announced by him in his paper 'Ueber die Reibung der Flüssigkeiten', *Pogg. Ann.* **113** (1861), 55–86, 193–238, and 383–425.

‖ Brush, op. cit., p. 250.

ing the structure of matter.† Atoms were considered by these critics as unobservable, metaphysical entities, the typical product of a crude mechanical view of nature; they were to be shunned together with all theories in which their existence was assumed.

But the vortex theories did not prevail, and the deep philosophical criticisms too seem to have been rejected by the majority of practising scientists,‡ who from about 1860 became increasingly committed to the kinetic theory as now developed by Clausius and Maxwell. In December 1863 Thomas Graham went so far as to state that the theory was 'now generally received',§ and a similar opinion was expressed by Balfour Stewart some three years later in the light of Maxwell's experiments on viscosity.¶ Thereafter the kinetic theory of gases steadily gained in strength.‖ There still lay ahead a lengthy and by no means uninteresting process of refinement and extension, a process that was to occupy some of the most distinguished physicists of the period and was to culminate in the assimilation of the theory into the body of early twentieth-century quantum physics. But the truly creative phase was over before 1870. By the time such men as Ludwig Boltzmann, J. D. van der Waals, H. A. Lorentz, and Lord Rayleigh entered the field in the last three decades of the nineteenth century the theory, for all its defects, had become accepted as a part of 'normal science', and the work of its leading exponents can be characterized with fair accuracy as 'puzzle-solving', albeit 'puzzle-solving' of a very high order.††

† See, for example, Dr Brush's introduction to L. Boltzmann, *Lectures on gas theory* (Berkeley and Los Angeles, 1964), pp. 13–14 and (for a useful list of references) 464–74.

‡ Although in the 1890s Ludwig Boltzmann, one of the most ardent supporters of the theory, feared that the philosophical criticisms of his opponents in the school of 'energetics' might succeed in obliterating his achievements; see Brush, *Kinetic theory*, vol. 2, p. 7.

§ Graham, *Phil. Mag.*, 4th ser. **26** (1863), 410.

¶ Stewart, *An elementary treatise on heat* (Oxford, 1866), p. 367.

‖ On the impressive support for the theory that was expressed in the 1860s see Brush, *Ann. Sci.* **14** (1958), 249–55. It is interesting that Tyndall maintained at least a show of agnosticism in his *Heat considered as a mode of motion* (1863), although he treated the theory fully and very favourably on pp. 62–4.

†† Here again I use the terminology of Professor Kuhn's *Structure of scientific revolutions*; see especially pp. 10–42.

9

CONCLUSION: A SCIENCE IN DECLINE

To a large extent the history of the caloric theory of gases must be a story of French science, a story in which such names as Lavoisier, Laplace, Gay-Lussac, and Sadi Carnot have pride of place and in which the Institute's prize competition of 1812 stands as possibly the most important single event. This is not to deny, of course, that in the early years of the nineteenth century the caloric theory was accepted as standard doctrine in Britain almost as widely as it was in France. Yet there was always something peculiarly French about the caloric theory. Indeed, how could it be otherwise when the theory formed so essential a part of that characteristically French style of science which reached maturity in the Napoleonic period and to which I have earlier applied the term Laplacian?

So it is that I have chosen to devote my concluding remarks to France, and more particularly to some reflections on one specific problem in the history of nineteenth-century French science. Briefly the problem is to identify the nature, extent, and causes of the marked decline in the quality of French work in the physical sciences, or at least in physics, which is generally thought to have occurred after about 1830.† The absence of any major French contributions to thermodynamics, the kinetic theory of gases, and electromagnetism between the 1840s and the 1870s, when such important developments were taking place in all these fields elsewhere in Europe, has recently been

---

† The literature on this decline is considerable, but of special interest are: F. Klein, *Vorlesungen über die Entwicklung der Mathematik im 19. Jahrhundert* (Berlin, 1926), vol. 1, pp. 63–82 and 87–8; T. J. Rainoff, 'Wave-like fluctuations of creative productivity in the development of west-European physics in the eighteenth and nineteenth centuries', *Isis* **12** (1929), 287–319; M. Yuasa, 'Center of scientific activity: its shift from the 16th to the 20th century', *Jap. Stud. Hist. Sci.* **1** (1962), 57–75; J. W. Herivel, 'Aspects of French theoretical physics in the nineteenth century', *Br. J. Hist. Sci.* **3** (1966–7), 109–32. The point with regard to medical work is made well in J. Ben-David, 'Scientific productivity and academic organization in nineteenth century medicine', *Am. soc. Rev.* **25** (1960), 828–43.

CONCLUSION: A SCIENCE IN DECLINE

cited, with considerable cogency, as evidence that such a decline did in fact occur.† Statistics of scientific 'productivity' also serve to confirm the point,‡ although they are incomplete and somewhat unsatisfactory on a number of counts. In a history of the caloric theory of gases, however, it must surely be the sterile experimenting of Victor Regnault that has pride of place as evidence of the decline; indeed, by any standards Regnault's failure to play a significant part in the development of thermodynamics deserves more than the passing comment it has received in earlier studies of our problem.§

Regnault, we should recall, was a man of outstanding ability, and by the early 1840s he had the familiarity with steam-engine operation that seems to have been so important to most of the pioneers of energy conservation.¶ Moreover, thanks to the French Government, he had been provided with assistants, equipment, and a laboratory that would have been the envy of his contemporaries both in France and elsewhere.‖ With all this, how could he have failed? A partial answer may well lie in the stultifying effect of the Government commission, but it is doubtful whether this need have been quite as stultifying as it became in Regnault's hands. Far more relevant to the problem is surely the widespread mood of scepticism in French science which was noted in the last chapter and which had shown itself in some cases in the form of true positivism, in others merely in an extreme caution in the use of such hypothetical concepts as atoms or caloric. It was into this latter category that Dumas's philosophy fell, and Dumas, as I have argued, was close to Regnault in the late 1830s. There is much in Regnault's work to suggest that the association was a crucial one and that

---

† Herivel, op. cit., pp. 114–32. The point was made earlier, with special reference to thermodynamics, by Charles Fabry in G. A. A. Hanotaux (ed.), *Histoire de la nation française* (1924), vol. 14, p. 329. Here Fabry described the Second Empire as a period of 'eclipse' for French physics.

‡ See especially the papers by Rainoff and Yuasa cited in the note on p. 314.

§ Although Mr Herivel for one clearly sees the style of Regnault's work as significant evidence; see Herivel, op. cit., p. 129.

¶ See Professor T. S. Kuhn's comments in Clagett, *Critical problems*, pp. 329–36.

‖ For example, it would have compared very favourably with the inadequate facilities that were available in France to such men as François Magendie, Claude Bernard, and Louis Pasteur about the middle of the century. Also Regnault's laboratory had sufficient renown to attract a stream of students from outside France, including the young William Thomson in 1845.

Regnault, like Dumas, did not consider the search for causes as worthless or impossible but was merely opposed to premature theorizing. This view is supported by Dumas himself, who recalled of Regnault that whereas he could enthusiastically accept the interconvertibility of heat and work once this was proved beyond reasonable doubt, nevertheless:

Bold speculations concerning the unity of matter or the unity of forces held no attraction for him. The conversion of light into heat, of magnetism into electricity, and of these four forces into one another had not concerned him. He accepted such conversion as an ingenious idea and not as a definite result.†

If this comment does in fact represent Regnault's attitude, and it certainly seems consistent with his work as a whole, then the validity of the single term 'positivist' which has been used to describe the state of French physics about the middle of the century‡ must be accepted with reserve. Regnault was working in the tradition not of Fourier and Comte but rather of Dumas, a vigorous advocate of the accumulation of facts, though only as a preliminary to the establishment of true causes. Regnault, as it happened, hardly ever progressed beyond the first and surer part of Dumas's programme, since, by the standards that he set himself, the problems involved even at this stage were simply too great. Others, less sensitive perhaps to the mood of the time, could continue with their theorizing, even in France, but the backgrounds of these men were distinctly unconventional and so, in terms of education, contacts, and standing in the scientific community, were in sharp contrast to that of Regnault. It is surely significant in this respect that of the only Frenchmen to play any part at all in the establishment of the principle of energy conservation— Marc Seguin, Émile Verdet, and Adolph Hirn—not one was a full academician and only Verdet, first as examiner and then as professor of physics from 1862 until his death in 1866, had any close contact with the École Polytechnique, where the influences of Dumas and Lamé, for example, were presumably most strongly felt.

† Dumas, 'Éloge de Regnault', *Mém. Acad. Sci. Inst. Fr.* **43** (1883), pp. lvii–lviii.

‡ Herivel, op. cit., pp. 121–5. It is stressed that Mr Herivel himself defines the sense in which he uses the term 'positivist' (or 'analytico-positivistic') with great care.

CONCLUSION: A SCIENCE IN DECLINE

Regnault is an important figure in any study of the decline of French physical science precisely because his background, training, and general intellectual environment were so conventional. He certainly towered above his contemporaries in the quality of his work, yet his general approach clearly reflects the norm in French experimental physics about the middle of the century.† For this reason a comparison between Regnault's precise but quite unimaginative experiments and the highly speculative conjectures of the Laplacian school is all the more instructive. The Laplacian science may well have been wrong and ill-conceived in its basic principles; but it was often fertile, and the unadventurous style of science that had taken its place in France by the 1850s and of which Regnault was a master was decidedly not that.

A full discussion of the changes that came about between the scientifically brilliant Napoleonic period and the age of Regnault would obviously be out of place in a study devoted to the history of one particular theory. Yet it may not be unduly rash to invoke the evidence accumulated in this book in suggesting that the symptoms of the decline can be seen, as early as the 1820s, in the growing disenchantment with theory and a consequent preoccupation with the accumulation of data and with mathematical treatments in which the discussion of causes was eliminated or made irrelevant. By then the task of filling the vacuum left by the sudden rejection of Laplacian science was already proving difficult, far more difficult than the group of talented young physicists who came to prominence after the Bourbon restoration—Arago, Dulong, Fresnel, and Petit— could ever have imagined, or feared. These men were not positivists by any stretch of the imagination and they must surely have viewed the extreme caution of such men as Fourier and Dumas and the emergence of the Comtian philosophy of science as an indication of their own failure.

Sadly this brilliant group had virtually ceased to exist by the early 1830s. With the deaths of two of its most gifted members, Petit in 1820 and Fresnel in 1827, it had suffered important, even crucial losses, while of those who remained Arago, now permanent secretary of the Académie des Sciences, had largely

† I would argue that the work of Léon Foucault and Armand Hippolyte Louis Fizeau reflects precisely the same norm.

withdrawn from scientific work and Dulong, beset by illness and with the best years of his life already wasted on a demanding government commission, was experiencing growing frustration in what little science he was still able to pursue. So within less than two decades the impetus of the exciting years immediately after 1815 had been lost, and it may well be in this loss that we should look for the real tragedy of French physical science during the nineteenth century.

APPENDIX

A letter from Justus Liebig to Jules Pelouze, in the archives of the Académie des Sciences, Paris (dossier 'Pelouze'). Reproduced by courtesy of MM. les Secrétaires Perpétuels of the Académie. Liebig's French is uncorrected.

Giessen ce 14 Oct. 1838

Mon cher Pelouze. Permettez moi d'introduire aupres de vous Mr. Winkelblech professeur de chimie à Marburg (vous savez c'est 6 lieux [sic] de Giessen) mon ami et ancien élève. Vous le connaissez tres bien par ses beaux travaux sur les oxides de cobalt. Il m'a temoigne le plus vive desir de faire votre connaissance, c'est un excellent caractere que j'aime beaucoup.

Nous avons bien regretté de ce que vous n'etiez pas de notre compagnie a Freybourg et pendant notre voyage dans l'Alsace. Jamais mon cher Pelouze je n'ai fait un voyage plus agreable, jamais je n'ai pensé aussi souvent à vous et regrette plus amerement votre absence. Pensez, Wohler, Magnus, H. Rose, L. Gmelin et moi nous etions toujours ensemble, nous ne nous separions pas une seule quart heure pendant trois semaines. Nous etions ensemble comme 5 etudians joyeux, nous abandonnant à toutes les folies, sans jalousie l'un contre l'autre, causant science, formant des projets pour l'avenir je ne peux oublier cette heureuse journée, c'était vous qui nous manquait nous allions toujours à train de poste, Wohler avait emmené une excellent voiture, deux de nous etaient toujours sur le siege du cocher, trois dans l'interieur et il y avait encore une bien belle place pour vous, que jai regretté que ce n'était pas vous qui le remplissait. Vous etiez pendant ce tems en Angleterre et vous avez vu bien de choses utiles, mais je suis sur vous n'aviez pas eu un seul instant de notre bonheur.

Le resultat le plus immediat de nos discussions et de nos entretiens sera je crois de quelque influence. Nous cinq nous sommes convenu d'abandonner tout à fait les nombres d'atomes qui expriment le rapport de volumes, nous sommes decides à adopter dorenavant comme poids d'atome le poids des equivalens et pour nombres d'atomes les nombres equivalens. Ce ne seront que ceux qui ne sont soumis à aucun changement dans l'avenir pendant que tous ceux qui exprimeront les idees regnantes du jour devraient etre changé a chaque moment. Suivons nous la loi de Mr. Dulong sur les chaleurs specifiques nous aurons d'autres nombres qu'en admettant pour guide la theorie des volumes ou l'isomorphisme, toute ces doctrines sont basés sur des hypotheses; mais pour un emploi aussi simple il ne nous faut pas des hypotheses, pourquoi trainer nos idees sur l'état moleculaire des corps dans notre mode d'exprimer la composition des corps, d'ecrire nos formules. On ne voit pas une seule raison qui puisse justifier un tel emploi. Et particulierement pour l'enseignement les nombres equivalens ont une utilité réel, facile à saisir et a comprendre ce qu'ils signifient il ne troubleront jamais l'esprit de ceux qui suivent les cours de chimie.

Nous desirons bien ardemment que quelques chimistes francais se joigneraient à nous de manière qu'il soit adopté en meme tems dans plusieurs pays. Nous avons l'intention d'engager Mr. Berzelius de se mettre à la téte, si cela vient d'etre prononcé le premier par lui, nous menagerons le mieux possible sa susceptibilité. Tous nos formules deviennent plus simples et seront à jamais degage de tout ce qui ne peut etre prononcé comme vrai. Dites moi votre opinion la dessus.

Vous aurez appris que ma fille ainée Agnes souffre beaucoup depuis 9 mois c'est une maladie scrophuleuse qui vient affecter les os de la main droite, qui nous fait infiniment de chagrins, elle a deux plaies ouvertes et nous craignons beaucoup qu'elle ne perdra l'usage du bras droit. Quelle remede emploie-t-on à Paris dans cette cruelle maladie ? Parlez moi de Mad. Pelouze vos enfans et de votre santé. Ma femme et nous autres nous nous trouvons assez bien. J'ai a vous dire mille, mille choses de leur part. Adieu mon meilleur ami je vous aime de tout mon cœur.

J. LIEBIG

TABLES

TABLE A

Specific heats of gases determined at constant pressure

The figures given are volume specific heats on the scale where the volume specific heat of air is unity. Where the authority concerned gave a value for the specific heat by weight this value has been added in brackets, but the scale in this case is such that the value for water is unity

Authority	Date	Air	Oxygen	Nitrogen	Hydrogen	Carbon monoxide	Nitrous oxide	Carbon dioxide	Ethylene	Water vapour	Page ref. in text
Lavoisier and Laplace	1793	1·00† (0·33031)	2·17† (0·65)								35
Leslie	1804	1			1						108–9
Delaroche and Bérard	1812	1·0000 (0·2669)	0·9765 (0·2361)	1·000‡ (0·2754)	0·9033 (3·2936)	1·0340 (0·2884)	1·3503 (0·2369)	1·2583 (0·2210)	1·5530 (0·4207)	1·96 (0·8470)	138–41
Delaroche and Bérard (alternative method)	1812	1·000	0·974	1·000‡	0·893	0·983	1·315	1·311	1·680		138–41
Haycraft	1823	1	1	1	1	1					256–7
Dulong (indirect method)	1828	1	1		1		1·16	1·175	1·531		253–5
de la Rive and Marcet	1835							1·222			259–60
Apjohn	1835	1·000	1·0000	0·9887 (0·3135)	1·8948 (6·1892)	1·0808 (0·3123)	1·1652 (0·2240)	1·0146 (0·2124)	1·5309		258–9
Suerman	1836	1·0000 (0·3046)	0·9954 (0·2750)	1·0005	1·5979	0·9923	1·1229	1·0655			259
Apjohn	1837	1·000	0·808	1·048	1·459	0·996	1·193	1·195			258–9
Regnault	1853	1·0000 (0·23741)	1·0125 (0·21751)	0·9970 (0·24380)	0·9933 (3·40900)	0·9978 (0·2450)	1·4514 (0·2262)	1·3923 (0·2169)	1·7287 (0·4040)	1·2583 (0·4805)	297–8

† indicates that the figure given has been calculated from the specific heat by weight with the aid of contemporary values for specific gravity.
‡ indicates an estimated figure.

TABLE B

Specific heats of gases determined at constant volume

The figures given are volume specific heats on the scale where the volume specific heat of air is unity. Where the authority concerned gave a value for the specific heat by weight this value has been added in brackets, but the scale in this case is such that the value for water is unity

Authority	Date	Air	Oxygen	Nitrogen	Hydrogen	Carbon monoxide	Nitrous oxide	Carbon dioxide	Ethylene	Water vapour	Page ref. in text
Crawford	1779	1·00† (18·670)	5·14† (87·000)		1·16† (281)			0·02† (0·270)			33–4
Crawford	1788	1·00† (1·7900)	2·92† (4·7490)	0·43† (0·7936)	0·92† (21·400)			0·88† (1·0454)		0·53† (1·5500)	36–8
Dalton (theoretical calculation)	1808	1·000† (1·759)	0·834† (1·333)	1·029† (1·866)	0·427† (9·382)	0·840† (0·777)	0·499† (0·549)	0·424† (0·491)	1·477† (1·555)	0·410† (1·166)	115
Clément and Desormes	1812	1·000	1·000	1·000	0·6640			1·5000			142–7
de la Rive and Marcet	1827	1	1	1	1	1	1	1	1		257–8
Dulong (indirect method)	1828	1·000	1·000		1·000	1·000	1·227	1·249	1·754		253–5
de la Rive and Marcet	1829	1			1		1	1	1		258

† indicates that the figure given has been calculated from the specific heat by weight with the aid of contemporary values for specific gravity.

TABLE C
The expansion of air by heat

Authority	Date	Vol. (pressure) at f.p. of ice	Vol. (pressure) at room temp.	Vol. (pressure) at b.p. of water	Vol. (pressure) at other temperatures	Expansion (pressure increment) per °C, expressed as a fraction of volume (pressure) at f.p. of ice	Page ref. in text
Amontons	1699 and 1702		1000	App. 1333		(1/250)	60–1
	1703	1000		1417		(1/240)	61
Nuguet	1705		1000	App. 16000		(1/6.7)	62
la Hire	1708		1000	1295		(1/283)	62
Hauksbee	1709	1000			1099 at 130° on Hauksbee's scale (= app. 102 °F)	(1/394)	62 n.
Cruquius	1724–5	1000		1411		(1/243)	62
Poleni	1731–2	1000		1334		(1/299)	62
Musschenbroek	1762	1000		1500		(1/200)	62 n.
De Luc	1772					(1/250)	63
Luz	1774	1000		1377.5		(1/265)	66
Roy	1777	1000		1484		(1/207)	63
Shuckburgh	1777					1/228	63

TABLES

Lambert	1779	1000	1370	(1/270)	66
Saussure	1783			1/295	64 n.
Mayer	1786			1/266	71
Trembley	1786			(1/229)	63–4
Vandermonde, Berthollet, and Monge	1786			1/231	64–5
Guyton de Morveau and Prieur	1789	1000	1937	(1/107)	65
Volta	1793	1000	1370	1/270	66–7
Dalton	1801	1000	1325	(1/256)	69
Gay-Lussac	1802	1000	1375	(1/267)	69–70
Rudberg	1837	1000	1364(·5)	(1/275)	299 n.
Regnault (i) pressure coeff.	1842	1000	1366·5	(1/273)	298
(ii) volume coeff.	1842	1000	1367·0	(1/273)	298

Except in the case of Regnault no attempt has been made in this table to distinguish between experiments designed to measure pressure increments and those in which expansion at constant pressure was measured.

The figures in the column headed Expansion (pressure increment) per °C which appear in brackets have been calculated from the data given by the authority concerned and do not appear in the original work. They are given only as a guide for comparisons and should not be taken to imply that volume or pressure were supposed to increase uniformly with temperature.

TABLE D

The expansion by heat of gases other than air

	Volumes (pressures) at the boiling-point of water on a scale such that the volume (pressure) at the freezing-point of ice is 1000				
	Vandermonde, Berthollet, and Monge (1786)	Guyton de Morveau and Prieur (1789)	Gay-Lussac (1802)	Regnault (1842)	
				At constant volume	At constant pressure
Hydrogen	1443†	1392		1366·7	1366·1
Oxygen		5478	1375		
Nitrogen		6942	1375	1366·8	1367·0
Nitrous oxide			1375	1367·6	1371·9
Nitric oxide		1606	1375		
Carbon monoxide				1366·7	1366·9
Carbon dioxide		2009	1375	1368·8	1371·0
Sulphur dioxide			1375	1384·5	1390·3
Ammonia		6801	1375		
Page ref. in text	64–5	65	69–70	298	298

† In the original paper it is stated simply that hydrogen expands by 1/181·02 of its initial volume for every °R rise in temperature. It is assumed here that the original volume referred to is that at the melting-point of ice.

TABLE E

A table of affinities for caloric, etc.

	Vol. specific heats, as determined experimentally by Delaroche and Bérard (*Annls Chim.* **85** (1813), 157)	Vol. specific heats calc. by Avogadro (for method see pp. 209–10) (Avogadro, *Biblioteca italiana* **4** (1816), 489–91, and **5** (1817), 78 and 83)	Attractive power for caloric	Affinity for caloric	Refractive power, as determined experimentally by Biot and Arago (*Mém. phys. Inst. Fr.* **7** (1806), 320)	Refractive power calc. by Avogadro (for method see text, p. 213) (*Memorie Mat. Fis. Soc. ital. Sci.* **18** (1820), memorie fis. 161–2. Read 14 Oct. 1817)	Affinity for caloric calc. by Avogadro (for method see text, pp. 213 and 217) (*Memorie Mat. Fis. Soc. ital. Sci.* **19** (1823), memorie fis. 128. Dated 17 Jan. 1822. Also *Memorie Accad. Sci. Torino* **28** (1824), 62)	Vol. specific heats calculated from the affinities in previous column (*Memorie Mat. Fis. Soc. ital. Sci.* **19** (1823), memorie fis. 129, and *Memorie Accad. Sci. Torino* **18** (1824), 39 n.)
Oxygen	0·9765†	0·9765†	0·95355	0·8640	0·8616	0·902	0·8500	0·9706
Hydrogen	0·9033†	0·9033†	0·8160	11·1460	6·6144‡	6·614‡	10·2573	0·8401
Nitrogen	1·0058†	1·0058†	1·0116	1·0438	1·0341	1·031	1·0454	1·0075
Carbon vapour		1·0085	1·1828	1·4216			1·4296	
Carbon monoxide	1·0340†	1·0340†	1·0692	1·1047			1·0984	1·0321
Carbon dioxide	1·2583	1·2434	1·5460	1·0174	1·0048		1·0081	1·2396
Nitrous oxide	1·3503	1·2200	1·4884	0·9786		1·012	0·9744	1·2191
Nitric oxide			0·9826	0·9481				
Nitrogen trioxide				0·9304				
Nitrogen peroxide			2·91875	0·9189				
Nitrogen pentoxide				0·91075				
Ethylene			2·8148	2·8769			2·6831	1·6123
Ammonia	1·5530	1·6777	1·7298	2·9103	2·1685	2·2114	2·6602	
Water vapour			1·2928	2·0685			1·8886	
Ammonium carbonate				1·8499				
Ammonium bicarbonate				1·5510				
Column number	1	2	3	4	5	6	7	8

In all columns figures are given on a scale such that the relevant value for air is unity.

† and ‡ indicate that the figures given are experimental figures taken as the basis for the calculation of the theoretical values.

BIBLIOGRAPHY

Unless the contrary is stated, all references to books are to first editions, and references to journals are to first series.

PRIMARY SOURCES

(A) MANUSCRIPTS

Académie des Sciences, Paris:
   Laboratory notebooks of Lavoisier in 13 volumes; especially volume 8, for 25 March 1783 to February 1784 (Lavoisier papers, box 24).
   N. Clément and C. B. Desormes, 'Détermination expérimentale du zéro absolu de la chaleur'—their entry for the 1812 prize competition (unnumbered file marked '1813, Grand Prix des Sciences Physiques').
   V. F. da Olmi, 'Mémoire sur la détermination de la chaleur spécifique des fluides élastiques'—an entry for the 1814 prize competition (unnumbered file marked '1813. Grand Prix des Sciences Physiques').
   Letter of Justus Liebig to Jules Théophile Pelouze, 12 October 1838 (unnumbered file marked 'Pelouze').

Conservatoire National des Arts et Métiers, Paris:
   Notes taken at Nicolas Clément's lectures on applied chemistry between 1824 and 1828 by J. M. Baudot, bound in 3 volumes (MS. 8° Fa 40 (2)).

Darwin Papers, Down House, Kent:
   Commonplace book of Erasmus Darwin.

École Nationale Supérieure des Beaux-Arts, Paris:
   Notes taken at Nicolas Clément's lectures on applied chemistry in 1823–4 by Louis Benjamin Francœur (MS. 407).

École Polytechnique, Paris:
   Record cards of Pierre Dulong, Sadi Carnot, and Victor Regnault.
   Sadi Carnot, 'Recherche d'une formule à représenter la puissance motrice de la vapeur d'eau'—an nudated manuscript paper probably written about 1827 (unnumbered file marked 'Lazare et Sadi Carnot. Promotion 1812').
   Minute book of the *Conseil de Perfectionnement*; especially volumes 3 and 4, for the years 1806 to 1815.

Edinburgh University Library:
   'Students in the College of Chemistry', a class register kept by William Cullen between 1755 and 1765 (MS. Da. 3).
   Volume 1 of 'Notes of Doctor Black's Lectures on Chymistry'—notes taken by an unnamed student in 1775 (MS. Dc. 3. 11).

Institut de France, Paris:

Jacques Alexandre César Charles's notes for a lecture course on physics given by him about 1806 (File 2104).

Letter of Pierre Dulong to an unknown correspondent, 4 November 1828 (File 2220).

Académie des Sciences, Belles-Lettres et Arts de Lyon:

Joseph Mollet, 'Mémoire sur un nouveau fait physique'—manuscript paper read to the Lyons Academy on 14 December 1802 (MS. 230, ff. 69–74).

Manuscript paper presented by Jean Antoine Saissy to the First Class of the French Institute in Paris on 11 November 1811 (MS. 230, ff. 221–32).

Maison d'Auguste Comte, Paris:

Notes taken at Alexis Thérèse Petit's lectures on physics at the École Polytechnique in 1814–15 by Auguste Comte.

University of Manchester Institute of Science and Technology, Department of the History of Science and Technology:

Draft for a lecture dating from between June 1845 and April 1847 (Joule MSS., notebook 2).

Museum of the History of Science, Oxford:

Minute book of the Chapter Coffee House Society (MS. Gunther 4).

(B) PRINTED SOURCES

G. ADAMS, *Lectures on natural and experimental philosophy, considered in its present state of improvement* (1st edn, 5 vols., London, 1794; 2nd edn, 5 vols., London, 1799).

[P. A. ADET], *Supplément à la seconde édition d'histoire naturelle et de chimie* (Paris, 1789). The author is identified in A. F. Fourcroy, *Élémens d'histoire naturelle et de chimie* (3rd edn, Paris, 1789), vol. 1, p. vi.

F. U. T. AEPINUS, *Tentamen theoriae electricitatis et magnetismi* (St. Petersburg, n.d. [1759]).

G. B. AIRY, 'The Astronomer Royal's remarks on Professor Challis's theoretical determination of the velocity of sound', *Phil. Mag.*, 3rd ser. **32** (1848), 339–43.

G. AMONTONS, 'Moyen de substituer commodement l'action du feu à la force des hommes et des chevaux pour mouvoir les machines', *Hist. Acad. Sci.* (1699), pp. 112–26. Dated 20 June 1699.

—— 'Discours sur quelques proprietez de l'air, et le moyen d'en connoître la temperature dans tous les climats de la terre', *Mém. Acad. Sci.* (1702), pp. 161–80. Dated 28 June 1702.

—— 'Le thermometre réduit à une mesure fixe et certaine . . .', *Mém. Acad. Sci.* (1703), pp. 50–6. Dated 18 April 1703.

—— 'Que les nouvelles experiences que nous avons du poids et du ressort de l'air, nous font connoître qu'un degre de chaleur mediocre peut reduire l'air dans un état assez violent pour causer seul de très-grands tremblements et bouleversement sur le globe terrestre', *Mém. Acad. Sci.* (1703), pp. 101–8.

A. M. AMPÈRE, *Théorie mathématique des phénomènes électro-dynamiques, uniquement déduite de l'expérience* (Paris, 1826).

—— *Correspondance du grand Ampère*, ed. L. de Launay (3 vols., Paris, 1936–43).

—— 'Lettre . . . sur la détermination des proportions dans lesquelles les corps se combinent . . .', *Annls Chim.* **90** (1814), 43–86.

—— 'Essai d'une classification naturelle pour les corps simples', *Annls Chim. Phys.* **1** (1816), 295–308 and 373–94, and **2** (1816), 5–32 and 105–25.

—— 'Note de M. Ampère sur la chaleur et sur la lumière considérées comme résultant de mouvemens vibratoires', *Annls Chim. Phys.* **58** (1835), 432–44. This paper had appeared in a slightly modified form in *Bibliothèque universelle* **49** (1832), 225–35.

—— *See also* DULONG *et al.*; DUPIN *et al.*; FOURIER, AMPÈRE, and ARAGO.

J. APJOHN, 'On the specific heats of the permanently elastic fluids', *Notices of communications to the British Association . . . 1835* (London, 1836), pp. 30–2.

—— 'Upon a new method of investigating the specific heats of the gases', *Trans. R. Ir. Acad.* **18** (1838), 1–16. Read 16 March 1837.

D. F. J. ARAGO, *Œuvres complètes de François Arago*, ed. J. A. Barral (17 vols., Paris and Leipzig, 1854–62).

—— 'Note sur un phénomène remarquable qui s'observe dans la diffraction de la lumière', *Annls Chim. Phys.* **1** (1816), 199–202.

—— 'Résultats des expériences faites par ordre du Bureau des Longitudes, pour la détermination de la vitesse du son dans l'atmosphère', *Annls Chim. Phys.* **20** (1822), 210–23.

—— 'Éloge historique de T. Young', *Mém. Acad. Sci. Inst. Fr.* **13** (1835), 57–105. Read 26 November 1832.

—— 'Éloge historique de Joseph Fourier', *Mém. Acad. Sci. Inst. Fr.* **14** (1838), 69–138. Read 18 November 1833.

—— and A. T. PETIT, 'Sur les puissances réfractives et dispersives de certains liquides et des vapeurs qu'ils forment', *Annls Chim. Phys.* **1** (1816), 1–9. Read to the First Class of the Institute on 11 December 1815.

—— *See also* BIOT and ARAGO; FOURIER, AMPÈRE, and ARAGO.

L. A. VON ARNIM, 'Anweisung zum Gebrauche des Areometers von Say ohne Barometerbeobachtungen . . .', *Gilb. Ann.* **2** (1799), 238–45.

J. C. ARNOLD, *De thermometri sub campana antliae pneumaticae suspensi variationibus* (Erlangen, n.d. [1759]). J. C. Poggendorf, in his *Biographisch-literarisches Handwörterbuch . . .* (Leipzig, 1863), vol. 1, p. 63, follows J. G. Meusel, *Lexikon der vom Jahr 1750 bis 1800 verstorbenen teutschen Schriftseller* (Leipzig, 1802), vol. 1, p. 105, in dating Arnold's paper to 1757. Since Arnold was not made professor of physics until

1759 (Meusel, op. cit., vol. 1, p. 105), the date of 1759 seems in little doubt.

N. ARNOTT, *Elements of physics or natural philosophy* (4th edn, 2 vols., London, 1829).

A. AVOGADRO, *Opere scelte di Amedeo Avogadro pubblicate dalla R. Accademia delle Scienze di Torino* (Turin, 1911).

—— *Fisica de' corpi ponderabili ossia trattato della costituzione generale de' corpi* (4 vols., Turin, 1837–41).

—— 'Idées sur l'acidité et l'alcalinité', *J. Phys.* **69** (1809), 142–8.

—— 'Essai d'une manière de déterminer les masses relatives des molécules élémentaires des corps, et les proportions selon lesquelles elles entrent dans ces combinaisons', *J. Phys.* **73** (1811), 58–76.

—— 'Réflexions sur la théorie électro-chimique de M. Berzelius', *Annls Chim.* **87** (1813), 286–92.

—— 'Mémoire sur les masses relatives des molécules des corps simples', *J. Phys.* **78** (1814), 131–56. Dated January 1814.

—— 'Memoria sul calore specifico de' gaz composti parragonato a quello de' loro gaz componenti', *Bibltca ital.* **4** (1816), 478–91, and **5** (1817), 73–87.

—— 'Memoria sopra la relazione che esiste tra i calori specifici e i poteri refringenti delle sostanze gasose', *Memorie Mat. Fis. Soc. ital. Sci.* **18** (1820), memorie di fisica, 153–73. Dated 14 October 1817.

—— 'Nuove considerazioni sulle affinità de' corpi pel calorico calcolato per mezzo de' loro calori specifici, e de' loro poteri refringenti allo stato gasoso', *Memorie Mat. Fis. Soc. ital. Sci.* **19** (1823), memorie di fisica, part 1, 83–137. Dated 17 January 1822.

—— 'Mémoires sur l'affinité des corps pour le calorique et sur les rapports d'affinité qui en résultent entre eux', *Memorie Accad. Sci. Torino* **28** (1824), 1–122, and **29** (1825), 79–162. Read 12 January and 4 May 1823.

—— Extracts of the last four papers appeared also in *G. Fis.* Decade II, **7** (1824), 427–37, and **8** (1825), 1–9, 108–17, 160–4, and 313–30.

—— 'Osservazioni sopra un articolo del Bolletino delle Scienze del sig. B. de Ferussac . . .', *G. Fis.* Decade II, **8** (1825), 432–8.

—— 'Sur la densité des corps solides et liquides comparée avec la grosseur de leurs molécules, et avec leurs nombres affinitaires', *Memorie Accad. Sci. Torino* **30** (1826), 81–154, and **31** (1827), 1–94. Read 7 March and 20 June 1820. Both papers were summarized in *Bull. Sci. math.* **9** (1828), 33–48 and 327–39.

—— 'Comparaison des observations de Mr Dulong sur les pouvoirs réfringens des corps gazeux avec les formules de la relation entre les pouvoirs et les affinités pour le calorique déduites des chaleurs spécifiques', *Memorie Accad. Sci. Torino* **33** (1829), 49–111. Read 26 November 1826.

—— 'Sur les nombres affinitaires, ou déterminations des rapports électrochimiques des corps', *Bull. Sci. math.* **7** (1827), 129–42.

—— 'Mémoire sur les pouvoirs neutralisans des différens corps simples déduits de leurs proportions en poids dans les composés neutres qui en sont formés', *Memorie Accad. Sci. Torino* **34** (1830), 146–216. Read

7 December 1828. The paper was summarized in *Bull. Sci. math.* **12** (1829), 42–8.

A. AVOGADRO, 'Note sur la relation entre les chaleurs spécifiques des gaz composés, et celles de leurs gaz composans, qui résulte des observations de M. Dulong', *Bull. Sci. math.* **13** (1830), 211–17.

—— 'Memoria sui calori specifici de' corpi solidi e liquidi', *Memorie Mat. Fis. Soc. ital. Sci.* **20** (n.d. [1833]), memorie di fisica, part 2, 451–576. Dated 8 May 1832, with an additional note on pp. 577–86 dated 28 July 1828. The paper was summarized in *Annls Chim. Phys.* **55** (1833), 80–111.

—— 'Nouvelles recherches sur la chaleur spécifique des corps solides et liquides', *Annls Chim. Phys.* **57** (1834), 113–48.

—— 'Mémoire sur les volumes atomiques et sur leur relation avec le rang que les corps occupent dans la série électro-chimique', *Memorie Accad. Sci. Torino*, 2nd ser. **8** (1846), 129–93. Read 17 December 1843.

—— 'Mémoire sur les volumes atomiques des corps composés', *Memorie Accad. Sci. Torino*, 2nd ser. **8** (1846), 293–532. Read 13 April 1845.

—— 'Mémoire sur les volumes atomiques . . .', *Memorie Accad. Sci. Torino*, 2nd ser. **11** (1851), 231–318. Read 25 February 1849.

—— 'Mémoire sur les volumes atomiques des corps liquides à leur température d'ébullition . . .', *Memorie Accad. Sci. Torino*, 2nd ser. **12** (1852), 39–122. Read 2 June 1850.

—— The last four papers were summarized in *Annls Chim. Phys.*, 3rd ser. **14** (1845), 330–68, **29** (1850), 248–52, and **36** (1852), 96–102.

P. H. AZAÏS, *Essai sur le monde* (Paris, 1806).

—— *Mémoire sur le mouvement moléculaire et sur la chaleur; présenté à la Classe des Sciences Physiques et Mathématiques de l'Institut, le 15 septembre 1815: suivi d'un coup d'œil général sur le Système du Monde, et d'une lettre à M. de la Place* (Paris, 1806).

A. N. BAILLET, 'Lettre . . . sur la glace produite par l'expansion de l'air comprimé', *J. Phys.* **48** (1799), 166–7.

—— 'Remarques sur les expériences précédentes' (following a translation of Dalton's paper on the evacuated receiver experiments which had appeared in *Nicholson's Journal*), *J. Phys.* **76** (1802–3), 267–9.

E. BARRUEL, *La physique réduite en tableaux raisonnés* (Paris, 1806).

A. E. BAUDRIMONT, *Introduction à l'étude de la chimie par la théorie atomique* (Paris, 1833–4). The date 1833 appears on the title-page; 1834 appears on the paper cover.

—— *Traité de chimie générale et expérimentale, avec les applications aux arts, à la médecine et à la pharmacie* (2 vols., Paris, 1844–6).

J. E. BÉRARD, see DELAROCHE and BÉRARD.

T. O. BERGMAN, *Opuscula physica et chemica* (6 vols., Stockholm, Uppsala, and Abo, 1779–90).

—— 'Disquisitio de attractionibus electivis', *Nova Acta R. Soc. Scient. upsal.* **2** (1775), 159–248.

D. BERNOULLI, *Hydrodynamica, sive de viribus et motibus fluidorum commentarii* (Strasbourg, 1738).

C. L. BERTHOLLET, *Recherches sur les lois de l'affinité* (Paris, an 9 [1801]).

C. L. BERTHOLLET, *Essai de statique chimique* (2 vols., Paris, 1803).
—— Review of Chaptal's *Élémens de chymie* in *Annls Chim.* **6** (1790), 197–203.
—— 'Sur la chaleur produite par le choc et la compression', *Mém. Phys. Chim. Soc. Arcueil* **2** (1809), 441–8.
—— See also GUYTON DE MORVEAU, LAVOISIER, BERTHOLLET, and FOURCROY.

J. J. BERZELIUS, *Essai sur la théorie des proportions chimiques et sur l'influence chimique de l'électricité* (Paris, 1819): a translation of pp. 1–122 of volume 3 of his *Lärbok i Kemien* (Stockholm, 1818).
—— *Lehrbuch der Chemie*, trans. F. Wöhler (4 vols., Dresden, 1825–31); also the French translation, by A. J. L. Jourdan (for volume 1) and Madame Esslinger (for volumes 2–8), which appeared as *Traité de chimie* (8 vols., Paris, 1829–33).
—— *Briefwechsel zwischen J. Berzelius und F. Wöhler im Auftrage der Königl. Gesellschaft der Wissenschaften zu Göttingen*, ed. O. Wallach with commentary by J. von Braun (2 vols., Leipzig, 1901).
—— *Jac. Berzelius Bref*, ed. H. G. Söderbaum (6 vols. in 15 parts, Uppsala, 1912–35).
—— 'Versuch, die bestimmten und einfachen Verhältnisse aufzufinden, nach welchen die Bestandtheile der unorganischen Natur mit einander verbunden sind', *Gilb. Ann.* **37** (1811), 249–334 and 415–72. A partial translation appeared in *Annls Chim.* **78** (1811), 5–37.
—— 'De l'influence de l'électricité sur les affinités. Base d'une théorie électro-chimique', *Annls Chim.* **86** (1813), 146–74 and 225–61, and **87** (1813), 50–97 and 113–52; translated from J. S. C. Schweigger's *Neues Journal für Chemie und Physik* **6** (1812), 119–76 and 284–322.
—— 'Untersuchungen über die Zusammensetzung der Phosphorsäure, der phosphorigen Säure und ihre Salze', *Gilb. Ann.* **53** (1816), 393–446, and **54** (1816), 31–55; also the translation in *Annls Chim. Phys.* **2** (1816), 151–76, 217–40, and 329–39.
—— *Jahres-Bericht über die Fortschritte der physischen Wissenschaften*, trans. F. Wöhler, especially vol. 14 (Tübingen, 1834).
—— See also DULONG and BERZELIUS.

J. B. BIOT, *Traité de physique expérimentale et mathématique* (4 vols., Paris, 1816).
—— *Mélanges scientifiques et littéraires* (3 vols., Paris, 1858).
—— 'Sur la théorie du son', *J. Phys.* **55** (1802), 173–82.
—— 'Note sur la formation de l'eau par la seule compression, avec des réflexions sur la nature de l'étincelle électrique', *Annls Chim.* **53** (1805), 321–7. Read to the First Class of the Institute 25 March 1805. An earlier report of the same work had appeared in *Bull. Soc. philomath. Paris* for *frimaire an 13* (November–December 1804), **3**, 259.
—— 'Supplément à la théorie de l'action capillaire (extrait)', *J. Phys.* **65** (1807), 88–95.
—— 'Expériences sur la production du son dans les vapeurs', *Mém. Phys. Chim. Soc. Arcueil* **2** (1809), 94–103. Read to the First Class of the Institute 12 October 1807.

J. B. BIOT, 'Notice historique sur M. Petit', *Annls Chim. Phys.* **16** (1821), 327–35; also in *J. Phys.* **92** (1821), 241–8. Read to the Société Philomathique 15 February 1821.

—— and D. F. J. ARAGO, 'Mémoire sur les affinités des corps pour la lumière, et particulièrement sur les forces réfringentes des différens gaz', *Mém. Sci. math. phys. Inst. Fr.* **7** (1806), 301–87. Read 24 March 1806.

J. BLACK, *Dissertatio medica inauguralis, de humore acido a cibis orto, et magnesia alba* (Edinburgh, 1754). Presented 11 June 1754.

—— *Lectures on the elements of chemistry, delivered in the University of Edinburgh*, ed. J. Robison (2 vols., Edinburgh, 1803).

—— 'Experiments upon magnesia alba, quicklime, and some other alcaline substances', *Essays and observations, physical and literary. Read before a society in Edinburgh, and published by them* (Edinburgh, 1756), vol. 2, pp. 157–225. Dated 5 June 1755. In the second edition of the *Essays* (Edinburgh, 1770) the paper was on pp. 172–248.

H. BOERHAAVE, *Elementa chemiae, quae anniversario labore docuit, in publicis, privatisque, scholis, Hermannus Boerhaave* (2 vols., Leiden, 1732).

L. BOLTZMANN, *Lectures on gas theory*, trans. and ed. S. G. Brush (Berkeley and Los Angeles, 1964).

A. R. BOUVIER, 'Note sur les machines à vapeur, et description d'une de ces machines propre à produire immédiatement le mouvement de rotation', *Annls Chim. Phys.* **3** (1816), 177–92.

R. BOYLE, *New experiments physico-mechanicall, touching the spring of the air and its effects* (1st edn, Oxford, 1660; 2nd edn, Oxford, 1662).

—— *New experiments and observations touching cold, or an experimental history of cold, begun* (London, 1665).

W. T. BRANDE, *A manual of chemistry; containing the principal facts of the science, arranged in the order in which they are discussed and illustrated at the Royal Institution of Great Britain* (1st edn, London, 1819; 3rd edn, 2 vols., London, 1830).

A. BRAVAIS, 'Note sur la vitesse du son', *Annls Chim. Phys.*, 3rd ser. **34** (1852), 82–9.

[ANON.] 'Briquet pneumatique par Dumotiez', *J. Phys.* **62** (1806), 189.

C. H. D. BUIJS-BALLOT (or BUYS-BALLOT), 'Ueber die Art von Bewegung, welche wir Wärme und Electricität nennen', *Pogg. Ann.* **103** (1858), 240–59.

'C', A review of W. Morgan's *Examination of Dr. Crawford's theory*, in *The critical review* **51** (1781), 212–16.

N. L. S. CARNOT, *Réflexions sur la puissance motrice du feu et sur les machines propres à développer cette puissance* (Paris, 1824). The text was reprinted in *Annls scient. Éc. norm. sup., Paris*, 2nd ser. **1** (1872), 393–457, and again (with a biographical sketch by Hippolyte Carnot and the manuscript notes) in an edition of 1878 published by Gauthier Villars in Paris.

CASSINI DE THURY, 'Sur la propagation du son', *Mém. Acad. Sci.* (1738), pp. 128–46. Dated 16 April 1738.

T. CAVALLO, *A treatise on the nature and properties of air and other permanently elastic fluids* (London, 1781).

—— *The elements of natural or experimental philosophy* (4 vols., London, 1803).

H. CAVENDISH, 'Three papers, containing experiments on factitious air', *Phil. Trans. R. Soc.* **56** (1766), 141–84. Read 29 May, 6 November, and 13 November 1766.

J. CHALLIS, 'Theoretical determination of the velocity of sound', *Phil. Mag.*, 3rd ser. **32** (1848), 276–84. Dated 17 March 1848.

J. A. C. CHAPTAL, *Élémens de chymie* (3 vols., Montpellier, 1790).

B. P. É. CLAPEYRON, 'Mémoire sur la puissance motrice de la chaleur', *J. Éc. polytech.* **14**, cahier 23 (1834), 153–90.

R. J. E. CLAUSIUS, 'Ueber die bewegende Kraft der Wärme und die Gesetze, welche sich daraus für die Wärmelehre selbst ableiten lassen', *Pogg. Ann.* **79** (1850), 368–97 and 500–24.

—— 'Ueber die Art der Bewegung, welche wir Wärme nennen', *Pogg. Ann.* **100** (1857), 353–80. Dated 5 January 1857.

—— 'Ueber die mittlere Länge der Wege, welche bei der Molecularbewegung gasförmiger Körper von den einzelnen Molecülen zurückgelegt werden; nebst einigen anderen Bemerkungen über die mecanische Wärmetheorie', *Pogg. Ann.* **105** (1858), 239–58. Dated 14 August 1858.

W. CLEGHORN, *Disputatio physica inauguralis, theoriam ignis complectens* (Edinburgh, 1779). Presented 12 September 1779. A translation, with commentary, has since appeared in D. McKie and N. H. da V. Heathcote, 'William Cleghorn's *De igne*', *Ann. Sci.* **14** (1958), 1–82.

N. CLÉMENT-DESORMES (formerly CLÉMENT), 'Tableau relatif à la théorie générale de la puissance mécanique de la vapeur', *Nouv. Bull. Soc. philomath. Paris*, N.S. **2** (1826), 50–3.

—— and C. B. DESORMES, 'Détermination expérimentale du zéro absolu de la chaleur et du calorique spécifique des gaz', *J. Phys.* **89** (1819), 321–46 and 428–55.

—— See also DESORMES and CLÉMENT.

C. COLDEN, *The principles of action in matter, the gravitation of bodies, and the motion of the planets, explained from these principles* (London, 1751).

I. A. M. F. X. COMTE, *Cours de philosphie positive* (6 vols., Paris, 1830–42).

—— *Correspondance inédite d'Auguste Comte* (4 vols., Paris, 1903–4).

J. M. COUPÉ, 'De la chaleur qui précède l'arrivée d'un vent froid, en masse, dans certains points de l'atmosphère, et réciproquement', *J. Phys.* **53** (1801), 262–4.

A. CRAWFORD, *Experiments and observations on animal heat, and the inflammation of combustible bodies* (1st edn, London, 1779; 2nd edn, London, 1788).

N. CRUQUIUS, 'Observationes . . .', *Phil. Trans. R. Soc.* **33** (1724–5), 4–7.

W. CULLEN, 'Of the cold produced by evaporating fluids and of some other means of producing cold', *Essays and observations, physical and literary. Read before a society in Edinburgh, and published by them* (Edinburgh, 1756), vol. 2, pp. 145–56. Dated 1 May 1755. In the second edition of the *Essays* (Edinburgh, 1770) the paper was on pp. 159–71.

J. L. N. F. CUVIER, Annual reports on the work of the First Class of the Institute and the Académie des Sciences, especially those in *Mém. Sci. math. phys. Inst. Fr.* **13**, part 2 (1812), pp. lxxxi–cxxxii, and *Mém. Acad. Sci. Inst. Fr.* **4** (1819–20), pp. lxxxi–cxxvi.

J. DALTON, *A new system of chemical philosophy* (2 vols., Manchester, 1808–27); also a second edition of part 1 of the first volume (London, 1842).

—— 'Experiments and observations on the heat and cold produced by the mechanical condensation and rarefaction of air', *Mem. Proc. Manch. lit. phil. Soc.* **5**, part 2 (1802), 515–26. Read 27 June 1800.

—— 'On the force of steam or vapour from water and various other liquids both in a vacuum and in air', *Mem. Proc. Manch. lit. phil. Soc.* **5**, part 2 (1802), 550–74.

—— 'On the expansion of elastic fluids by heat', *Mem. Proc. Manch. lit. phil. Soc.* **5**, part 2 (1802), 595–602. This and the last item listed were two of four essays read before the Manchester Literary and Philosophical Society on 2, 16, and 30 October 1801.

—— 'Letter . . . concerning the determination of the zero of heat, the thermometrical graduation, and the law by which dense or non-elastic fluids expand by heat', *Nicholson's Journal*, 2nd ser. **5** (1803), 34–6. Dated 20 April 1803.

—— 'Remarks on Mr. Gough's two essays on the doctrine of mixed gases; and on Professor Schmidt's experiments on the expansion of dry and moist air by heat', *Mem. Proc. Manch. lit. phil. Soc.*, 2nd ser. **1** (1805), 425–36. Read 4 October 1805.

J. F. DANIELL, *Introduction to the study of chemical philosophy: being a preparatory view of the forces which concur to the production of chemical phenomena* (1st edn, London, 1839; 2nd edn, London, 1843).

E. DARWIN, *The botanic garden* (London, 1791).

—— 'Frigorific experiments on the mechanical expansion of air', *Phil. Trans. R. Soc.* **78** (1788), 43–52. Read 13 December 1787.

H. DAVY, *The collected works of Sir Humphry Davy, Bart. LL.D. F.R.S.*, ed. J. Davy (9 vols., London, 1839–40).

—— *A syllabus of a course of lectures on chemistry, delivered at the Royal Institution of Great Britain* (London, 1802).

—— *Elements of chemical philosophy* (London, 1812).

—— *Six discourses delivered before the Royal Society at their anniversary meetings, on the award of the Royal and Copley medals* (London, 1827).

—— 'An essay on heat, light, and the combinations of light', in *Contributions to physical and medical knowledge, principally from the West of England, collected by Thomas Beddoes, M.D.* (Bristol, 1799), pp. 5–147.

—— 'The Bakerian lecture, on some chemical agencies of electricity', *Phil. Trans. R. Soc.* **97** (1807), 1–56. Read 20 November 1806.

—— 'The Bakerian lecture. An account of some new analytical

researches on the nature of certain bodies, particularly the alkalies, phosphorus, sulphur, carbonaceous matter, and the acids hitherto undecompounded; with some general observations on chemical theory', *Phil. Trans. R. Soc.* **99** (1809), 34–104. Read 15 December 1808.

J. B. J. DELAMBRE, Annual reports on the work of the First Class of the Institute, especially that in *Mém. Sci. math. phys. Inst. Fr.* **7**, part 1 (1806), 1–42 of the *Histoire* for 1806.

J. C. DELAMÉTHERIE, 'Note sur un froid considérable produit par la sortie prompte de l'air atmosphérique fortement comprimé', *J. Phys.* **47** (1798), 186.

F. DELAROCHE, 'Observations sur le calorique rayonnant', *J. Phys.* **75** (1812), 201–28. Read to the First Class of the Institute 3 June 1811.

—— and J. É. BÉRARD, 'Mémoire sur la détermination de la chaleur spécifique des différens gaz', *Annls Chim. Phys.* **85** (1813), 72–110 and 113–82.

J. A. DE LUC, *Recherches sur les modifications de l'atmosphère* (2 vols., Geneva, 1772).

—— *Idées sur la météorologie* (2 vols., London, 1786–7).

—— *Introduction à la physique terrestre par les fluides expansibles, précédée de deux mémoires sur la nouvelle théorie chimique, considérée sous différens points de vue* (2 vols., Paris and Milan, 1803).

—— 'To the conductors of the Edinburgh Review', a letter dated 18 April 1805 in *Edinb. Rev.* **6** (1805), 502–15.

W. DERHAM, 'Experimenta et observationes de soni motu, aliisque ad id attinentibus', *Phil. Trans. R. Soc.* **26** (1708), 2–35.

J. T. DESAGULIERS, *A course of experimental philosophy* (2 vols., London, 1734–44).

C. B. DESORMES and N. CLÉMENT, 'Mémoire sur la théorie des machines à feu (extrait)', *Bull. Soc. philomath. Paris*, N.S. **6** (1819), 115–18. The complete paper was read before the Académie des Sciences on 16 and 23 August 1819.

—— See also CLÉMENT and DESORMES.

C. M. DESPRETZ, *Traité élémentaire de physique* (1st edn, Paris, 1825; 2nd edn, Paris, 1827; 4th edn, Paris, 1836).

J. P. DESSAIGNES, 'Mémoire sur les phosphorescences', *J. Phys.* **68** (1809), 444–67, and **69** (1809), 5–35.

—— 'Mémoire sur la propriété lumineuse de tous les corps de la nature par la compression', *J. Phys.* **73** (1811), 41–53.

R. V. DIXON, *A treatise on heat* (Dublin, 1849).

M. DONOVAN, *A treatise on chemistry* (4th edn, London, 1839).

P. L. DULONG, 'Recherches sur la décomposition mutuelle des sels insolubles et des sels solubles', *Annls Chim.* **82** (1812), 273–308. Read to the First Class of the Institute 29 July 1811.

—— 'Mémoire sur les combinaisons du phosphore avec l'oxigène', *Mém. Phys. Chim. Soc. Arcueil* **3** (1817), 405–52; also the summary in *Annls Chim. Phys.* **2** (1816), 141–50. Read to the Académie des Sciences 1 and 15 July 1816.

P. L. Dulong, 'Recherches sur les pouvoirs réfringens des fluides élastiques', *Annls Chim. Phys.* **31** (1826), 154–81. Read to the Académie des Sciences 10 October 1825.

—— 'Recherches sur la chaleur spécifique des fluides élastiques', *Annls Chim. Phys.* **41** (1829), 113–59. Read to the Académie des Sciences 18 May 1828.

—— 'Mémoire sur la chaleur animale', *Annls Chim. Phys.*, 3rd ser. **1** (1841), 440–55. Read to the Académie des Sciences 2 December 1822.

—— and J. J. Berzelius, 'Nouvelles déterminations des proportions de l'eau et de la densité de quelques fluides élastiques', *Annls Chim. Phys.* **15** (1820), 386–95.

—— and A. T. Petit, 'Recherches sur les lois de dilatation des solides, des liquides et des fluides élastiques, et sur la mesure exacte des températures', *Annls Chim. Phys.* **2** (1816), 240–63. Read to the First Class of the Institute 29 May 1815.

—— 'Recherches sur la mesure des températures et sur les lois de la communication de la chaleur', *Annls Chim. Phys.* **7** (1818), 113–54, 225–64, and 337–67.

—— *et al.* (on behalf of a commission consisting of Prony, Ampère, Girard, and himself), 'Rapport supplémentaire concernant les mesures de sûreté relatives à l'emploi des machines à feu', *Annls Chim. Phys.* **27** (1824), 95–101.

—— 'Exposé des recherches faites par ordre de l'Académie . . . pour déterminer les forces élastiques de la vapeur d'eau à de hautes températures', *Annls Chim. Phys.* **43** (1830), 74–111. Read to the Académie des Sciences 30 November 1829.

—— See also Gay-Lussac and Dulong; Petit and Dulong.

J. B. A. Dumas, *Traité de chimie appliquée aux arts* (8 vols., Paris and Brussels, 1828–46).

—— *Leçons sur la philosophie chimique. Professées au Collège de France. Recueillies par M. Binau* (Paris, 1837). This edition was published by Bechet Jeune. Another edition, with identical pagination but undated, was published in Paris by Ébrard. In the latter the name of the recorder of the lectures was given as M. Bineau.

—— 'Mémoire sur quelques points de la théorie atomistique', *Annls Chim. Phys.* **33** (1826), 337–91.

—— 'Sur la densité de la vapeur du phosphore', *Annls Chim. Phys.* **49** (1832), 210–14.

—— 'Dissertation sur la densité de la vapeur de quelques corps simples', *Annls Chim. Phys.* **50** (1832), 170–81.

—— 'Éloge historique de Henri-Victor Regnault', *Mém. Acad. Sci. Inst. Fr.* **42** (1883), pp. xxxvii–lxxv. Read 14 March 1881.

F. P. C. Dupin, *Géométrie et mécanique des arts et métiers et des beaux-arts* (3 vols., Brussels, 1825–6). A Paris edition was published in 1826, also in three volumes.

—— *et al.* (on behalf of a commission consisting of Laplace, Prony, Ampère, Girard, and himself), 'Rapport sur les machines à vapeur à haute pression', *Procès-verbaux*, vol. 7, pp. 469–79 (14 April 1823).

J. B. EMMETT, 'On the chemical phenomena of heat', *Ann. Phil.* **9** (1817), 421–30.

J. C. P. ERXLEBEN, *Anfangsgründe der Naturlehre entworsen . . . Fünfte Auflage. Mit Zusatzen von G. C. Lichtenberg* (Göttingen, 1791).

—— 'Legem vulgarem, secundum quam calor corporum certo temporis intervallo crescere vel decrescere dicitur, ad examen revocat J. C. P. Erxleben', *Novi commentarii societatis regiae scientiarum Gottingensis*, commentationes physicae et mathematicae classis, **8** (1778), 74–95.

L. EULER, *Leonhardi Euleri opera omnia*, especially 2nd ser. vol. 13 (Zürich, 1956), and 3rd ser. vol. 1 (Leipzig and Berlin, 1926).

[ANON.], 'Expériences du Docteur Black, sur la marche de la chaleur dans certaines circonstances', in Rozier's *Introduction aux observations sur la physique* **2** (1777), 428–31; first published in September 1772.

M. FARADAY, 'Experimental researches in electricity—seventh series', *Phil. Trans. R. Soc.* **124** (1834), 77–122. Dated 31 December 1833.

E. G. FISCHER, *Physique mécanique*, trans. J. B. Biot (3rd edn, Paris, 1819; 4th edn, Paris, 1830).

J. D. FORBES, *A review of the progress of mathematical and physical science in more recent times, and particularly between the years 1775 and 1850* (Edinburgh, 1858).

A. F. FOURCROY, *Leçons élémentaires d'histoire naturelle et de chimie* (2 vols., Paris, 1782); also the second edition of this work, entitled *Élémens d'histoire naturelle et de chimie* (2 vols., Paris, 1786), and the third edition (5 vols., Paris, 1789).

—— *Système des connaissances chimiques, et de leurs applications aux phénomènes de la nature et de l'art* (6 vols., Paris, an 9–an 10 [1801–2]).

—— See also GUYTON DE MORVEAU, LAVOISIER, BERTHOLLET, and FOURCROY.

J. B. J. FOURIER, *Œuvres de Fourier*, ed. G. Darboux (2 vols., Paris, 1888–90).

—— *Théorie analytique de la chaleur* (Paris, 1822).

—— 'Éloge historique de M. le Marquis de Laplace', *Mém. Acad. Sci. Inst. Fr.* **10** (1831), pp. lxxi–cii. Read 15 June 1829.

—— A. M. AMPÈRE and D. F. J. ARAGO, 'Rapport fait à l'Académie sur un mémoire de Fresnel, relatif à la double réfraction', *Annls Chim. Phys.* **20** (1822), 337–44.

A. J. FRESNEL, *Œuvres complètes d'Augustin Fresnel*, ed. É. Verdet (3 vols., Paris, 1866–70).

—— 'Mémoire sur la diffraction de la lumière, où l'on examine particulièrement le phénomène des franges colorées que présentent les ombres des corps éclairés par un point lumineux', *Annls Chim. Phys.* **1** (1816), 239–81.

J. GADOLIN, *Dissertatio chymico-physica, de theoria caloris corporum specifici* (Abo, 1784).

—— 'Versuche und Bemerkungen über der Körper absolute Wärme', *Der Konigl. Schwedischen Akademie der Wissenschaften neue Abhandlungen* **5** (1784; published Leipzig, 1786), 222–39.

J. GADOLIN, 'Disquisitio de theoria caloris corporum specifici', *Nova Acta R. Soc. Scient. upsal.* **5** (1792), 1–49.

T. GARNETT, *Outlines of a course of lectures on chemistry* (Liverpool, 1797).

—— *Outlines of a course of lectures on chemistry: delivered at the Royal Institution of Great Britain, 1801* (London, 1801)

—— *Popular lectures on zoonomia, or the laws of animal life in health and disease* (London, 1804).

J. L. GAY-LUSSAC, 'Recherches sur la dilatation des gaz et des vapeurs', *Annls Chim.* **43** (1802), 137–75. Read to the First Class of the Institute 31 January 1802.

—— 'Relation d'un voyage aérostatique fait par M. Gay-Lussac le 29 fructidor an 12', *Annls Chim.* **52** (1804), 75–94. Read to the First Class of the Institute 1 October 1804.

—— 'Premier essai pour déterminer les variations de température qu'éprouvent les gaz en changeant de densité, et considérations sur leur capacité pour le calorique', *Mém. Phys. Chim. Soc. Arcueil* **1** (1807), 180–203. Read to the First Class of the Institute 15 September 1806.

—— 'Mémoire sur la combinaison des substances gazeuses, les unes avec les autres', *Mém. Phys. Chim. Soc. Arcueil* **2** (1809), 207–34. Read to the Société Philomathique 31 December 1808.

—— 'Extrait d'un mémoire sur la capacité des gaz pour le calorique', *Annls Chim.* **81** (1812), 98–108. Read to the First Class of the Institute 10 January 1812.

—— 'Note sur la capacité des fluides élastiques pour le calorique', *Annls Chim.* **83** (1812), 106–8.

—— 'Sur le calorique des combinaisons', *Annls Chim. Phys.* **1** (1816), 214–16.

—— 'Sur le froid produit par la dilatation des gaz', *Annls Chim. Phys.* **9** (1818), 305–10.

—— 'Sur le calorique du vide', *Annls Chim. Phys.* **13** (1820), 304–8.

—— 'Extrait d'un mémoire sur le froid produit par l'évaporation des liquides', *Annls Chim. Phys.* **21** (1822), 82–92. Read to the First Class of the Institute 6 March 1815.

—— and P. L. DULONG, 'Rapport fait à l'Académie ... sur un mémoire de M. Dumas qui a pour objet plusieurs points de la théorie atomistique', *Annls Chim. Phys.* **34** (1827), 326–31.

—— and J. J. WELTER, 'Sur la dilatation de l'air', *Annls Chim. Phys.* **19** (1822), 436–7. Read to the Académie des Sciences 29 April 1822.

—— See also HUMBOLDT and GAY-LUSSAC.

S. GERMAIN, *Œuvres philosophiques de Sophie Germain suivies de pensées et de lettres inédites et précédées d'une notice sur sa vie et ses œuvres par Hte Stupuy* (Paris, 1879).

D. GILBERT (formerly GIDDY), 'On the expediency of assigning specific names to all such functions of simple elements as represent definite physical properties; with the suggestion of a new term in mechanics; illustrated by an investigation of the machine moved by recoil, and also by some observations on the steam engine', *Phil. Trans. R. Soc.* **117** (1827), 25–38. Read 25 January 1827.

P. S. GIRARD, Review of Sadi Carnot's *Réflexions sur la puissance motrice du feu*, in *Revue encyclopédique* **23** (1824), 411–14.

—— See also DULONG *et al.*; DUPIN *et al.*

L. GMELIN, *Handbuch der Chemie* (10 vols., Heidelberg, 1843–70).

J. GORHAM, *The elements of chemical science* (2 vols., Boston, 1819–20).

T. GRAHAM, *Elements of chemistry, including the applications of the science in the arts* (1st edn, London, 1842; 2nd edn, 2 vols., London, 1850–8).

—— 'On the molecular mobility of gases', *Phil. Mag.*, 4th ser. **26** (1863), 409–34.

W. J. S. VAN 'SGRAVESANDE, *Physices elementa mathematica, experimentis confirmata: sive introductio ad philosophiam Newtonianam* (2 vols., Leiden, 1720–1).

—— *Mathematical elements of natural philosophy, confirmed by experiments; or, an introduction to Isaac Newton's philosophy*, trans. J. T. Desaguliers (1st edn, 2 vols., London, 1720–1; 2nd edn, 2 vols., London, 1726; 5th edn, 2 vols., London, 1737).

—— *Philosophiae Newtonianae institutiones, in usus academicos* (Leiden, 1723).

G. GREGORY, *The economy of nature explained and illustrated on the principles of modern philosophy* (1st edn, 3 vols., London, 1796; 2nd edn, 3 vols., London, 1798; 3rd edn, 3 vols., London, 1804).

F. A. C. GREN, *Systematisches Handbuch der gesammten Chemie* (4 vols., Halle, 1787–96).

A. GUENYVEAU, *Essai sur la science des machines. Des moteurs, des roues hydrauliques, des machines à colonne d'eau, du bélier hydraulique, des machines à vapeur et des animaux* (Lyons, 1810).

L. B. GUYTON DE MORVEAU, 'Essai sur la dilatabilité de l'air et des gaz par la chaleur, et la nécessité de la déterminer avec exactitude pour perfectionner la méthode de réduction des volumes de ces fluides aux volumes qu'ils auroient à une température donnée', *Annls Chim.* **1** (1790), 256–99.

—— A. L. LAVOISIER, C. L. BERTHOLLET, and A. F. FOURCROY, *Méthode de nomenclature chimique* (Paris, 1787); also the English translation, *Method of chymical nomenclature*, by James St John (London, 1788).

J. N. P. HACHETTE, *Correspondance sur l'École Impériale [Royale] Polytechnique* (3 vols., Paris, 1808–16). This was edited by Hachette.

—— *Programmes d'un cours de physique; ou précis de leçons sur les principaux phénomènes de la nature, et sur quelques applications des mathématiques à la physique* (Paris, 1809).

—— *Traité élémentaire des machines* (1st edn, Paris, 1811; 2nd edn, Paris, 1819; 4th edn, Paris, 1828). A version of the first edition which was published in 1811 in Paris and St Petersburg also exists. The pagination is identical to that of the Paris edition.

—— 'Méthode pratique pour comparer les effets des machines à vapeur', *Bull. Soc. Encour. Ind. natn.*, XVIIe année (1818), 169–74.

—— 'Sur les machines à vapeur', *Bull. Soc. Encour. Ind. natn.*, XVIIIe année (1819), 254–5.

A. HALDAT, 'Recherches sur la chaleur produite par le frottement', *J. Phys.* **65** (1807), 213–22.

S. HALES, *Vegetable staticks: or, an account of some statical experiments on the sap in vegetables* (2 vols., London, 1727).

E. HALLEY, 'A discourse of the rule of the decrease of the height of the mercury in the barometer, according as places are elevated above the surface of the earth . . .', *Phil. Trans. R. Soc.* **16** (1686), 104–16.

J. B. DU HAMEL, *Regiae scientarum academiae historia* (Paris, 1698).

S. HAUGHTON, 'Remarks on Professor Potter's theory of sound', *Phil. Mag.*, 4th ser. **1** (1851), 332–4. Dated 22 March 1851.

F. HAUKSBEE, *Physico-mechanical experiments on various subjects. Containing an account of several surprizing phenomena touching light and electricity, producible on the attrition of bodies* (London, 1709).

R. J. HAÜY, *Traité élémentaire de physique* (1st edn, 2 vols., Paris, 1803; 2nd edn, 2 vols., Paris, 1806; 3rd edn, 2 vols., Paris, 1821).

W. T. HAYCRAFT, 'On the specific heat of the gases', *Trans. R. Soc. Edinb.* **10** (1824), 195–216. Read 3 November 1823.

H. VON HELMHOLTZ, *Über die Erhaltung der Kraft, eine physikalische Abhandlung* (Berlin, 1847). Read before the Physikalischen Gesellschaft of Berlin 23 July 1847.

W. HENRY, *An epitome of chemistry* (1st edn, London, 1801; 5th edn, London, 1808; 6th edn, Edinburgh, 1806). The fifth and sixth editions were entitled *An epitome of experimental chemistry*.

—— 'A review of some experiments which have been supposed to disprove the materiality of heat', *Mem. Proc. Manch. lit. phil. Soc.* **5**, part 2 (1802), 603–21. Read 5 June 1801.

—— 'Experiments on the quantity of gases absorbed by water at different temperatures, and under different pressures', *Phil. Trans. R. Soc.* **93** (1803), 29–42. Read 23 December 1802.

—— 'Illustrations of Mr. Dalton's theory of the constitution of mixed gases. In a letter from Mr. Wm. Henry of Manchester, to Mr. Dalton', *Nicholson's Journal*, 2nd ser. **8** (1804), 297–301. Dated 20 June 1804.

J. HERAPATH, *Mathematical physics; or the mathematical principles of natural philosophy: with a development of the causes of heat, gaseous elasticity, gravitation, and other great phenomena of nature* (2 vols., London, 1847).

—— 'Observations on M. Laplace's communication to the Royal Academy of Sciences, "Sur l'attraction des sphères, et sur la répulsion des fluides élastiques" ', *Phil. Mag.* **62** (1823), 61–2 and 136–9. Dated 19 July and 19 August 1823.

—— 'On the caloric of gases and vapours, by M. Poisson . . . with observations by John Herapath, Esq.', *Phil. Mag.* **62** (1823), 328–38. Dated 15 October 1823.

—— 'On the velocity of sound and variation of temperature and pressure in the atmosphere', *Q. Jl Sci.*, 2nd ser. **7** (1830), 167–75.

HERO (of Alexandria), *Heronis Alexandrini opera quae supersunt omnia*, ed. and trans. W. Schmidt (5 vols., Leipzig, 1899–1914).

R. HERON, *Elements of chemistry: comprehending all the most important facts and principles in the works of Fourcroy and Chaptal with the addition of the more recent chemical discoveries which have been made known in Britain and on the Continent* (London, 1800).

A. M. HÉRON DE VILLEFOSSE, *De la richesse minérale. Considérations sur les mines, usines et salines des différens états, présentées comparativement* (3 vols. and atlas, Paris, 1819). An earlier printing of the first volume, dating from 1810, also exists.

J. F. W. HERSCHEL, 'Light', in *Encyclopaedia metropolitana*, vol. 4 (2nd edn, London, 1845), pp. 341–85.

W. HERSCHEL, 'Experiments on the solar, and on the terrestrial rays that occasion heat; with a comparative view of the laws to which light and heat, or rather the rays which occasion them are subject, in order to determine whether they are the same, or different', *Phil. Trans. R. Soc.* **90** (1800), 293–326 and 437–538. Read 15 May and 6 November 1800.

B. HIGGINS, *A syllabus, of chemical and philosophical enquiries composed for the use of the noblemen and gentlemen who have subscribed to the proposals made, for the advancement of natural knowledge* (London, n.d. [1775]).

—— *A syllabus of the discourses and experiments, with which the meetings of the subscribers are to be opened, after the course of chemistry is concluded* (London, n.d. [1775]).

—— *A philosophical essay concerning light* (London, 1776), in which the syllabus described in the previous entry is reproduced (on pp. ix–liii).

—— *Experiments and observations relating to acetous acid, fixable air, dense inflammable air, oils, and fuel; the matter of fire and light, metallic reproduction, combustion, fermentation, putrefaction, respiration, and other subjects of chemical philosophy* (London, 1786).

W. HIGGINS, *A comparative view of the phlogistic and antiphlogistic theories. With inductions. To which is annexed, an analysis of the human calculus, with observations on its origin, etc.* (2nd edn, London, 1791).

P. DE LA HIRE, 'Expériences et remarques sur la dilatation de l'air par l'eau boüillante', *Mém. Acad. Sci.* (1708), pp. 274–88.

C. H. A. HOLTZMANN, 'On the heat and elasticity of gases and vapours', *Taylor's scientific memoirs*, vol. 4 (1846), pp. 189–217; translated from a pamphlet published at Manheim in 1845.

J. G. HOYAU, 'Description de la machine à vapeur à haute pression, importée en France par M. Humphrey Edwards', *Bull. Soc. Encour. Ind. natn.*, XXVIIe année (1818), 365–84.

A. HUMBOLDT and J. L. GAY-LUSSAC, 'Expériences sur les moyens eudiométriques et sur la proportion des principes constituants de l'atmosphère', *J. Phys.* **60** (1805), 129–68. Read to the First Class of the Institute 21 January 1805.

C. HUTTON, *A mathematical and philosophical dictionary: containing an explanation of the terms, and an account of the several subjects, comprised under the heads mathematics, astronomy, and philosophy both natural and experimental* . . . (1st edn, 2 vols., London, 1795–6; 2nd edn,

2 vols., London, 1815). In the second edition the title was changed to *A philosophical and mathematical dictionary*. . . .

J. IMISON, *Elements of science and art: being a familiar introduction to natural philosophy and chemistry. Together with their application to a variety of elegant and useful arts*, a new edition by Thomas Webster (2 vols., London, 1808).

W. IRVINE, *Essays, chiefly on chemical subjects, by the late William Irvine, M.D. F.R.S.Ed., and by his son William Irvine, M.D.* (London, 1805).

W. IRVINE (Jr), 'A letter . . . concerning the late Dr. Irvine of Glasgow, his doctrine, which ascribes the disappearance of heat, without increase of temperature, to a change of capacity in bodies, and that of Dr. Black, which supposes caloric to become latent by chemical combination with bodies, with particular remarks on the mistakes of Dr. Thompson, in his accounts of these doctrines', *Nicholson's Journal*, 2nd ser. **6** (1803), 25–31.

—— 'Experimental determinations of the latent heat of spermaceti, bee's wax, tin, bismuth, lead, zinc, and sulphur', *Nicholson's Journal*, 2nd ser. **9** (1804), 45–52. Dated 24 August 1804.

—— 'Letter of inquiry respecting the late Dr. Irvine's fundamental experiment on the relative capacities for heat of ice and water. With an answer by Mr. Irvine', *Nicholson's Journal*, 2nd ser. **11** (1805), 50–2.

J. IVORY, 'Remarks on the gradation of heat in the atmosphere', *Phil. Mag.* **58** (1821), 24–31. Dated 5 July 1821.

—— 'On the laws of the condensation and dilatation of air and the gases, and the velocity of sound', *Phil. Mag.* **66** (1825), 3–13. Dated 4 July 1825.

—— 'Investigation of the heat extricated from air when it undergoes a given condensation', *Phil. Mag.*, 2nd ser. **1** (1827), 89–94. Dated 4 January 1827.

—— 'Continuation of the subject relating to the absorption and extrication of heat in a mass of air that changes its volume', *Phil. Mag.*, 2nd ser. **1** (1827), 165–70. Dated 5 February 1827.

—— 'Answer to an article by Mr. Henry Meikle published in No. vii of the Quarterly Journal of Science', *Phil. Mag.*, 2nd ser. **4** (1828), 321–6. Dated 13 October 1828.

—— 'Some observations on Mr. Meikle's reply, published in the last number of the Quarterly Journal of Science', *Phil. Mag.*, 2nd ser. **5** (1829), 104–6. Dated 13 January 1829.

G. JARS, 'Description d'une nouvelle machine exécutée aux mines de Schemnitz en Hongrie, au mois de Mars 1755', *Mém. prés. div. Sav. Acad. Sci.* **5** (1768), 67–71.

J. JOLY, 'On the specific heats of gases at constant volume. Part I. Air, carbon dioxide, and hydrogen', *Proc. R. Soc.* **48** (1890), 440–1. Received 2 September 1890.

J. P. JOULE, *The scientific papers of James Prescott Joule* (2 vols., London, 1887).

—— 'On a new class of magnetic forces', in William Sturgeon's *The annals of electricity, magnetism, and chemistry; and guardian of experimental science* **8** (1842), 219–24. Read at the Royal Victoria Gallery, Manchester, 16 February 1841.

J. P. JOULE, 'On the calorific effects of magneto-electricity, and on the mechanical value of heat' *Phil. Mag.*, 3rd ser. **23** (1843), 263–76, 347–55, and 435–43.

—— 'On the changes of temperature produced by the rarefaction and condensation of air', *Phil. Mag.*, 3rd ser. **26** (1845), 369–83. Dated June 1844.

—— 'On the heat evolved during the electrolysis of water', *Mem. Proc. Manch. lit. phil. Soc.*, 2nd ser. **7** (1846), 87–112. Dated 24 January 1843, with an appendix dated 20 February 1844.

—— 'On the theoretical velocity of sound', *Phil. Mag.*, 3rd ser. **31** (1847), 114–15.

—— 'Expériences sur l'identité entre le calorique et la force mécanique. Détermination de l'équivalent par la chaleur dégagée pendant la friction du mercure', *C.r. hebd. Séanc. Acad. Sci., Paris* **25** (1847), 309–11. Read 23 August 1847.

—— 'Some remarks on heat, and the constitution of elastic fluids', *Mem. Proc. Manch. lit. phil. Soc.*, 2nd ser. **9** (1851), 107–14. Read 3 October 1848. Later republished in *Phil. Mag.*, 4th ser. **14** (1857), 211–16.

—— and W. THOMSON, 'On the thermal effects experienced by air in rushing through small apertures', *Phil. Mag.*, 4th ser. **4** (1852), 481–92. Read at the British Association for the Advancement of Science at Belfast 3 September 1852.

P. KELLAND, *Theory of heat* (Cambridge, 1837).

G. KNIGHT, *An attempt to demonstrate, that all the phaenomena in nature may be explained by two simple active principles, attraction and repulsion wherein the attractions of cohesion, gravity, and magnetism, are shewn to be one and the same; and the phaenomena of the latter are more particularly explained* (London, 1748).

A. K. KRÖNIG, 'Grundzüge einer Theorie der Gase', *Pogg. Ann.* **99** (1856), 315–22. Dated June 1856.

J. L. LAGRANGE, *Méchanique analitique* (Paris, 1788).

—— 'Nouvelles recherches sur la nature et la propagation du son', *Mélang. Phil. Math. Soc. R. Turin* **2** (1760–1), 11–172.

J. H. LAMBERT, *Pyrometrie oder vom Maasse des Feuers und der Wärme* (Berlin, 1779).

—— 'Tentamen de vi caloris, qua corpora dilatat, eiusque dimensione', *Acta helvet.* **2** (1755), 172–242.

—— 'Sur la vitesse du son', *Hist. Acad. Sci. Berlin* **24** (1768), 70–9.

G. LAMÉ, *Cours de physique de l'École Polytechnique* (2 vols., Paris, 1836–7).

M. LANDRIANI, 'Dissertation sur la chaleur latente', *J. Phys.* **26** (1785), 88–100 and 197–207.

P. S. LAPLACE, *Exposition du système du monde* (1st edn, 2 vols., Paris, 1796; 2nd edn, 2 vols., Paris, 1799).

—— *Traité de mécanique céleste* (5 vols., Paris, 1799–1825).

—— 'Sur la vitesse du son dans l'air et dans l'eau', *Annls Chim. Phys.* **3** (1816), 238–41. Read to the Académie des Sciences 23 December 1816.

P. S. LAPLACE, 'Sur l'attraction des sphères, et sur la répulsion des fluides élastiques', *Connaissance des tems . . . pour l'an 1824* (Paris, 1821), pp. 328–43. Read to the Académie des Sciences 10 September 1821. A shortened version appeared in *Annls Chim. Phys.* **18** (1821), 181–90.

——— 'Éclaircissemens de la théorie des fluides élastiques', *Annls Chim. Phys.* **18** (1821), 273–80.

——— 'Considérations sur l'attraction des corps sphériques, et sur la répulsion des fluides élastiques', *J. Phys.* **94** (1822), 84–90.

——— 'Développement de la théorie des fluides élastiques, et application de cette théorie à la vitesse du son', *Connaissance des tems . . . pour l'an 1825* (Paris, 1822), pp. 219–27. Presented to the Bureau des Longitudes 12 December 1821. A continuation of this memoir and an addition to it appeared on pp. 302–23 and 386–7 of the same volume.

——— 'Sur la vitesse du son', *Connaissance des tems . . . pour l'an 1825* (Paris, 1822), pp. 371–2.

——— *See also* DUPIN *et al.*; LAVOISIER and LAPLACE.

A. L. LAVOISIER, *Traité élémentaire de chimie. Présenté dans un ordre nouveau et d'après les découvertes modernes* (2 vols., Paris, 1789).

——— *Mémoires de chimie*, ed. Mme Lavoisier (Paris, 1805). Only volumes 1 and 2 and part of volume 4 were published.

——— 'De la combinaison de la matière du feu avec les fluides évaporables, et de la formation des fluides élastiques aëriformes', *Mém. Acad. Sci.* (1777), pp. 420–32. Dated 5 September 1777.

——— 'Mémoire sur la combustion en général', *Mém. Acad. Sci.* (1777), pp. 592–600. Dated 5 September 1777.

——— and P. S. LAPLACE, Mémoire sur la chaleur', *Mém. Acad. Sci.* (1780), pp. 355–408. Read 18 June 1783.

——— 'Troisième mémoire servant de supplément au précédent, et contenant les expériences faites sur la chaleur, pendant l'hiver de 1783 à 1784', in Lavoisier, *Mémoires de chimie*, vol. 1, pp. 121–47.

——— *See also* GUYTON DE MORVEAU, LAVOISIER, BERTHOLLET, and FOURCROY.

G. L. LE SAGE, *Essai de chymie méchanique* (Geneva, n.d. [1758]).

J. LESLIE, *An experimental inquiry into the nature, and propagation, of heat* (London, 1804).

——— 'Dissertation fourth exhibiting a general view of the progress of mathematical and physical science chiefly during the eighteenth century', *Encyclopaedia Britannica*, vol. 1 (7th edn, Edinburgh, 1842), pp. 573–677. The dissertation also appeared in the first volume of the eighth edition (Edinburgh, 1860).

[G. B. I. T. LIBRI], 'Lettres à un Américain sur l'état des sciences en France', *Revue des deux mondes*, 4th ser. **21** (1840), 789–818, **22** (1840), 532–54, and **23** (1840), 410–37.

J. LIEBIG, *Animal chemistry, or organic chemistry in its applications to physiology and pathology*, ed. W. Gregory (London, 1842).

——— *Chemische Briefe* (2nd edn, Heidelberg, 1845).

J. LIND, 'Description and use of a portable wind gage', *Phil. Trans. R. Soc.* **65** (1775), 353–65.

F. LUNN, 'Heat', in *Encyclopaedia metropolitana*, vol. 4 (2nd edn, London, 1845), pp. 225–340.

J. F. LUZ, *Vollständige und auf Erfahrung gegründete Beschreibung* (Frankfurt and Leipzig, 1784).

P. J. MACQUER, *Élémens de chymie-théorique* (Paris, 1749).

—— *Dictionnaire de chymie, contenant la théorie et la pratique de cette science* (1st edn, 2 vols., Paris, 1766; 2nd edn, 4 vols., Paris, 1778).

J. H. DE MAGELLAN, *Essai sur la nouvelle théorie du feu élémentaire, et de la chaleur des corps* (London, 1780); reprinted, with only minor modifications, in Rozier's *Observations sur la physique* **17** (1781), 375–86 and 411–22.

G. MAGNUS, 'Ueber die Ausdehnung der Gase durch die Wärme', *Pogg. Ann.* **55** (1842), 1–27. Read to the Berlin Academy of Science 25 November 1840.

—— 'Ueber die Ausdehnung der atmosphärischen Luft in höheren Temperaturen', *Pogg. Ann.* **57** (1842), 177–99.

J. J. D. DE MAIRAN, 'Nouvelles recherches sur la cause générale du chaud en été et du froid en hiver, en tant qu'elle se lie à la chaleur interne et permanente de la terre . . .', *Mém. Acad. Sci.* (1765), pp. 143–266.

F. MARCET, see DE LA RIVE and MARCET.

E. MARIOTTE, *Œuvres de Mr. Mariotte, de l'Académie Royale des Sciences . . . comprenant tous les traitez de cet auteur, tant ceux qui n'avoient pas encore ete publiez* (2 vols. in 1, Leiden, 1717).

G. MARTINE, *Essays medical and philosophical* (London, 1740).

J. C. MAXWELL, 'Illustrations of the dynamical theory of gases', *Phil. Mag.*, 4th ser. **19** (1860), 19–32, and **20** (1860), 21–37. Read at the British Association for the Advancement of Science at Aberdeen in September 1859.

—— 'The Bakerian lecture, on the viscosity or internal friction of air and other gases', *Phil. Trans. R. Soc.* **96** (1866), 249–68. Read 8 February 1866 (Received 23 November and 7 December 1865, and 6 February 1866).

J. R. MAYER, *Kleinere Schriften und Briefe von Robert Mayer. Nebst Mittheilungen aus seinem Leben*, ed. J. J. Weyrauch (Stuttgart, 1893).

—— 'Bemerkungen über die Kräfte der unbelebten Natur', *Annalen der Chemie und Pharmacie* (eds. F. Wöhler and J. Liebig) **42** (1842), 233–40.

J. T. MAYER (the younger), *Physicalisch-mathematische Abhandlung über das Ausmessen der Wärme in Rucksicht und deren Anwendung auf das Höhenmessen vermittelst des Barometers* (Frankfurt and Leipzig, 1786).

—— 'Commentatio de apparentiis obiectorum terrestrium a refractione lucis in atmosphaera nostra pendentibus', *Commentationes societatis regiae scientiarum Gottingensis recentiores* **1**, commentationes mathematicae (1811).

—— 'Commentatio de lege vis elasticae vaporum', *Commentationes societatis regiae scientiarum Gottingensis recentiores* **1**, commentationes mathematicae (1811). Read 10 June 1809.

H. MEIKLE, 'On the theory of the air-thermometer', *Edinb. new phil. J.* **1** (1826), 332–41.

BIBLIOGRAPHY

H. MEIKLE, 'On the law of temperature', *Ann. Phil.*, 2nd ser. **12** (1826), 366–9. Dated 10 September 1826.

—— 'Refutation of Mr. Ivory's new law of the heat extricated from air by condensation', *Edinb. new phil. J.* **3** (1827), 149–57.

—— 'On Mr Ivory's investigations of the velocity of sound', *Q. Jl Sci.*, 2nd ser. **4** (1828), 124–35.

—— 'Reply to Mr. James Ivory's answer in No. XXIII of the Philosophical Magazine and Annals of Philosophy', *Q. Jl Sci.*, 2nd ser. **4** (1828), 315–19.

—— 'On the velocity of sound', *Edinb. new phil. J.* **6** (1828), 26–32.

—— 'On the relation between the density, pressure, and temperature of air; and on experiments regarding the theory of clouds, rain, etc.; and with a conjecture about thunder and lightning', *Q. Jl Sci.*, 2nd ser. **5** (1829), 56–75.

—— 'Refutation of the charges contained in Mr. Ivory's remarks, etc. in the Philosophical Magazine and Annals for February last', *Q. Jl Sci.*, 2nd ser. **5** (1829), 109–13.

M. MERSENNE, *De l'utilité de l'harmonie* (Paris, 1636); a work forming part of Mersenne's *Harmonie universelle*. Consulted by me in the Cambridge University Library copy.

—— *Ballistica et acontismologia* (Paris, 1644).

S. L. METCALFE, *Caloric. Its mechanical chemical and vital agencies in the phenomena of nature* (2 vols., London, 1843).

O. E. MEYER, 'Ueber die Reibung der Flüssigkeiten', *Pogg. Ann.* **113** (1861), 55–86, 193–238, and 383–425.

E. MITSCHERLICH, 'Ueber die Krystallisation der Salze, in denen das Metall der Basis mit zwei Proportionen Sauerstoff verbunden ist', *Abhandlungen der physikalischen Klasse der Königlich-Preussischen Akademie der Wissenschaften* (1818–19), pp. 427–37. Read 9 December 1819.

—— 'Sur le rapport de la densité des gaz à leur poids atomique', *Annls Chim. Phys.* **55** (1833), 5–41.

C. F. MOHR, 'Ansichten über die Natur der Wärme', *Annalen der Pharmacie* (ed. J. Liebig) **24** (1837), 141–7.

J. MOLLET, *Mémoire sur deux faits nouveaux, l'inflammation des matières combustibles, et l'apparition d'une vive lumière, obtenues par la seule compression de l'air* (Lyons, 1811). Read to the Académie des Sciences, Belles-Lettres et Arts de Lyon 27 March 1804.

—— 'De la constitution intime des gaz et de leur capacité pour le calorique', *J. Phys.* **90** (1820), 113–30. Read to Académie des Sciences, Belles-Lettres et Arts de Lyon 17 June 1817.

G. MONGE, 'Théorie de la chaleur', *Journal gratuit* **1** (1790), x^e classe (physique), 26–32, 41–4, 49–53, 65–7, and 81–3; reproduced with additions and modifications in *Encyclopédie méthodique. Physique*, vol. 2 (Paris, 1816), pp. 170–4.

—— 'Mémoire sur la cause des principaux phénomènes de la météorologie', *Annls Chim.* **5** (1790), 1–71.

W. MORGAN, *An examination of Dr. Crawford's theory of heat and combustion* (London, 1781).

J. MURRAY, *Elements of chemistry* (2 vols., Edinburgh, 1801).

—— *A system of chemistry* (1st edn, 4 vols., Edinburgh, 1806–7; 2nd edn, 4 vols., Edinburgh, 1809; 3rd edn, 4 vols., Edinburgh, 1814).

P. VAN MUSSCHENBROEK, *Introductio ad philosophiam naturalem* (2 vols., Leyden, 1762). The pagination is continuous.

C. L. M. H. NAVIER, 'Sur la variation de température qui accompagne les changemens de volume des gaz', *Bull. Soc. philomath. Paris*, N.S. **7** (1820), 97–101; also in *Annls Chim. Phys.* **17** (1821), 372–9.

—— 'Note sur l'action mécanique des combustibles', *Annls Chim. Phys.* **19** (1821), 357–72.

F. E. NEUMANN, 'Untersuchung über die specifische Wärme der Mineralien', *Pogg. Ann.* **23** (1831), 1–39.

I. NEWTON, *Philosophiae naturalis principia mathematica* (1st edn, London, 1687; 2nd edn, Cambridge, 1713; 3rd edn, London, 1726).

—— *Opticks: or, a treatise of the reflections, refractions, inflections and colours of light* (1st edn, London, 1704; 4th edn, London, 1730).

W. NICHOLSON, *An introduction to natural philosophy* (1st edn, 2 vols., London, 1782; 2nd edn, 2 vols., London, 1787; 3rd edn, 2 vols., London, 1790; 4th edn, 2 vols., London, 1796; 5th edn, 2 vols., London, 1805).

—— *The first principles of chemistry* (1st edn, London, 1790; 2nd edn, London, 1792; 3rd edn, London, 1796).

—— 'Flash from an air-gun', *Nicholson's Journal*, 2nd ser. **4** (1803), 280.

J. A. NOLLET, *Leçons de physique expérimentale* (5 vols., 1745–55). The first three volumes were published in Amsterdam, the last two in Paris.

—— 'Sur la vapeur qu'on aperçoit dans le récipient d'une machine pneumatique, lorsqu'on commence à raréfier l'air qu'il contient', *Mém. Acad. Sci.* (1740), pp. 243–53.

—— 'Mémoire sur les instruments qui sont propres aux expériences de l'air', *Mém. Acad. Sci.* (1740), pp. 385–432, 567–85, and 338–62. Dated 16 June 1741.

T. NORTHMORE, 'Experiments on the remarkable effects which take place in the gases, by change in their habitudes, or elective attractions, when mechanically compressed', *Nicholson's Journal*, 2nd ser. **12** (1805), 368–74. Dated 17 December 1805.

[ANON.] 'Nouveaux thermomètres métalliques de MM. Bréguet', *Annls Chim. Phys.* **5** (1817), 312–15.

L. NUGUET, 'Nouvelles expériences pour déterminer de combien précisément l'air se dilate par la chaleur de l'eau boüillante, avec des remarques sur le thermometre de Mr. Amontons de l'Académie des Sciences', *Mém. Hist. Sci.* (1705), pp. 1790–1807.

'II'. 'Miscellaneous information . . . on the supposed determination of the real zero of heat', *Nicholson's Journal*, 2nd ser. **4** (1803), 220–4. Dated 17 February 1803.

J. C. É. PÉCLET, *Traité élémentaire de physique* (3rd edn, 2 vols., Paris, 1838).

A. T. PETIT, 'Théorie mathématique de l'action capillaire', *J. Éc. polytech.* **9**, cahier 16 (1813), 1–40.

A. T. PETIT, 'Sur l'emploi du principe des forces vives dans le calcul de l'effet des machines', *Annls Chim. Phys.* **8** (1818), 287–305.

—— and P. L. DULONG, 'Sur quelques points importans de la théorie de la chaleur', *Annls Chim. Phys.* **10** (1819), 395–413. Read to the Académie des Sciences 12 April 1819.

—— *See also* ARAGO and PETIT; DULONG and PETIT.

M. A. PICTET, *Essais de physique* (Geneva, 1790).

J. PLAYFAIR, *Outlines of natural philosophy, being heads of lectures delivered in the University of Edinburgh* (2 vols., Edinburgh, 1812–14).

S. D. POISSON, *Théorie mathématique de la chaleur* (Paris, 1835).

—— 'Mémoire sur la théorie du son', *J. Éc. polytech.* **7**, cahier 14 (1808), 319–92. Read to the First Class of the Institute 17 August 1807.

—— 'Sur la vitesse du son', *Connaissance des tems . . . pour l'an 1826* (Paris, 1823), pp. 257–77. Part of the article also appeared in *Annls Chim. Phys.* **23** (1823), 5–16.

—— 'Sur la chaleur des gaz et des vapeurs', *Annls Chim. Phys.* **23** (1823), 337–52.

—— 'Mémoire sur les équations générales de l'équilibre et du mouvement des corps solides élastiques et des fluides', *J. Éc. polytech.* **13**, cahier 20 (1831), 1–174. Read to the Académie des Sciences 12 October 1829.

G. POLENI, 'Epistola qua continetur summarium observationum meteorologicarum per sexennium Patavii habitarum', *Phil. Trans. R. Soc.* **37** (1731–2), 201–16.

R. POTTER, 'The solution of the problem of sound, founded on the atomic constitution of fluids', *Phil. Mag.*, 4th ser. **1** (1851), 101–4. Dated 28 December 1850.

C. S. M. POUILLET, *Éléments de physique expérimentale et de météorologie* (2 vols., Paris and Brussels, 1829–30).

P. PRÉVOST, 'Discussion de quelques expériences relatives à l'influence de la densité sur la chaleur spécifique des gaz', *Mém. Soc. Phys. Hist. nat. Genève* **4** (1828), 255–64. Read 4 September 1828.

J. PRIESTLEY, *Experiments and observations on different kinds of air* (3 vols., London, 1774–7).

—— *Experiments and observations relating to various branches of natural philosophy; with a continuation of the observations on air* (3 vols., 1779–86). The first volume was published in London, the other two in Birmingham.

—— *Scientific correspondence of Joseph Priestley*, ed. H. C. Bolton (New York, 1892).

—— 'Observations on different kinds of air', *Phil. Trans. R. Soc.* **62** (1772), 147–264. Read 5, 12, 19, and 26 March 1772.

—— 'An account of further discoveries in air . . . in letters to Sir John Pringle, Bart. F.R.S. and the Rev. Dr. Price, F.R.S.', *Phil. Trans. R. Soc.* **65** (1775), 384–94. The letters are dated 15 March, 1 April, and 25 May 1775.

See also SCHOFIELD in list of secondary sources.

R. DE PRONY, *Nouvelle architecture hydraulique, contenant l'art d'élever l'eau au moyen de différentes machines, de construire dans ce fluide, de le diriger, et généralement de l'appliquer, de diverses manières, aux besoins de la société* (2 vols., Paris, 1790–6).

—— See also DUPIN et al.

W. PROUT, *Chemistry meteorology and the function of digestion considered with reference to natural theology* (London, 1834); the eighth Bridgewater Treatise.

—— 'On the relation between the specific gravities of bodies in the gaseous state and the weights of their atoms', *Ann. Phil.* **6** (1815), 321–30, and **7** (1816), 111–13; also the extract by P. L. Dulong in *Annls Chim. Phys.* **1** (1816), 411–16.

W. J. M. RANKINE, 'On Laplace's theory of sound', *Phil. Mag.*, 4th ser. **1** (1851), 225–7. Dated February 1851.

—— 'On the centrifugal theory of elasticity, as applied to gases and vapours', *Phil. Mag.*, 4th ser. **2** (1851), 509–42.

—— 'On the thermo-dynamic theory of steam-engines with dry saturated steam, and its applications to practice', *Phil. Trans. R. Soc.* **149** (1859), 177–92. Dated 27 December 1858. Read 27 January 1859.

F. REECH, *Machine à air d'un nouveau système déduit d'une comparaison raisonnée des systèmes de MM. Ericsson et Lemoine* (Paris, 1854).

H. V. REGNAULT, *Relation des expériences entreprises par ordre de Monsieur le ministre des travaux publics, et sur la proposition de la commission centrale des machines à vapeur, pour déterminer les principales lois et données numériques qui entrent dans le calcul des machines à vapeur* (3 vols., Paris, 1847–70); also published, with identical pagination, as *Mém. Acad. Sci. Inst. Fr.* **21** (1847), 1–767; **26** (1862), pp. iii–x and 3–928; **37**, part 1 (1868), 3–575; and **37**, part 2 (1870), 579–968.

—— 'Recherches sur la chaleur spécifique des corps simples et composés', *Annls Chim. Phys.* **73** (1840), 5–72, and 3rd ser. **1** (1841), 129–207.

—— 'Recherches sur la dilatation des gaz', *Annls Chim. Phys.*, 3rd ser. **4** (1842), 5–63, and **5** (1842), 52–83.

—— 'Note sur la comparaison du thermomètre à air et du thermomètre à mercure', *Annls Chim. Phys.*, 3rd ser. **6** (1842), 370–80.

—— 'Recherches sur les chaleurs spécifiques', *Annls Chim. Phys.*, 3rd ser. **9** (1843), 322–49.

É. RITTER, 'Note sur la constitution physique des fluides élastiques', *Mém. Soc. Phys. Hist. nat. Genève* **11** (1846–8), 99–114. Read 19 June 1845.

J. W. RITTER, *Beweiss dass ein beständiger Galvanismus den Lebensprocess in dem Thierreiche begleite. Nebst neuen Versuchen und Bemerkungen über den Galvanismus* (Weimar, 1798).

—— *Die Begründung der Elektrochemie und Entdeckung der ultravioletten Strahlen von Johann Wilhelm Ritter*, ed. A. Hermann [Ostwalds Klassiker der exacten Wissenschaften, new series, volume 2] (Frankfurt, 1968).

BIBLIOGRAPHY

A. DE LA RIVE and F. MARCET, 'Expériences relatives au froid produit par l'expansion des gaz', *Bibliothèque universelle* **22** (1823), 265–82; also the shortened version in *Annls Chim. Phys.* **23** (1823), 209–16. Read to the Société de Physique et d'Histoire Naturelle de Genève 17 April 1823.

—— 'Recherches sur la chaleur spécifique des gaz', *Annls Chim. Phys.* **35** (1827), 5–34. Read to the Société de Physique et d'Histoire Naturelle de Genève 19 April 1827.

—— 'Nouvelles recherches sur la chaleur spécifique des gaz', *Annls Chim. Phys.* **41** (1829), 78–92. Read to the Société de Physique et d'Histoire Naturelle de Genève 16 April 1829.

—— 'Quelques recherches sur la chaleur spécifique', *Annls Chim. Phys.* **75** (1840), 113–44. Read to la Société de Physique et d'Histoire Naturelle de Genève 17 June 1835.

W. ROY, 'Experiments and observations made in Britain, in order to obtain a rule for measuring heights with the barometer', *Phil. Trans. R. Soc.* **67** (1778), 653–788.

F. RUDBERG, 'Ueber die Ausdehnung der trocknen Luft zwischen 0° und 100 °C', *Pogg. Ann.* **41** (1837), 271–93, and **44** (1838), 119–23.

COUNT RUMFORD (Benjamin Thompson), 'An inquiry concerning the source of heat which is excited by friction', *Phil. Trans. R. Soc.* **88** (1798), 80–102. Read 25 January 1798.

—— 'An inquiry concerning the weight ascribed to heat', *Phil. Trans. R. Soc.* **89** (1799), 179–94. Read 2 May 1799.

—— 'Inquiries concerning the heat developed in combustion, with a description of a new calorimeter', *Nicholson's Journal*, 2nd ser. **32** (1812), 105–25. Read to the First Class of the (French) Institute 24 February 1812.

H. B. DE SAUSSURE, *Voyages dans les Alpes, précédés d'un essai sur l'histoire naturelle des environs de Genève* (4 vols., Neuchâtel, 1779–96).

—— *Essais sur l'hygrométrie* (Neuchâtel, 1783). Two editions, differing in page size and pagination, were published at Neuchâtel in 1783, both by Samuel Fauche. I have used the quarto edition throughout (pp. xviii+367).

C. W. SCHEELE, *The collected papers of Carl Wilhelm Scheele*, trans. L. Dobbin (London, 1931).

A. SEGUIN, 'Observations générales sur le calorique et ses différens effets, et réflexions sur la théorie de MM. Black, Crawford, Lavoisier et de Laplace, sur la chaleur qui se dégage pendant la combustion; avec un résumé de tout ce qui a été fait et écrit jusqu'à ce moment sur ce sujet', *Annls Chim.* **3** (1789), 148–242.

—— 'Second mémoire sur le calorique', *Annls Chim.* **5** (1790), 191–271; also the slightly abbreviated version which appeared in *Encyclopédie méthodique. Chimie*, vol. 2, pp. 699–741.

—— 'Réponse . . . à la lettre de M. De Luc . . .', *J. Phys.* **36** (1790), 417–21.

M. SEGUIN, *De l'influence des chemins de fer et de l'art de les tracer et de les construire* (Paris, 1839).

M. Seguin, 'Note à l'appui de l'opinion émise par M. Joule sur l'identité du mouvement et du calorique', *C.r. hebd. Séanc. Acad. Sci., Paris* **25** (1847), 420–2 (20 September 1847.)

G. Shuckburgh, 'Observations made in Savoy, in order to ascertain the height of mountains by means of the barometer . . .', *Phil. Trans. R. Soc.* **67** (1778), 513–97. Read 8 and 15 May 1777.

—— 'Comparison between Sir George Shuckburgh and Colonel Roy's rules for the measurement of heights with the barometer; in a letter to Col. Roy, F.R.S.', *Phil. Trans. R. Soc.* **68** (1778), 681–8. Dated 20 April 1778. Read 7 May 1778.

A. F. Silvestre, *Rapport général des travaux de la Société Philomathique de Paris depuis le 23 frimaire an VI jusqu'au 30 nivôse an VII* (Paris, 1799).

J. M. Socquet, *Essai sur le calorique* (Paris, 1801).

M. Somerville, *On the connexion of the physical sciences* (London, 1834).

B. Stewart, *The conservation of energy, being an elementary treatise on energy and its laws* (3rd edn, London, 1874).

G. G. Stokes, 'An examination of the possible effect of the radiation of heat on the propagation of sound', *Phil. Mag.*, 4th ser. **1** (1851), 305–17. Dated 6 February and 11 March 1851.

A. K. W. Suerman, 'Responsio ad quaestionem physicam, ab ordine nobilissimo matheseos et philosophiae naturalis in academia Lugduno-Batava, A. MDCCCXXIX propositam: Exponantur ac dijudicentur variae cum observandi tum computandi rationes, quibus quantum vaporis aquei in atmosphaera vel aere quocunque contineatur, determinari possit', in *Annales Lugduno-Batavae, 1829–30* (Leiden, 1831).

—— 'Expériences sur la chaleur spécifique des gaz et de l'air à pressions différentes', *Annls Chim. Phys.* **63** (1836), 315–32.

P. G. Tait, *Sketch of thermodynamics* (1st edn, Edinburgh, 1868; 2nd edn, Edinburgh, 1877).

—— *Heat* (London, 1884).

L. J. Thenard, *Traité de chimie élémentaire, théorique et pratique* (1st edn, 4 vols., Paris, 1813–16; 4th edn, 5 vols., Paris, 1824).

—— 'Sur la lumière produite par la décharge du fusil à vent', *Annls Chim. Phys.* **22** (1823), 436–9.

—— 'Observations sur la lumière qui jaillit de l'air et de l'oxigène par compression', *Annls Chim. Phys.* **44** (1830), 181–8.

T. Thomson, *A system of chemistry* (1st edn, 4 vols., Edinburgh, 1802; 2nd edn, 4 vols., Edinburgh, 1804; 4th edn, 5 vols., Edinburgh, 1810).

—— *An outline of the sciences of heat and electricity* (London, 1830).

W. Thomson (Lord Kelvin), 'On vortex atoms', *Proc. R. Soc. Edinb.* **6** (1866–9), 94–105 (18 February 1867); also in *Phil. Mag.*, 4th ser. **34** (1867), 15–24.

—— 'On the dissipation of energy', *Fortnightly review*, 2nd ser. **51** (1892), 313–21.

—— 'Heat', in *Encyclopaedia Britannica* (9th edn, Edinburgh, 1880), vol. 11, pp. 554–89.

—— See also Joule and Thomson.

A. TILLOCH, 'A brief examination of the received doctrines respecting heat or caloric', *Phil. Mag.* **8** (1800), 70–8, 119–26, and 211–21. Read to the Askesian Society in December 1799.

—— 'An attempt to prove that the matter of heat, like other substances, possesses not only volume but gravity; being a second essay on caloric', *Phil. Mag.* **9** (1801), 158–67. Read to the Askesian Society in November 1800.

T. S. TRAILL, 'Memoir of Dr. Thomas Charles Hope, late Professor of Chemistry in the University of Edinburgh', *Trans. R. Soc. Edinb.* **16** (1849), 419–34. Read 6 June 1847.

T. TREDGOLD, *The steam engine* (London, 1827).

E. TURNER, *Elements of chemistry* (1st edn, London, 1827; 6th edn, London, 1837).

J. TYNDALL, *Fragments of science: a series of detached essays, addresses, and reviews* (6th edn, 2 vols., London, 1879).

A. URE, *A dictionary of chemistry* (London, 1820).

A. VANDERMONDE, C. L. BERTHOLLET, and G. MONGE, 'Mémoire sur le fer considéré dans ses différens états métalliques', *Mém. Acad. Sci.* (1786), pp. 132–200. Dated May 1786.

É. VERDET, *Œuvres de É. Verdet*, ed. A. de la Rive (8 vols., Paris, 1868–73).

A. VOLTA, *Le opere di Alessandro Volta* [Edizione Nazionale] (7 vols., Milan, 1918–29).

—— 'Della uniforme dilatazione dell'aria', *Annali Chim.* **4** (1793), 227–94.

—— 'Neue Abhandlung über die thierische Electrizität. In Briefen an Herrn Abbe Anton Maria Vassali', *Neues Journal der Physik* (ed. F. A. C. Gren) **2** (1795), 141–50 and 151–72. Dated June and August 1794.

J. J. WATERSTON, 'On the physics of media that are composed of free and perfectly elastic molecules in a state of motion', *Phil. Trans. R. Soc.* **183** A (1893), 1–79. Received 11 December 1845. Read 5 March 1846. This version includes a brief introduction by Lord Rayleigh and some other information concerning the paper.

R. WATSON, *Chemical essays* (5 vols., Cambridge, 1781–7).

T. WEDGWOOD, 'Experiments and observations on the production of light from different bodies, by heat and by attrition', *Phil. Trans. R. Soc.* **82** (1792), 28–47 and 270–82. Read 22 December 1791 and 10 May 1792.

J. J. WELTER. See GAY-LUSSAC and WELTER.

J. C. WILCKE, 'Über die specifische Menge des Feuers in festen Körpern, und derselben Abmessung', *Der Königl. Schwedischen Akademie der Wissenschaften neue Abhandlungen* **2** (1781; published Leipzig, 1784), 48–79.

N. M. WOLFE, 'Descriptio fontis Hieronis in metallifodinis Chemnicensibus in Hungaria, anno 1756 extructi', *Phil. Trans. R. Soc.* **52** (1762), 547–54. Read 17 June 1762.

W. H. WOLLASTON, 'A synoptic scale of chemical equivalents', *Phil. Trans. R. Soc.* **104** (1814), 1–22. Read 4 November 1813.

T. Young, *A syllabus of a course of lectures on natural and experimental philosophy* (London, 1802).
—— *A course of lectures on natural philosophy and the mechanical arts* (2 vols., London, 1807).

SECONDARY SOURCES

(A) THESES

J. H. Brooke, 'The role of analogical argument in the development of organic chemistry' (University of Cambridge Ph.D. thesis, 1969).

G. R. Talbot, 'Origins and solutions of some problems in heat in the eighteenth century' (University of Manchester Ph.D. thesis, 1967).

(B) PRINTED SOURCES

G. Bachelard, *Étude sur l'évolution d'un problème de physique. La propagation thermique dans les solides* (Paris, 1928).

H. Balfour, 'The fire-piston', in Thomas, *Anthropological essays presented to E. B. Tylor*, pp. 17–49.

J. Ben-David, 'Scientific productivity and academic organization in nineteenth century medicine', *Am. soc. Rev.* **25** (1960), 828–43.

P. E. M. Berthelot, *La révolution chimique. Lavoisier* (Paris, 1890).

J. L. F. Bertrand, *Éloges académiques* (Paris, 1890).

E. Blanc and L. Delhoume, *La vie émouvante et noble de Gay-Lussac* (Paris, 1950).

F. Bourquelot, see Louandre and Bourquelot.

W. H. Brock, 'William Prout and barometry', *Notes Rec. R. Soc. Lond.* **24** (1969), 281–94.

C. Brunold, *L'entropie. Son rôle dans le développement historique de la thermodynamique* (Paris, 1930).

S. G. Brush, *Kinetic theory* (2 vols., Oxford, 1965–6).

—— 'The development of the kinetic theory of gases. I. Herapath', *Ann. Sci.* **13** (1957), 188–98.

—— 'The development of the kinetic theory of gases. II. Waterston', *Ann. Sci.* **13** (1957), 273–82.

—— 'The development of the kinetic theory of gases. III. Clausius', *Ann. Sci.* **14** (1958), 185–96.

—— 'The development of the kinetic theory of gases. IV. Maxwell', *Ann. Sci.* **14** (1958), 243–55.

—— 'Development of the kinetic theory of gases. V. The equation of state', *Am. J. Phys.* **29** (1961), 593–605.

—— 'John James Waterston and the kinetic theory of gases', *Am. Scient.* **49** (1961), 202–14.

—— 'Development of the kinetic theory of gases. VI. Viscosity', *Am. J. Phys.* **30** (1962), 269–81.

—— 'The Royal Society's first rejection of the kinetic theory of gases (1821), John Herapath versus Humphry Davy', *Notes Rec. R. Soc. Lond.* **18** (1963), 161–80.

G. BUCHDAHL, 'Sources of scepticism in atomic theory', *Br. J. Phil. Sci.* **10** (1959), 120–34.

D. S. L. CARDWELL, 'Science and technology in the eighteenth century', *History of science* **1** (1962), 30–43.

—— 'Power technologies and the advance of science, 1700–1825', *Technology Cult.* **6** (1965), 188–207.

—— (ed.), *John Dalton & the progress of science* (Manchester, 1968).

M. CERMENATI, *Alessandro Volta, alpinista* (Turin, 1899).

M. CLAGETT (ed.), *Critical problems in the history of science* (Madison, 1957).

I. B. COHEN, *Franklin and Newton. An inquiry into speculative Newtonian experimental science and Franklin's work in electricity as an example thereof* (Philadelphia, 1956).

—— see also KOYRÉ and COHEN.

N. G. COLEY, 'The physico-chemical studies of Amedeo Avogadro', *Ann. Sci.* **20** (1964), 195–210.

P. COSTABEL, 'Le "calorique du vide" de Clément et Desormes (1812–1819)', *Archs int. Hist. Sci.* **21** (1968), 3–14.

C. A. COULSON, *Waves. A mathematical account of the common types of wave motion* (7th edn, Edinburgh and London, 1955).

M. P. CROSLAND, *The Society of Arcueil. A view of French science at the time of Napoleon I* (London, 1967).

—— 'The origins of Gay-Lussac's law of combining volumes of gases', *Ann. Sci.* **17** (1961), 1–26.

E. E. DAUB, 'Atomism and thermodynamics', *Isis* **58** (1968), 293–303.

J. B. DUMAS, *Histoire de l'Académie Royale des Sciences, Belles-Lettres et Arts de Lyon* (2 vols., Lyons, 1839).

A. M. DUNCAN, 'Some theoretical aspects of eighteenth-century tables of affinity', *Ann. Sci.* **18** (1962), 177–94 and 217–32.

A. FERGUSON, 'Minutes of the life and character of Joseph Black, M.D.', *Trans. R. Soc. Edinb.* **5**, part 3 (1805), 101–17.

B. S. FINN, 'Laplace and the speed of sound', *Isis* **55** (1964), 7–19.

A. FOURCY, *Histoire de l'École Polytechnique* (Paris, 1828).

R. FOX, 'Dalton's caloric theory', in Cardwell, *Dalton & science*, pp. 187–201.

—— 'The background to the discovery of Dulong and Petit's law', *Br. J. Hist. Sci.* **4** (1968–9), 1–22.

—— 'The fire piston and its origins in Europe', *Technology Cult.* **10** (1969), 355–70.

—— 'Watt's expansive principle in the work of Sadi Carnot and Nicolas Clément', *Notes Rec. R. Soc. Lond.* **24** (1969), 233–53.

—— 'The intellectual environment of Sadi Carnot: a new look', *Actes du XIIe Congrès International d'Histoire des Sciences. Paris, 25–31 August 1968* (in press).

—— 'The rejection of Laplacian physics: a turning-point in the history of the physical sciences in France', to be published shortly.

R. FRIC, 'Contribution à l'étude de l'évolution des idées de Lavoisier sur la nature de l'air et sur la calcination des métaux', *Archs int. Hist. Sci.* **12** (1959), 137–68.

W. A. GABBEY and J. W. HERIVEL, 'Un manuscrit inédit de Sadi Carnot', *Revue Hist. Sci. Applic.* **19** (1966), 151–66.

C. C. GILLISPIE, *The edge of objectivity* (London, 1960).

J. GIRARDIN and C. LAURENS, *Dulong de Rouen. Sa vie et ses ouvrages* (Rouen, 1854).

H. GUERLAC, *Lavoisier—the crucial year. The background and origin of his first experiments on combustion in 1772* (Ithaca, N.Y., 1961).

—— 'A lost memoir of Lavoisier', *Isis* **50** (1959), 125–9.

N. H. DE V. HEATHCOTE, see MCKIE and HEATHCOTE.

W. C. HENRY, *Memoirs of the life and scientific researches of John Dalton* (London, 1854).

J. W. HERIVEL, 'Aspects of French theoretical physics in the nineteenth century', *Br. J. Hist. Sci.* **3** (1966–7), 109–32.

—— See also GABBEY and HERIVEL.

W. HOUGH, *Fire as an agent in human culture*, United States National Museum Bulletin 139 (Washington, 1926).

R. JENKINS, 'A Cornish engineer: Arthur Woolf, 1766–1837', *Trans. Newcomen Soc.* **13** (1932–3), 55–73.

R. KARGON, 'The decline of the caloric theory of heat: a case study', *Centaurus* **10** (1964), 35–9.

F. KLEIN, *Vorlesungen über die Entwicklung der Mathematik im 19. Jahrhundert* (2 vols., Berlin, 1926–7).

A. KOYRÉ and I. B. COHEN, 'Newton's "electric & elastic spirit" ', *Isis*, **51** (1960), 337.

E. KRAUSE, *The life of Erasmus Darwin by Charles Darwin, being an introduction to an essay on his scientific works by Ernst Krause* (2nd edn, London, 1887).

T. S. KUHN, *The structure of scientific revolutions* (Chicago, 1962).

—— 'The caloric theory of adiabatic compression', *Isis* **49** (1958), 132–40.

—— 'Energy conservation as an example of simultaneous discovery', in Clagett, *Critical problems*, pp. 321–56.

—— 'Engineering precedent for the work of Sadi Carnot', *Actes du IXe Congrès International d'Histoire des Sciences. Barcelona-Madrid, 1–7 Septembre 1959* (Barcelona and Paris, 1960), pp. 530–5.

—— 'Sadi Carnot and the Carnot engine', *Isis* **52** (1961), 567–74.

P. LEMAY and R. E. OESPER, 'Pierre Dulong, his life and work', *Chymia* **1** (1948), 171–90.

E. M. LÉMERAY, *L'éther actuel et ses précurseurs* (Paris, 1922).

S. LILLEY, 'Social aspects of the history of science', *Archs int. Hist. Sci.* **2** (1948–9), 376–443.

C. LOUANDRE and F. BOURQUELOT, *La littérature française contemporaine, 1827–1844. Continuation de la France littéraire* (6 vols., Paris, 1848).

E. MAINDRON, *Les fondations de prix à l'Académie des Sciences. Les lauréats de l'Académie 1714–1880* (Paris, 1881).

D. MCKIE and N. H. DE V. HEATHCOTE, *The discovery of specific and latent heats* (London, 1935).

E. MENDOZA, 'Contributions to the study of Sadi Carnot and his work', *Archs int. Hist. Sci.* **12** (1959), 377–96.

BIBLIOGRAPHY

E. MENDOZA, 'A sketch for a history of early thermodynamics', *Physics today* **14** (February 1961), 32–42.

—— 'A sketch for a history of the kinetic theory of gases', *Physics today* **14** (March 1961), 36–9.

—— 'The surprising history of the kinetic theory of gases', *Mem. Proc. Manch. lit. phil. Soc.* **105** (1962–3), 15–28.

—— 'Sadi Carnot and the Cagnard engine', *Isis* **54** (1963), 262–3.

J. T. MERZ, *A history of European thought in the nineteenth century* (4 vols., Edinburgh and London, 1896–1914).

H. METZGER, *Newton, Stahl, Boerhaave et la doctrine chimique* (Paris, 1930).

W. E. K. MIDDLETON, *A history of the thermometer and its use in meteorology* (Baltimore, 1966).

A. and J. K. MOILLIET, *Sketch of the life of James Keir, Esq., F.R.S., with a selection from his correspondence* (London, n.d.).

A. DE MONTET, *Dictionnaire biographique des Genevois et des Vaudois* (2 vols., Lausanne, 1877–8).

J. P. MUIRHEAD, *The origin and progress of the mechanical inventions of James Watt* (3 vols., London, 1854).

—— *The life of James Watt, with selections from his correspondence* (London, 1858).

L. K. NASH, 'The origins of Dalton's chemical atomic theory', *Isis* **47** (1956), 101–16.

N. J. T. M. NEEDHAM, *Science and civilisation in China*, vol. 4, part 2 (Cambridge, 1965).

'T. H. N.' (T. H. NORTON), Obituary of Victor Regnault, in *Nature, Lond.* **17** (1878), 263–4.

R. E. OESPER, *see* LEMAY and OESPER.

A. J. PACEY, *see* TALBOT and PACEY.

J. R. PARTINGTON, *A history of chemistry*, vols. 3 and 4 (London, 1962–3).

—— and W. B. SHILLING, *The specific heats of gases* (London, 1924).

J. PAYEN, 'Une source de la pensée de Sadi Carnot', *Archs int. Hist. Sci.* **21** (1968), 15–37.

G. PINET, *Histoire de l'École Polytechnique* (Paris, 1887).

J. PLAYFAIR, 'Biographical account of the late Dr. James Hutton, F.R.S. Edin.', *Trans. R. Soc. Edinb.* **5** (1805), part 3, 39–99.

T. J. RAINOFF, 'Wave-like fluctuations of creative productivity in the development of West-European physics in the eighteenth and nineteenth centuries', *Isis* **12** (1929), 287–319.

H. E. ROSCOE and A. HARDEN, *A new view of the origins of Dalton's atomic theory* (London, 1896).

L. ROSENFELD, 'La genèse des principes de la thermodynamique', *Bull. Soc. R. Sci. Liège* **10** (1941), 197–212.

R. E. SCHOFIELD, *The Lunar Society of Birmingham. A social history of provincial science and industry in eighteenth-century England.* (Oxford, 1963).

—— (ed.), *A scientific autobiography of Joseph Priestley (1733–1804)*, (Cambridge, Mass., and London, 1966).

W. B. SHILLING, see PARTINGTON and SHILLING.

R. H. SILLIMAN, 'William Thomson: smoke rings and nineteenth-century atomism', *Isis* **54** (1963), 461–74.

W. A. SMEATON, *Fourcroy, chemist and revolutionary, 1755–1809* (Cambridge, 1962).

S. SMILES, *Lives of Boulton and Watt. Principally from the original Soho MSS* (2nd edn, London, 1866).

G. R. TALBOT and A. J. PACEY, 'Some early kinetic theories of gases: Herapath and his predecessors', *Br. J. Hist. Sci.* **3** (1966–7), 133–49.

F. S. TAYLOR, 'The origin of the thermometer', *Ann. Sci.* **5** (1942), 129–56.

A. W. THACKRAY, 'Documents relating to the origins of Dalton's chemical atomic theory', *Mem. Proc. Manch. lit. phil. Soc.* **108** (1965–6), 21–42.

—— 'The origin of Dalton's chemical atomic theory: Daltonian doubts resolved', *Isis* **57** (1966), 35–55.

—— 'The emergence of Dalton's chemical atomic theory: 1801–08', *Br. J. Hist. Sci.* **3** (1966–7), 1–23.

—— 'Quantified chemistry—the Newtonian dream', in Cardwell, *Dalton & science*, pp. 92–108.

N. W. THOMAS (ed.), *Anthropological essays presented to Edward Burnett Tylor in honour of his 75th birthday. October 2 1907* (Oxford, 1907), pp. 17–49.

J. THOMSON, *An account of the life, lectures, and writings of William Cullen, M.D.* (2nd edn, 2 vols., Edinburgh and London, 1859).

C. A. TRUESDELL, *Essays in the history of mechanics* (Berlin, Heidelberg, and New York, 1968).

W. WHEWELL, *A history of the inductive sciences* (3 vols., London, 1837).

—— *Philosophy of the inductive sciences* (2 vols., London, 1840).

—— 'Comte and positivism', *Macmillan's magazine*, **13** (1866), 353–62.

E. T. WHITTAKER, *A history of the theories of aether and electricity. I. The classical theories* (2nd edn, London, 1951).

L. P. WILLIAMS, 'The physical sciences in the first half of the nineteenth century: problems and sources', *History of science* **1** (1962), 1–15.

M. YUASA, 'Center of scientific activity: its shift from the 16th to the 20th century', *Jap. Stud. Hist. Sci.* **1** (1962), 57–75.

INDEX

Absolute heat, 26, 33, 34, 36–7
Absolute zero:
  first defined, 61
  interest revived by Irvinists, 62
  Irvinist calculation described, 27–8, Plate 1; used, 32, 36, 75, 107, 117, 140, 146, 153; criticized, 32, 39, 105–6, 140; abandoned, 155
  calculated from data on gas expansion, 61–2, 146–7, 153
  Dalton's calculations, 74–5, 106, 109
  scepticism concerning, 61–2 n., 151–3, 192, 239
Académie des Sciences, Paris, 29, 30, 61, 63–4, 69, 137, 179, 184, 191, 243, 266, 267, 273, 292, 316
  Lavoisier at, 9–10; with Laplace at, 29, 64
  Laplace at, 157, 165, 167 n.; with Lavoisier at, 29, 64
  Fourier at, 234–5, 271–2
  Dulong at, 228, 239, 250–2, 255–6, 261, 267, 268
  Arago at, 261, 271–2, 295, 317–18
  Regnault at, 295, 297, 300–1
  Biot's absence from, 271 n., 273
  prize competition of 1817 and 1818, 236–9; of 1819, 234–5; of 1823, 249; of 1842, 294–5
  organizes research on velocity of sound, 84; on vapour pressures, etc., 249–50
  elections harming Laplacian interests, 270–3
  see also Institute, First Class of the French
Académie Française, 272
Accum, Friedrich Christian, 95 n., 121
Adams, George (the younger), 54 n., 158 n.
Adiabatic phenomena:
  discovered, 41, 44–5; conditions for discovery, 43–4
  'adiabatic' first used, 39 n.; use in this book, xv
  work of Boyle, 41–5; of Cullen, 40–2, 44–6, 48, 49, 50–1, 56–7; of Arnold, 45, 47, 50; of Cleghorn, 42, 46–7, 108, 117; of Lambert, 40, 47–8, 50–2; of Magellan, 49, 58 n.; of Bergman, 41–2, 45; of Saussure, 40, 45, 48, 50–1; of De Luc, 48 n., 49, 52–3, 100, 158; of Crawford, 37–8, 49–50, 53–4, 58 n., 60, 80, 88, 107; of Darwin, 42, 44 n., 49, 54–60, 80, 100, 119 n., 193; of Dalton, 42, 48, 53, 68, 86–9, 100 n., 107, 117–18, 129–30, 144, 148, 159; of Leslie, 47, 108–9, 117, 130, 158; of Biot, 82–6, 94–6, 100 n., 118, 128; at Lyons Academy, 80 n., 89–98; of Gay-Lussac, 47, 89 n., 98 n., 130–1, 148, 151–4; of Poisson, 86, 118, 140, 144, 151, 175–6, 185–6, 189, 193, 275; of Haüy, 128; of Young, 117–18; of Southern, 86 n.; of Clément and Desormes, 151–4, 170; of Bréguets, 152; of Laplace, 41, 49, 52, 82, 85, 107, 125–7, 157, 161–5, 168–71, 195; of Navier, 189 n., 192; of Dulong, 245–6, 253–5; of de la Rive and Marcet, 143, 154, 275; of Joule, 39–40, 253, 306 n.
  attributed to mechanical deformation, 41–2, 45, 58, 89; to friction, 43, 45, 89; to evaporation, 44–5, 50–1, 58, 79, 89
  known in Lunar circle, 50; at Chapter Coffee House, 50, 53 n., 58 n., 60; to Higgins, 55 n.
  little known at Royal Society, 54; in France, 40–1, 44 n., 49–50, 52, 67, 68, 79–81
  and chemical combination, 94–5
  and caloric theory, 40, 81, 98–9, 103, 107–8, 119–20, 142–3
  and energy conservation, 39–40, 130, 253–5, 306 n.
  and hygrometry, 48–9; mist in receivers, 43–4; snow on Schemnitz engine, 58–9, 179–80
  and meteorology, 48–9, 59, 80

362 INDEX

Adiabatic phenomena (cont.)
  and specific heats, 88, 108–9, 154, 159, 294
  and velocity of sound, 68–9, 81–6, 161–75, 193–4
  accompanied by emission of light, 90–1, 96–8
  relationship between P, V, and T, 175–6, 188–9, 193; between temperature change and specific heat, 192
  to attain low temperatures, 59, 151
Aepinus, Franz Ulrich Theodor, 16, 18–19
Affinity:
  in eighteenth-century chemistry, 197, 198
  in Berthollet's chemistry, 125, 196–7, 199–202, 204, 206, 210, 218
  in Avogadro's chemistry, 210–12, 214, 215, 217–18, 225
  tables of, 198, 199, 211, 212, 218, 221
  for light, 197–202, 204, 210, 213–19
  for caloric, 205–19, 221, 222, 224, 225
  see also Affinity numbers
Affinity numbers, 218
Airy, Sir George Biddell, 194
Alembert, Jean Lerond d', 82
Alfort veterinary school, 236
Altitude measurement:
  barometric method, 62–3, 70–1
  incentive from mountaineering, 63, 66, 71; from ballooning, 63, 66, 71
  stimulates work on gas expansion, 63–4, 70–1
Amontons, Guillaume, 60–2, 84, 146–7
Ampère, André Marie, 215, 249 n., 250 n.
  on classification of elements, 204
  his gas hypothesis, 207, 260 n.
  on nature of heat, 3 n., 277–8
  on imponderable fluids, 234–5, 262 n.
  and Laplacian physics, 234–5, 271
  positivistic strain in, 262 n.
Animal heat, 134
  Crawford on, 33–4, 36–7, 106, 135, 249, 275
  Lavoisier on, 249

Delaroche on, 136
Dulong on, 249
Annales de chimie et de physique:
  Fresnel's papers in, 234
  Prout's papers in, 247, 248
  change of editor, 234, 270
Apjohn, James, 258–9
Arago, Dominique François Jean, 296
  on refraction, 78, 95 n., 166 n., 197–204, 208, 213, 217
  supports wave theory, 233–5; against Biot, 235
  on velocity of sound, 171
  critic of Laplacians, 227–9, 235, 271, 272–3, 274, 317–18
  and government commission, 249–50
  and Annales de chimie et de physique, 234, 247 n., 270
  at Académie des Sciences (Institute), 166, 179, 202, 261, 271, 272, 274, 295
  at Arcueil, 229
  at École Polytechnique, 228–9
Arcueil:
  Society of, 132, 136, 229, 233, 262
  Berthollet's laboratory, 71–2, 129, 236, 241
  Laplace at, 72, 129
  encouragement of young men, 72, 132–4, 136, 236
  experiments at, 133, 241, 247
Arden, John, 58 n.
Arnim, Ludwig Achim von, 81
Arnold, Johann Christian, 45, 47, 50
Arnott, Neil, 276
Askesian Society, London, 109
Atomic theory:
  Higgins on, 11
  Dalton's theory, 11, 110–15, 246, 282, 290
  early French opposition, 72, 236, 246–8
  supported by Dulong, 239, 247–8, 261–2, 263; by Dulong and Petit, 240, 246, 265, 282; by Berzelius, 248
  caution of Wollaston, 266, 290; of Davy, 266; of Dumas, 261, 264, 285–92; of Liebig, 291–2, 319–20; of others in 1830s, 223, 261, 263–4, 290–5

INDEX

positivist attitude to, 265–6, 290, 312–13
see also Atomic weights
Atomic weights:
of Dalton, 281; of Wollaston, 290; of Berzelius, 225, 282, 283, 284, 288; of Dumas, 283–9
Liebig on, 292, 319–20
Attractive power, 208–10, 221, 224
Avogadro, Lorenzo Romano Amedeo Carlo, 1, 156, 197, 278–9, 304
on heat, 196, 205–26, 293–4, 297
see also Avogadro's hypothesis
Avogadro's hypothesis:
used by Avogadro, 196, 206–26 *pass.*, 293
Dalton on, 78, 110–12, 114–15
neglected, 196, 207, 222, 226
partially accepted by Dulong and Petit, 207, 215, 216–17, 221, 283–4 n.; by Berzelius, 207, 215, 221; by Dumas, 221, 283–8
accepted by Dulong, 252, 260, 284, 292; at Karlsruhe, 196
Ampère's version, 207, 260 n.

Babbage, Charles, 95–6 n.
Bacon, Francis, 14
Baillet, Arsène Nicolas, 79–80, 87, 89
Ballooning, 63, 66, 71
Barruel, Étienne, 97
Baudot, J. M., 180 n.
Baudrimont, Alexandre Édouard, 260–1, 292
Beaumont, Jean Baptiste Armand Louis Léonce Élie de, 296.
Becquerel, Antoine César, 295
Bérard, Jacques Étienne, 204, 293
in 1812 competition, 131–42, 147–50, 154–5, 160, 203, 240, 258
results used by Laplace, 162, 164–5, 172–4, 299; by Poisson, 179, 299; by Carnot, 186–7, 299; by Clapeyron, 190–1; by Navier, 192; by Avogadro, 207–9, 217, 219–21
results criticized by Dulong, 254, 256; discredited by Regnault, 298–9
friendships at Arcueil, 132, 133, 134, 136–7
Bergman, Torbern Olof, 37 n., 123 n., 125
on adiabatic phenomena, 41–2, 45

Bernard, Claude, 315 n.
Bernoulli, Daniel, 82, 302–3
Berthelot, Pierre Eugène Marcellin, 290
Berthollet, Claude Louis, 77 n., 122 n., 134 n., 149, 271, 273
his chemistry, 196–7, 199–202, 204, 210, 218, 233, 239, 246–7, 263, 290
on phlogiston, 21
on caloric theory, 102 n., 124–6
on gas expansion, 64–5
interest in adiabatic phenomena, 96–8, 100 n.
as a patron, 2, 71–2, 96–8, 129, 133, 136–7, 205, 229, 236
Berzelius, Jöns Jacob, 94 n., 244, 246, 248, 292, 320
on electrochemical theory, 2–3, 212 n., 242–3
on nature of heat, 120, 241–3, 277 n.
on Avogadro's hypothesis, 207, 215, 221
on atomic weights, 225, 282, 283, 284, 288
Billardière, Houtou de la, 228 n.
Biot, Jean-Baptiste, 2, 72–3, 136–7, 230 n., 247 n., 271 n., 272, 273, 294
on nature of heat, 97
on heating by compression, 82–6, 94–6, 100 n., 118, 128
on velocity of sound, 82–6
on nature of light, 97, 198–202, 234–5
on refraction, 78, 95 n., 166, 197–204, 205, 208, 213, 217
Birmingham: *see* Lunar Society of Birmingham
Black, Joseph, 14, 40, 44 n., 47, 45–7, 55–7, 121
discovery of carbon dioxide, 20; of specific and latent heats, 21–2
on nature of heat, 23–5
on Irvinist doctrines, 25–6, 31, 105–7, 142
Blainville, Henri Marie Ducrotay de, 265 n., 266, 269, 273
Boerhaave, Herman, 7
on fire, 4, 12–14, 16, 17, 23 n.
on absolute zero, 62 n.
his *Elements of chemistry*, 12, 14
Boltzmann, Ludwig, 312

INDEX

Bonaparte, Napoleon: *see* Napoleon Bonaparte
Bosc, Louis Augustin Guillaume, 98 n., 134 n.
Boulton, Matthew, 190 n.
Bourbon restoration, 240
  and Laplacian science, 228–9, 240, 271–2
  and École Polytechnique, 230–1
Boussingault, Jean-Baptiste Joseph Dieudonné, 296
Boyle, Robert, 16, 20
  on structure of air, 8
  on nature of heat, 14
  on adiabatic phenomena, 41–5
  see also Boyle's law
Boyle's law, 84
  derived theoretically by Newton, 6–7, 74; by Bernoulli, 302; by Laplace, 168, 173
  used, 61, 64, 162, 182
  confirmed, 250
  discredited, 298–9
  and uniqueness of gases, 68, 78
Brande, William Thomas, 276
Bravais, Auguste, 194
Bréguet, Abraham Louis and Antoine, 152
British Association for the Advancement of Science, 291, 306, 306–7 n.
Brodie, Sir Benjamin Collins (the younger), 290
Buijs-Ballot (or Buys-Ballot), Christoph Hendrik Diederik, 311

Cabart, Charles François, 261
Calendar:
  republican and Gregorian forms, xv
  Positive, 265
Caloric:
  'caloric' first used, 6 n.; criticized, 6 n.
  two states of (sensible and latent), 10, 35–6, 38–9, 52, 91, 104–77 *pass.*, 193, 203, 230, 239–42, 275, 278–9
  see also Adiabatic phenomena; Heat, nature of; Irvinist doctrines; Latent heat
Calorimetry:
  origins, 21–2
  supports caloric theory, 22
  ice-calorimeter, 34–5, 145

Cambridge, University of:
  Plumian professor, 194
  Queens' College, 3 n.
  St. John's College, 56
  Trinity College, 81
Candolle, Augustin Pyramus de, 136
Capacity:
  meaning in this book, xv
  term first used, 26; criticized, 108
  see also Specific heat
Capillary action:
  treated by Laplace, 166, 231
  treated by Petit, 231
Carnot, Lazare Hippolyte (brother of N. L. S. Carnot), 183, 189
Carnot, Lazare Nicolas Marguerite, 178, 265
Carnot, Marie François Sadi (son of L. H. Carnot), 183 n.
Carnot, Nicolas Léonard Sadi, 39–40, 86, 94 n., 195, 245 n., 299, 314
  as a calorist, 177–91, 215
  supports vibrational theory, 3 n., 189, 277
  on heat engines, 177–91, 302 n., 308 n.
  influenced by Clément, 137, 179–84
  excluded from 'establishment', 184, 192, 274–5, 296
  see also Carnot cycle
Carnot cycle:
  used by Carnot, 182–9
  scepticism of Regnault, 301–2 n.
Cassini de Thury, 84–5
Cauchy, Augustin Louis, 264
Cavallo, Tiberius, 28, 59, 100 n.
Cavendish, The Hon. Henry, 16, 19 n., 20–1
Challis, James, 194
Chaptal, Jean Antoine Claude, Comte de Chanteloup, 49, 79, 98 n., 137
Chapter Coffee House Society, London, 50, 53 n., 58 n., 60
Charles, Jacques Alexandre César, 90 n., 91, 149 n.
  on gas expansion, 65–6, 69–70
Chauvain, Citizen, 90 n.
Chemical heat: *see* Reaction, heats of chemical
Chevreul, Michel Eugène, 247
Clapeyron, Benoît Paul Émile, 190–1, 275, 302 n.

INDEX 365

see also Clausius–Clapeyron (or Clapeyron) equation
Clausius, Rudolph Julius Emmanuel, 300, 310 n., 307–8
  on kinetic theory, 4–5, 305 n., 307 n., 311–13
  on nature of heat, 308 n., 310
  see also Clausius–Clapeyron (or Clapeyron) equation
Clausius–Clapeyron (or Clapeyron) equation, 191
Cleghorn, George, 9, 17
Cleghorn, William:
  on fire, 9, 11–12, 14, 24, 28
  on adiabatic phenomena, 42, 46–7, 108, 117
Clément, Nicolas:
  in 1812 competition, 88, 133, 136–8, 142–8
  work criticized by Gay-Lussac, 150–4
  results used by Poisson, 175; by Navier, 192
  on measurement of γ, 87, 170, 172
  on heat engines, 179–83
  influences Carnot, 137, 179–84
  see also Clément and Desormes's 'law'
Clément and Desormes's 'law':
  announced, 179
  used by Carnot, 179, 183; by Clément and Desormes, 181–2; by Dulong, 244–6
  discredited, 246, 298–9
Cold at high altitudes, 59, 193, 275
Colden, Cadwallader, 19
Collège de France, Paris, 82, 286
Column-of-water engine, 58–9, 79–80, 184 n.
Competitions:
  at Académie des Sciences, 134–51, 234, 236–8, 249, 294–5
  at French Institute, 86, 97, 107, 115, 131–2, 133–50, 158, 160, 179, 236–7, 314
  at University of Leiden, 259
  at Turin Academy, 293–5
Comte, Isidore Auguste Marie François Xavier, 232, 238, 262–70, 316
Conduction of heat, 236–7, 262, 272, 278
Conservatoire des Arts et Métiers, Paris, 66 n.

Carnot at, 137, 183
Clément at, 137, 154, 180 n., 183
Cooling, laws of:
  Newton's law, 87, 237–9, 265
  Dulong and Petit on, 236–9, 269–70
Coupé, Jacques Michel, 80
Crawford, Adair, 26 n., 72, 76, 106, 135, 249, 275–6
  and Irvinist doctrines, 10, 27 n., 28–9, 74, 75, 88, 117, 121, 138, 147, 154, 240, 275
  on determination of specific heats, 32–8, 113, 115, 138, 141, 157–9, 205
  on adiabatic phenomena, 37–8, 49–50, 53–4, 58 n., 60, 80, 88, 107
  on thermometry, 72, 76
Cruquius, Nicolaas Samuel, 62
Cullen, William, 40–2, 44–6, 48, 49, 50–1, 56–7
Cuvier, Jean Léopold Nicolas Frédéric (Georges), Baron, 91–2 n., 98, 100 n., 134 n., 149, 243, 272

Dalton, John, 1, 127, 206, 237 n., 276, 278, 304
  his atomic theory, 11, 110–15, 246, 282, 290
  as an Irvinist, 53, 88, 102 n., 105, 106–7, 109–15, 129–30, 154–6, 240
  on adiabatic phenomena, 42, 48, 53, 68, 86–9, 100 n., 107, 117–18, 129–30, 144, 148, 159
  on gas expansion, 68–79, 147
  on specific heats, 108 n., 110–15, 154–6, 157–9, 240
  see also Dalton's law of partial pressures
Dalton's law of partial pressures, 168, 173
Daniell, John Frederic, 276
Darwin, Charles (son of Erasmus), 55–6
Darwin, Erasmus:
  on adiabatic phenomena, 42, 44 n., 49, 54–60, 80, 100, 119 n., 193
  on heating by percussion and friction, 4 n., 49, 59 n., 100
Davy, Sir Humphry, 94 n., 95–6 n., 98
  critic of imponderable fluids, 17, 118–19

Davy, Sir Humphry (*cont.*)
  supports kinetic theory of heat, 99 n., 104, 115–16, 119–20, 243, 306, 311
  on Herapath's kinetic theory, 303
  on gas expansion, 298 n.
  on atomic theory, 290
  on electrochemical theory, 211–12, 242
Delambre, Jean-Baptiste Jean, 202
Delamétherie, Jean Claude, 79
Delaroche, François, 204, 293
  on radiant heat, 136
  in 1812 competition, 131–42, 147–50, 154–5, 160, 203, 240, 258
  results used by Laplace, 162, 164–5, 172–4, 299; by Poisson, 179, 299; by Carnot, 186–7, 299; by Clapeyron, 190–1; by Navier, 192; by Avogadro, 207–9, 217, 219–21
  results criticized by Dulong, 254, 256; discredited by Regnault, 198–9
  friendships at Arcueil, 132, 133, 134, 136–7
De Luc, Jean André, 29
  on fire, 52–3, 125, 158
  on adiabatic phenomena, 48 n., 49, 52–3, 100, 158
  on heating by percussion, 49, 52, 100
  on gas expansion, 62–5
  on thermometry, 72
  on Crawford, 34, 53
Derham, William, 82
Desaguliers, Jean Théophile, 7, 8, 16, 43 n.
Descartes, René, 8 n., 13 n., 14
Desfontaines, René Louiche, 134 n.
Desormes, Charles Bernard:
  in 1812 competition, 88, 133, 136–8, 142–8
  results used by Poisson, 175; by Navier, 192
  work criticized by Gay-Lussac, 150–4
  on heat engines, 179–83
  on measurement of γ, 87, 170, 172
  contacts in scientific community, 133, 137
  see also Clément and Desormes's 'law'

Despretz, César Manuel:
  on nature of heat, 276, 279
  on Clément and Desormes's 'law', 244, 246 n., 299 n.
Dessaignes, J. P., 97–8
Diffraction of light, 233–4
Diffusion of gases:
  Dalton's theories, 110–12, 206
  and kinetic theory, 311
Digby, Sir Kenelm, 16
Dixon, Robert Vickers, 276
Dobson, Matthew, 44 n.
Donovan, Michael, 276
Dublin: *see* Royal College of Surgeons of Ireland; Trinity College
Dubois, 93–4
Du Fay, Charles François de Cisternay, 16
Duhamel, Jean Marie Constant, 263, 268
Duhem, Pierre Maurice Marie, 312
Dulong, Pierre Louis, 235–6, 255–6
  on nature of heat, 99 n., 238–46, 264
  on absolute zero, 151, 239
  on animal heat, 249
  on Clément and Desormes's 'law', 244–6, 299 n.
  on expansion, thermometry, and cooling, 73, 77, 236–9, 269–70
  on measurement of specific heats, 160–1, 204–5, 216–17, 219–21, 253–6, 258, 259–61, 293–5; of γ, 220, 253–4
  on vapour pressures, etc., 249–50, 300
  on nature of light, 244, 252
  on refraction, 215–16, 219, 250–2
  on atomic theory, 239, 240, 246–8, 282–3
  on Avogadro's hypothesis, 207, 215, 216–17, 221, 252, 260, 283–4 n., 284, 292
  on Dalton, 155–6
  on Avogadro, 215–17
  on electrochemical theory, 241–4
  friendship with Berzelius, 241–4, 248
  critic of Laplacians, 227–9, 233, 239–48, 273, 317–18
  and positivism, 262–70
  see also Dulong and Petit's law
Dulong and Petit's law, 216

INDEX 367

announced, 239–44
approximate nature demonstrated, 282–3, 297, 299
scepticism of Dalton, 155–6; of Avogadro, 217
applied to gases, 216–17, 221, 253, 293–4, 297
possible extensions, 156, 250–5, 289
and atomic theory, 246, 248, 282–97
not positivistic, 265, 268
Dumas, Jean-Baptiste, 91
Dumas, Jean-Baptiste André:
on atomic theory, 261, 264, 283–92, 317
on Avogadro's hypothesis, 221, 283–8
on nature of heat, 276
influences Regnault, 295, 296–7, 315–16
Dumotiez, Louis Joseph and Pierre François, 93–4
Dupin, Pierre Charles François, Baron, 181 n., 249 n.
Durivau, 231

École des Mines, Paris, 79, 296
École des Ponts et Chaussées, Paris, 228–9 n.
École Normale, Paris, 236
École Polytechnique, Paris, 31, 71, 94, 137, 178, 184, 194, 229, 230, 236, 261, 271, 296, 316
textbooks at, 31, 230, 268–70
inadequate physics teaching, 231–2
political unrest at, 237
and Bourbon restoration, 230–1
Arago at, 228, 231, 233
Comte at, 232, 266–9
Dulong at, 235–6, 256
Hachette at, 178, 230
Lamé at, 232, 268–70
Petit at, 178, 228, 230–2, 233, 239
Regnault at, 281, 296
Edgeworth, Richard Lovell, 54
Edinburgh:
Philosophical Society, 41, 56, 57
Royal Society of, 256
see also Edinburgh, University of
Edinburgh, University of:
Cleghorn's dissertation, 9–12, 14, 17
Black at, 14, 23–4, 55
Cullen at, 56–7
Darwins at, 55–7

Hope at, 21, 279 n.
Leslie and Murray at, 107
Robison at, 23–4
Edwards, Humphrey, 178
Egyptian campaign, 71, 228, 271
Electric fluid:
as model for caloric, 15–17, 19, 102 n.
weaknesses of fluid theory, 15
accepted at École Polytechnique, 230 n.
in atoms of matter, 306
Electrochemical theory:
and overthrow of caloric, 2–3, 119, 241–4
Davy on, 211–12, 242
Avogadro on, 211–18
Berzelius on, 2–3, 212 n., 242–3
Dulong and Petit on, 243–4
Electrolysis, Faraday's laws of, 289–90
Electromagnetism, 281, 314–15
Emmett, John Barnes, 120 n.
Energy conservation, principle of:
emerges, 279, 281–2, 300–2, 305–6
and caloric theory, 3, 308–10
and kinetic theory, 99, 280, 305–8
and adiabatic phenomena, 39–40, 130, 253–5, 306 n.
early French attitudes to, 300–2, 314–16
Regnault's role in discovery, 282, 300–2, 306 n., 314–16
Engineering:
and energy conservation, 281–2, 315
stimulates work of Carnot, 177–84, 189, 191; of Dulong, 249–50; of Regnault, 282
see also Column-of-water engine; Steam-engine
Entropy, 308 n.
Equivalent weights, 290, 292, 319–20
Erlangen, University of, 45
Erxleben, Johann Christian Polykarp, 62 n., 237 n.
Ether:
Newton's, 18
luminiferous, 264, 277
in vortex theories, 309
Euler, Leonhard, 82
Exley, Thomas, 19 n.
Expansion coefficient of gases:
eighteenth-century determinations,

Expansion coefficient of gases (*cont.*)
  38, 60–7, 69–70, 71 n.; used by Crawford, 38; used by Biot, 84
  equality for all gases discovered, 69–70, 147; predicted by Laplace, 168, 173; confirms uniqueness of gases, 68–9, 77–9; disproved, 298–9
Eynard, Ennemond, 90, 98

Faculté des Sciences, Paris, 255–6
Faraday, Michael, 19 n., 225, 281, 289–90, 291
Faraday's laws of electrolysis, 289–90
Ferguson, Adam, 24 n.
Ferry, A., 231–2
Fire, doctrines of, 4, 8–14, 19, 23 n., 29–30, 46–8, 51–3, 104 n.
Fire piston, 92–4, 151, 153, Plate 2
Fischer, Ernst Gottfried, 276
Fletcher, Mr, 90
Forbes, James David, 274
Fourcroy, Antoine François de, 91
  rejects phlogiston, 21
  on nature of heat, 22–3, 30, 97
Fourier, Jean-Baptiste Joseph, Baron, 179
  on heat transfer, 236–7, 272, 278
  rejects Laplacian principles, 167 n., 227–9, 235, 236, 271–3
  his positivism, 229, 262–3, 264–5, 266, 268–9, 316
Fox, 59
Francoeur, Louis Benjamin, 180 n.
Franklin, Benjamin, 15–16, 18–19
Fresnel, Augustin Jean, 215
  on wave theory, 3, 202, 233–5, 273
  critic of Laplacian principles, 202, 227–9, 233–4, 271, 273, 317
Friction, heating by:
  explained by Boerhaave and Lambert, 4; by Darwin, 4, 59 n.
  Rumford on, 4, 99–103
  associated with compression, 99–103, 232
  weaknesses of calorist explanation, 29–30, 99–103, 116, 119

Gadolin, Johan, 22, 32
Galillei, Galileo, 60
Garnett, Thomas, 95 n., 276 n.

Gas laws:
  simplicity of, 78, 134–5, 199, 249, 250, 252, 298–9
  see also entries for individual laws
Gas particles:
  forces between, 2, 7–20 *pass.*, 77–8, 125–7, 166, 173–4, 397 n.
  size treated by Dalton, 74–7, 110–15, 206; by Mollet, 207
  structure treated by Avogadro, 206, 207, 213, 222–3; by Dumas, 283–90
  polyatomic elementary molecule, 222–3, 260, 283–90, 292
Gaudin, Marc Antoine Augustin, 260 n.
Gay-Lussac, Joseph Louis, 1, 136–7, 252, 255, 258, 273, 295, 296, 314
  critic of Irvinist doctrines, 89 n., 129, 138, 148, 150–4; of Clément and Desormes, 151–4, 179
  on absolute zero, 151, 152–3, 192
  on gas expansion, 67, 68–73, 77–9, 84, 146, 176, 245, 298–9
  on specific heats, 129, 131–5; determination of γ, 170–1, 174, 193
  on adiabatic phenomena, 47, 89 n., 98 n., 130–1, 148, 151–4
  on nature of light, 234
  on Avogadro's hypothesis, 284
  on simplicity of gas laws, 78, 134–5
  Arcueil influences, 70–2, 129, 247
  role in Institute's competitions, in 1812, 133–5, 136, 142–4, 148, 153; in 1814, 149
  see also Gay-Lussac's law of combining volumes
Gay-Lussac's law of combining volumes, 78, 134–5, 247
Gensoul, Joseph Ferdinand, 90
Germain, Sophie, 167 n.
Giessen, University of, 296, 320
Gilbert (formerly Giddy), Davies, 182 n., 303 n.
Gilbert, Ludwig Wilhelm, 75 n.
Gilbert, William, 15
Girard, Pierre Simon, 190 n., 249 n., 250 n.
Girardin, Jean Pierre Louis, 261 n.
Glasgow, University of:
  Black and Robison at, 23
  Irvine at, 25
  Watt at, 57

Gmelin, Leopold, 278, 291–2, 319
Gorham, John, 277 n.
Göttingen, Royal Academy of Science, 71
Graham, Thomas, 276, 279, 310 n., 313
'sGravesande, Willem Jakob Storm van, 7, 8, 13–14, 16
Green, George, 264
Gregory, George, 97 n.
Gren, Friedrich Albert Carl, 212
Guenyveau, André, 178
Guyton de Morveau, Louis Bernard, 137
    rejects phlogiston, 21
    on gas expansion, 64 n., 65
    on identity of heat and light, 96 n.

Hachette, Jean Nicolas Pierre, 31, 178, 182, 230
Haex (otherwise Haez or Haess), Thibaud, 96
Haldat, Alexandre, 101–2 n.
Hales, Stephen, 7, 8, 20
Hallé, Jean Noël, 149
Halley, Edmund, 63
Hamilton, Sir William Rowan, 291
Hassenfratz, Jean Henri, 31 n., 94–5, 230–2
Haughton, Samuel, 194
Hauksbee, Francis (the elder), 43 n., 62 n.
Haüy, René Just, 77 n., 87, 149
    on caloric, 127–9; on other imponderable fluids, 128
    on adiabatic phenomena, 128
    distinguishes c_p from c_v, 129, 131, 158–60
    on thermometry, 159–60
Haycraft, W. T., 217, 256–7, 275
Heat, nature of:
    caution in eighteenth century, 23–5, 28, 29–30, 51
    growing support for material theory, 22–5, 30–1
    strength of caloric c. 1800, 103, 104–5, 115, 230
    caloric supported by adiabatic phenomena, 40, 81, 98–9, 103, 107–8, 119–20; harmed by electrochemical theory, 2–3, 119, 241–4
    agnosticism in nineteenth century, 3–4, 105 n., 128, 191–2, 225, 263–4, 276–7
    diminishing interest in, 275–80, 281
    decline of caloric, 2–4, 151, 215, 240–6, 273–80, 304–13, 315
    usefulness of caloric, 25, 28, 192, 243 n., 279–80, 304
    analogy with light, 96–8, 104 n.
    see also Vibrational theory of heat
Helm, Georg, 312
Helmholtz, Hermann Ludwig Ferdinand von, 277, 281, 308 n., 309 n.
Henry, William, 102 n., 106, 116 n.
Herapath, John:
    on velocity of sound, 194
    on kinetic theory of gases, 277, 280, 303, 305 n., 307, 311
    critic of Laplace and Poisson, 174–5 n., 194
Hero of Alexandria, 60
Heron, Robert, 25 n., 105
Héron de Villefosse, Antoine Marie, 178
Herschel, Sir John Frederick William, 276
Herschel, Sir William, 104, 105 n.
Higgins, Bryan:
    on fire, 11, 12, 16, 19 n., 112 n.
    on phlogiston, 16–17 n.
    on adiabatic phenomena, 55 n.
Higgins, William, 11
Hire, Philippe de la, 62
Hirn, Gustave Adolphe, 282, 316
Holtzmann, Carl Heinrich Alexander, 282
Homberg, Wilhelm, 13
Hope, Thomas Charles, 21, 279 n.
Humboldt, Friedrich Heinrich Alexander, Baron von, 129
Hutton, Charles, 54 n., 104 n., 105 n.
Hutton, James, 19, 54–9
Hygrometry, 43–4, 48, 49

Ice-calorimeter, 34–5, 145
Imison, John, 277 n.
Imponderable fluids:
    in Newtonian tradition, 2, 18–19
    uniformity of properties, 17
    in Laplacian physics, 2, 229
    taught at École Polytechnique, 230, 232–3
    criticized by Davy, 118–19; by positivists, 229, 263

INDEX

Imponderable fluids (*cont.*)
  rejected, 2, 17, 205–6, 229
  see also Caloric; Electric fluid; Light; Magnetism, fluid theory of
Indicator diagram, 190
Institute, First Class of French, 69, 91, 118 n., 132, 135–6, 202, 237, 247, 272–3
  prize competition of 1809, 97; of 1811, 236–7; of 1812, 86, 107, 115, 131–2, 133–50, 158, 160, 179, 314; of 1814, 149–50
  papers on adiabatic phenomena, 90–1, 94, 97–8; on diffraction, 234
  organizes research, 198
  see also Académie des Sciences, Paris
Irvine, William (the elder), 28–9, 53, 117, 121, 135, 138, 147, 156, 240
  on nature of heat, 28
  doctrines described, 25–8, Plate 1
  determination of absolute zero, 27–8, 105–6, 107, 109, 140–1
  use of 'capacity', 26
  member of 'Scottish school', 10, 25–6
  influences Crawford, 32–3
  see also Irvinist doctrines
Irvine, William (the younger), 26, 27
  see also Irvinist doctrines
Irvinist doctrines:
  described, 25–8, Plate 1
  on adiabatic phenomena, 53, 88, 107–9, 148, 152
  supported by Crawford's experiments, 33–4, 36–7
  used by Crawford, 10, 27 n., 28–9, 74, 75, 88, 117, 121, 138, 147, 154, 240, 275; by Magellan, 26, 27 n., 29; by Wedgwood, 101; by vibrationalists, 101–3, 117–18, 120–1; by Dalton, 53, 88, 102 n., 105, 106–7, 109–15, 129–30, 154–6, 240; by Murray and Leslie, 74 n., 107; by Clément and Desormes, 138, 147–8, 152–4; by Haycraft, 275
  criticized by Black, 25–6, 31, 105–7, 142; by Lavoisier and Laplace, 10, 31–2, 35–6, 128, 135, 138, 142, 147; by Laplace, 129, 156; by Seguin, 25 n., 32, 38–9; by Heron, 105; by Thomson, 105–6; by Henry, 102 n., 106; by Gay-Lussac, 89 n., 129, 138, 148, 150–4; by Berzelius, 242; by Dulong and Petit, 155–6, 239, 240
  diminishing support for, 32, 106–7
  discredited in 1812 competition, 115, 140–2, 147–8, 150; generally by 1820s, 156
  see also Irvine William (the elder); Irvine, William (the younger); Reaction, heats of chemical
Isère, 228 n., 271
Isomorphism, 248, 283
Ivory, Sir James, 193–4

Jacobins, 80, 231
Jars, Gabriel, 79–80
Joly, John, 205 n.
Joule, James Prescott, 282, 300, 301 n.
  on adiabatic phenomena and energy conservation, 39–40, 253, 306 n.
  on kinetic theory of gases, 120, 277, 305 n., 306–8, 310, 311
  on velocity of sound, 195 n.
  see also Joule–Thomson effect
Joule–Thomson effect, 301

Kane, Sir Robert John, 291
Kantian tradition, 3 n.
Karlsruhe congress, 196
Kelland, Philip, 3 n., 276
Kelvin, Lord: *see* Thomson, William
Kinetic theory of gases:
  in eighteenth century, 4–5, 8–9 n., 302–3
  and adiabatic phenomena, 40, 81, 98–9, 103, 107–8, 119–20
  theories of Davy, 120, 243, 306, 311; of Herapath, 277, 280, 303, 305 n., 307, 311; of Waterston, 303–4; of Joule, 120, 277, 305 n., 306–8, 310, 311
  rejected at Royal Society, 303–4
  modern theory established, 1, 4–5, 99, 302–13
  velocity distribution introduced, 312
  vortex theories, 309–13
  lack of French contributions, 314

INDEX 371

Kirwan, Richard, 22, 27, 34, 37 n.
Knight, Gowin, 11–12 n., 18
Krönig, August Karl, 311

Lagrange, Joseph Louis, 82–5
Lambert, Johann Heinrich:
  on fire, 4, 47–8
  on adiabatic phenomena, 40, 47–8, 50–2
  on heating by friction, 4
  on gas expansion, 61, 66, 146
  on absolute zero, 61, 146
  on velocity of sound, 82
Lamé, Gabriel:
  on nature of heat, 263, 276
  as a positivist, 263, 268–70, 316
  admires Dulong and Petit, 232, 238, 269–70
Landriani, Marsilio, 25 n., 158 n.
Laplace, Pierre Simon, Count (later Marquis), 1, 98 n., 131, 136, 190, 191–2, 215, 249 n., 299, 304, 314
  in Newtonian tradition, 2, 126, 166–7, 196–9
  accepts imponderable fluids, 17, 229, 233–6
  supposed support for vibrational theory, 29–30
  critic of Irvinist doctrines, 10, 31–2, 35–6, 128, 129, 135, 138, 142, 147, 156, 240–1
  his definitive caloric theory, 165–77, 203; ignored, 191–2, 274–5; attitude of Poisson, 175–7
  on determination of specific heats, 34–6, 145; on distinction between c_p and c_v, 158, 159, 161; on γ, 157, 162–5, 169–74
  on adiabatic phenomena, 41, 49, 52, 82, 85, 107, 125–7, 157, 161–5, 168–71, 195
  on velocity of sound, 68–9, 81–2, 85, 157, 161–5, 170–1, 194–5, 253
  on gas expansion, 64, 70–3, 77 n., 160, 168, 173; on gas thermometry, 72–3, 160 n., 168
  opposes atomic theory, 239, 248
  influences Avogadro, 210; criticizes Fourier, 236 n.
  as a patron, 71–2, 82, 197–9, 205
  his influence declines, 229–31, 233–6, 270–5
  see also Laplacian physics

Laplacian physics:
  programme stated, 166–7
  Newtonian basis, 2, 166
  at École Polytechnique, 230–3
  scepticism of Comte, 263
  critics at Académie des Sciences, 270–3
  decline of, 2, 205, 227–9, 233–5, 239–48, 270–80, 317–18
  and decline of French physics, 314, 317–18
Latent heat:
  discovered, 24–6; determined, 25, 36–7
  Black on, 24–6
  discovery aids caloric, 24–5, 28
  explained by Irvinists, 25–8, 36–7; 107; by Thomson, 105; by Henry, 106; by Berthollet, 125
  latent caloric effecting expansion, 31–2, 52, 131, 158, 159, 168–9, 174, 193
  see also Caloric
Lavoisier, Antoine Laurent, 12, 13, 17, 51, 105, 136, 275–6, 303, 314
  on matter of fire, 9–11, 29–30
  on caloric as an element, 279
  on similarity of caloric and light, 10–11, 96
  on two states of caloric, 10, 31, 158, 240–1
  doubts concerning imponderables, 17
  debt to Boerhaave and Franklin, 16
  'Mémoire sur la chaleur', 29–32, 34–5, 64, 128
  critic of Irvinist doctrines, 10, 31–2, 35–6, 128, 135, 138, 142, 147
  on specific heats, 34–6, 125, 145; and distinction between c_p and c_v, 158, 159
  ignorant of adiabatic phenomena, 41, 49–50, 80, 158
  on gas expansion, 64, 158
  on animal heat, 249
  and the new chemistry, 21, 204, 212
Lefèvre-Gineau, Louis, 90 n.
Legendre, Adrien Marie, 167 n.
Legion of Honour, 299
Leiden, University of, 12, 259
Leiden jar, 15
Lemeray, Nicolas and Louis, 13

Length, units of, xv
Le Sage, Georges Louis, 198, 278 n.
Leslie, Peter Dugud, 19
Leslie, Sir John:
  rejects term 'caloric', 6 n.
  on Irvinist doctrines, 74 n., 107
  on absolute zero, 107
  on similarity of heat and light, 97 n.
  on adiabatic phenomena, 47, 108–9, 117, 130, 158
  on velocity of sound, 85 n.
  on heating by percussion, 100 n.
  on specific heats of gases, 108–9; ignores distinction between c_p and c_v, 158
Libri (Carrucci dalla Sommaia), Guglielmo Bruto Icilio Timoleone, Count, 234, 272
Liebig, Baron Justus von, 296
  on nature of heat, 277
  as pioneer of energy conservation, 282
  on atomic theory, 291–2, 319–20
Light:
  corpuscular theory in eighteenth century, 14–15, 197–8; in Laplacian programme, 166 n., 196–204; used by Biot and Arago, 198–202; supported by Biot, 97, 198–202, 234–5; used by Avogadro, 212–19; used by Dulong, 215–16, 250–1; taught at École Polytechnique, 230, 233; attacked 3, 151, 202, 204, 219, 233–4, 252
  wave theory of Knight, 18; of Fresnel, 2, 202, 233–5, 273; supported by Arago and Petit, 202, 204, 233–5; supported by Dulong, 244, 252; supported by Ampère, 233–4, 277; favoured by Fourier, 234–5; opposed by Laplacians, 234–5; and attitudes to caloric, 3, 116–17; difficulties of theory, 264
  Hassenfratz's teaching, 231–2
  conflict of Biot and Arago, 235
  analogy with heat, 96–98, 104 n., 212–13
  emitted from air gun, 89–90; in compression of gas, 90–1, 96–8
  see also Affinity; Diffraction of light; Molecular forces; Refracting force; Refraction; Refractive power
Lind, James, 55
London: see Askesian Society; Chapter Coffee House Society; Royal Society of London; Royal Institution; St. George's Hospital; University College
Lorentz, Hendrik Antoon, 313
Lorentz, Richard, 93–4
Lubbock, Sir John William, 304
Lunar Society of Birmingham, 54–5, 57–8, 60
Lunn, Francis, 276
Luz, Johann Friedrich, 66
Lycées, at Sorèze, 149; Henri IV, Paris, 250
Lyons:
  Académie des Sciences, 89–93, 97–8, 127
  École Centrale, 89
  Faculté des Sciences, 122
  Boussingault and Regnault at, 296

MacCullagh, James, 264
Mach, Ernst, 312
Macquer, Pierre Joseph:
  on nature of heat, 10 n., 22–3, 30
  accepts phlogiston, 21
Magellan, Jean Hyacinthe de:
  rejects vibrational theory, 28
  as an Irvinist, 26, 27 n., 29
  introduces term 'specific heat', 26
  on adiabatic phenomena, 49, 58 n.
Magendie, François, 315 n.
Magnetism, fluid theory of:
  modelled on theory of electricity, 16
  as model for caloric, 15–17, 19
  taught at École Polytechnique, 230 n.
Magnus, Heinrich Gustav, 291, 299 n., 319
Mairan, Jean Jacques Dortous de, 61
Malus, Étienne Louis, 166 n.
Manchester, 306
  Literary and Philosophical Society, 69, 73, 86, 106, 307
Marcet, François:
  on adiabatic phenomena, 143, 154, 275
  on specific heats of gases, 154, 217, 257–60, 275, 284, 289

supports Laplacian caloric theory, 257–8, 260
Marcet, Jane, 205 n.
Mariotte, Edme, 8 n.
Martine, George (the younger), 61, 146, 237 n.
Maxwell, James Clerk, 5, 305 n., 311–13
Mayer, Johann Tobias (the younger), 71, 147
Mayer, Julius Robert von, 39–40, 300, 301 n., 308
Mechanical equivalent of heat, 39–40, 302, 306–7 n.
Meikle, Henry, 193–4
Mersenne, Marin, 81
Metcalfe, Samuel, 278
Meteorology, 49, 55–6, 59, 80, 193, 275
Meyer, Oskar Emil, 312 n.
Michelson, Albert Abraham, 310 n.
Mitscherlich, Eilhard, 248, 283, 285, 286
Mohr, Carl Friedrich, 277, 306 n., 308 n.
Molecular forces:
  acting over short range, 166–7, 200
  analogy with gravitation, 196, 198
  in corpuscular theory of light, 197–202, 203–4, 212–19
  between gas particles, 2, 7–20 *pass.*, 77–8, 125–7, 166, 173–4, 307 n.
Mollet, Joseph:
  as a calorist, 91, 127–9, 274–5
  on similarity of heat and light, 97–8
  on adiabatic phenomena, 80 n., 89–98
  on heating by percussion, 100 n.
  invents fire piston, 93–4
  on specific heats of gases, 135, 154
  on structure of gases, 207
Monge, Gaspard, 178, 230
  as a calorist, 30–1
  on similarity of heat and light, 96–7
  ignorant of adiabatic phenomena, 44 n.
  on gas expansion, 64–5
  contacts in Paris, 29–31
Montpellier, 136
Morgan, William, 34
Mountaineering, 63, 66, 71
Murray, John, 101, 107–8, 158
Muséum d'Histoire Naturelle, Paris, 127

Musschenbroek, Peter van, 7, 8, 14, 43 n., 62 n.

Nairne, Edward, 53 n., 58 n.
Nancy, Académie des Sciences de, 101–2 n.
Napoleon Bonaparte, 128 n., 211 n., 228
Naturphilosophie, 281
Navier, Claude Louis Marie Henri, 267
  on specific heats of gases, 154, 197
  on adiabatic phenomena, 189 n., 192
  on wave theory of light, 264
  attitude to Comte, 266
  clashes with Poisson, 271, 273
  member of Académie des Sciences, 271; of Société Philomathique, 273
Neumann, Franz Ernst, 226, 289, 297
Neutralizing power, 218, 219
Newton, Sir Isaac, 24
  static theory of gases, 6–8, 74, 111, 126, 303
  derives Boyle's law, 6–8
  on vibrational theory of heat, 14, 23 n.
  his ether, 18
  on velocity of sound, 68, 81–2, 84, 161, 168, 175, 194
  on molecular forces, 166
  on refraction of light, 198–9
  see also Newton's law of cooling
Newton's law of cooling, 87, 237–9, 265
Nicholson, William, 54 n., 65 n., 90
Nollet, Jean Antoine, 15, 43 n., 44 n.
Northmore, Thomas, 95
Nuguet, Lazare, 62

Oersted, Hans Christian, 281
Olmi, Vincent Frédéric, 149–50
Ostwald, Wilhelm, 290, 312
Oxygenicity, 212

Paris: *see* Académie des Sciences; Académie Française; Collège de France; Conservatoire des Arts et Métiers; École des Mines; École des Ponts et Chaussées; École Normale; École Polytechnique; Faculté des Sciences; Institute,

Paris (*cont.*)
   First Class of French; Lycées; Muséum d'Histoire Naturelle; Société Philomathique
Pasteur, Louis, 315 n.
Péclet, Eugène, 276
Pelouze, Théophile Jules, 291, 319–20
Percussion, heating by:
   associated with decrease in volume, 48–9, 52–3, 68, 100–1, 103, 241
   analogy with adiabatic phenomena, 48–9, 52–3, 68
   analogy with friction, 100–1
   as evidence against caloric, 119
Petit, Alexis Thérèse:
   on nature of heat, 202, 204, 234
   on Irvinist doctrines, 155–6, 239, 240
   on absolute zero, 151, 239
   attitude to Dalton, 155–6
   on 'Laplacian' caloric theory, 240–1
   supports electrochemical theory, 241–4
   on expansion, thermometry, and cooling, 73, 77, 236–9, 269–70
   on heat engines, 178–9, 182, 230
   on Dulong and Petit's law, 239–40, 250, 282, 285
   on nature of light, 202, 204, 234
   on refraction in gases, 202, 204, 233–4
   Laplacian influence on, 231–3
   critic of Laplacians, 227–9, 239, 273, 317
   on Avogadro's hypothesis, 207, 215, 216–17, 221, 260, 283–4 n.
   friendship with Dulong, 235–7, 248–9
   admired by Lamé, 232, 238, 269–70
   at École Polytechnique, 178, 228, 230–2, 233, 239
   elected to Société Philomathique, 273
   see also Dulong and Petit's law
Pfaff, Christian Heinrich, 211–12
Philo of Byzantium, 60
Phlogiston, 118
   structure, 16–17 n.
   rejected, 21
   in Crawford's theory, 33
Picard, Jean, 85

Pictet, Marc Auguste:
   interests, 40, 48
   his theory of heat, 51, 119 n.
   on adiabatic phenomena, 48, 50–2, 60, 79, 90, 119 n.
   on heating by percussion, 100 n.
Playfair, John, 77 n.
Pneumatic chemistry, 20–1, 22
   stimulates work on gas expansion, 63–6, 70
Pneumatics, 42–4
Poinsot, Louis, 234
Poisson, Siméon Denis, 149 n., 167 n., 215, 299
   on caloric theory of gases, 175–7, 190
   late support for caloric, 191–2, 274
   on adiabatic phenomena, 86, 118, 140, 144, 151, 175–6, 185–6, 189, 193, 275
   on velocity of sound, 86, 175
   on theory of light, 234–5, 264
   as disciple of Laplace, 2, 175–6, 191–2, 234, 273–4
   clashes with Navier, 271, 273
   admires Petit and Dulong, 238
   at Société Philomathique, 273
Poleni, Giovanni, Marquis, 62
Poncelet, Jean Victor, 267
Positivism, 316 n.
   in work of Black, 23–5; of Fourier, 229, 262–3, 264–5, 266, 268–9, 316; of Ampère, 262 n.; of Dulong, 262–70; of Duhamel, 263, 268; of Lamé, 263, 268–70, 316; of Dumas, 264 n., 287, 315–16
   of Comte, 262–70
   and imponderable fluids, 229, 263
   and caloric, 229, 263, 311–12
   and atomic theory, 263–4, 265–6, 290, 312–13
   and kinetic theory of gases, 312–13
Positive Calendar, 265
Potter, Richard, 194
Pouillet, Claude Servais Mathias Marie Roland, 256, 278
Powell, Baden, 304 n.
Prévost, Pierre, 278
Priestley, Joseph:
   on imponderable fluids, 16 n.
   on gas expansion, 38, 64–5
   on adiabatic phenomena, 58

INDEX

on Davy's experiments, 116 n.
as a phlogistonist, 21
on pneumatic chemistry, 20–1, 34
Prieur Duvernois (otherwise Prieur de la Côte-d'Or), 65–6
Prony, Gaspard Clair François Marie Riche, Baron de, 65, 97, 167 n., 249 n., 250 n.
Proust, Joseph Louis, 271
Prout, William, 247, 278 n., 299 n.

Quantum physics, 313

Radiant heat, 121, 258
supports vibrational theory, 3, 116–17, 244, 277–8
Herschel on, 104, 105 n.
Socquet on, 121–3
Delaroche on, 136
Ampère on, 3, 277–8
molecular radiation, 167–9, 173
Rankine, William John Macquorn, 300
first to use 'adiabatic', 39 n.
on velocity of sound, 194
his vortex theory, 309, 311, 312
Rayleigh, Lord (John William Strutt), 304 n., 313
Reaction, heats of chemical, 129
Lavoisier on, 10
Lavoisier and Laplace on, 35–6, 240–1
Irvinist explanation, 26–7, 88, 113, 115, 121, 135, 141–2, 203, 240
in 1812 competition, 134–5, 141–2
in competition set in 1842, 295
electrochemical theory of, 2–3, 241–4
as evidence for caloric, 102, 116
Redtenbacher, Jacob Ferdinand, 312
Reech, Ferdinand, 301–2 n.
Refracting force, 198, 215, 252
Refraction:
astronomical, 64, 70–1, 193, 199
double, 235
Newton on, 198–9
Laplace on, 198–9
Biot and Arago on, 78, 95 n., 166, 197–204, 205, 208, 213, 217
Arago and Petit on, 202, 204, 233–4
Dulong on, 215–16, 219, 250–2
Refractive power, 197–202, 205, 208, 213–14, 215–17, 219, 250–2

Regnault, Henri Victor, 250, 252, 281–2
and Avogadro, 224 n., 225–6
on Boyle's law, 298–9
on Clément and Desormes's 'law', 298–9
on Dulong and Petit's law, 297, 299
on specific heats, 225–6, 297–9
on gas expansion, 298–9
and Carnot cycle, 182, 301–2 n.
and thermodynamics, 300–2
influenced by Dumas, 295, 296–7, 315–16
illustrates decline of French physics, 296, 300, 315–17
and government commission, 298, 299–300
Respiration, 32, 275
see also Animal heat
Revolution, French:
revolutionary calendar, xv
Revolutionary Tribunal, 35
Ritter, Élie, 305 n.
Ritter, Johann Wilhelm, 212
Rive, Auguste de la, 252
on adiabatic phenomena, 143, 154, 275
on specific heats, 154, 217, 257–60, 275, 284, 289
supports 'Laplacian' caloric theory, 257–8, 275
Robert, François and brother, 66
Roberval, Gilles Personne (or Personnier) de, 81
Robison, John, 23–4
Roemer, Ole Christian, 85
Roget, Peter Mark, 303 n.
Rose, Heinrich, 291, 319
Rouelle, Guillaume François, 10 n.
Roy, William, 38, 63, 65
Royal College of Surgeons of Ireland, 258
Royal Institution, London, 95, 111 n., 116, 276 n.
Royal Society of London, 80, 99, 100–1
and adiabatic phenomena, 54–60
and kinetic theory, 303–4
Davy at, 211 n., 242, 290, 303
Bakerian lectures, 211 n., 242, 312
medals, 290, 299

376 INDEX

Rumford, Count (Benjamin Thompson):
  supports vibrational theory, 4, 23, 99–103, 104, 115, 244, 246
  his Irvinist assumptions, 101–3
  on weight of heat, 109
  on calorimetry, 135–6, 139.

Sadler, James, 95 n.
Saigey, Jacques Frédéric, 214–15, 217
St. George's Hospital, London, 32
Saissy, Jean Antoine, 97–8
Santorii, Santorio (Sanctorius), 60
Saussure, Horace Bénédict de, 40, 45, 48, 50–1, 64 n.
Saussure, Nicolas Théodore de, 225 n.
Savary, Félix, 261
Scheele, Carl Wilhelm, 21
Schemnitz, 59–9, 79–80
Seguin, Armand, 25 n., 32, 38–9
Seguin, Marc, 282, 301, 308 n., 316
Sensible heat (or caloric): see Caloric
Shuckburgh, Sir George, 63, 65
Small, William, 57
Société Philomathique, Paris, 79, 172, 179, 272–3
Socquet, Joseph Marie, 122–4
Somerville, Mary, 103 n.
Sorèze, 149
Sound: see Velocity of sound
Southern, John, 86 n.
Specific heat:
  discovered, 21
  'specific heat' first used, 26
  symbols explained, xv
  determined by Wilcke, 22; by Gadolin, 22, 32; by Kirwan, 22, 27, 34; by Crawford, 32–8, 113, 115, 138, 141, 157–9, 205; by Lavoisier and Laplace, 34–6, 145; by Leslie, 108–9; by Gay-Lussac, 129, 131–5; by Mollet, 135; in 1812 competition, 88, 131–55, 160, 203, 240, 258; by Dulong and Petit, 238–40, 255; by Dulong, 160–1, 204–5, 219–21, 253–6, 258, 259–61, 293–5; by Haycraft, 217, 256–7; by de la Rive and Marcet, 154, 217, 257–60, 275, 284, 289; by Suerman, 259, 275; by Apjohn, 258–9; by Regnault, 225–6, 297–9

Dalton on, 108 n., 110–15, 154–6, 157–9, 240
Rumford on, 135–6, 139
Avogadro on, 205–26, 293–4, 297
competitions concerning, 134–50, 292–5
distinction between c_p and c_v, 37–8, 129, 138–9, 145, 157–61, 204–5, 220, 253
determination of γ, 37–8, 83, 87, 144, 148–9, 157, 159–60, 162–5, 169–74, 177, 193–5, 220, 253–4, 294
variation with density and pressure, 88–9, 131–3, 139–40, 148, 153, 162–5, 171–2, 186, 188, 190–1, 192, 193, 257, 259, 275, 298
and atomicity, 222–5, 260–1, 292–5
see also Capacity; Dulong and Petit's law
Steam-engine:
  treated by Petit, 178–9, 182, 230; by Hachette, 182; by Clément and Desormes, 179–82, Plate 8; by Carnot, 177–84; by Gilbert, 182 n.; by Regnault, 182 n., 297–8
  in France, 178–9, 181, 249–50, 297–8, 300
  expansive operation, 178–83
  see also Engineering
Stewart, Balfour, 313
Stokes, Sir George Gabriel, 194, 264
Strutt, John William: see Rayleigh, Lord
Suerman, Alexander Karel Willem, 259, 275
Symmer, Robert, 16 n.

Tait, Peter Guthrie, 99 n., 116, 310 n.
Thenard, Louis Jacques, Baron, 137, 149 n., 179, 236, 273, 283, 296
  knows of fire piston, 94 n.
  on identity of heat and light, 98 n.
  on heating by percussion, 100 n.
  on atomic theory, 247
Thermodynamics: see Carnot cycle; Energy conservation, principle of; Entropy; Mechanical equivalent of heat
Thermometry:
  liquid-in-glass, 62, 72–3, 76
  gas, 62, 66–7, 72–3, 77, 159–60, 168, 193, 237–8, 269–70

INDEX 377

temperature scales, xv, 76–7
Thompson, Benjamin: see Rumford, Count
Thomson, James, 301 n.
Thomson, Sir Joseph John, 309–10
Thomson, Thomas, 65–6 n., 69 n.
  misinterprets Black, 25 n., 105
  on nature of heat, 105, 276
  on heating by friction, 102 n.
  on Dalton, 75, 106
  on Irvinist doctrines, 101, 105–6
  on absolute zero, 75, 105–6
Thomson, Sir William (Lord Kelvin):
  on Davy's experiments, 99 n.
  at Regnault's laboratory, 299, 315 n.
  and thermodynamics, 300, 308
  discovers Joule–Thomson effect, 301
  his vortex theory, 309, 310 n., 312
Tilloch, Alexander, 89, 109
Traill, Thomas Stewart, 21 n., 278
Tredgold, Thomas, 303 n.
Trembley, Jean, 63–4
Trinity College, Dublin, 9, 194
Turin:
  Reale Accademia delle Scienze, 214, 217, 293–4
  University of, 122
Turner, Edward, 276, 279
Tyndall, John, 4, 99 n., 310, 313 n.

Universities: see entries for individual universities
University College, London, 194
Utrecht, University of, 259

Vacuum:
  entry of gas into, 40–60, 86–9, 130–1, 142–8, 152–4
  effect of volume change on, 130, 153
  heat capacity of, 48, 51–3, 88, 130, 142–6, 148, 152, 154
Vandermonde, Alexandre (or Alexis Théophile), 29, 64–5
Van der Waals, Johannes Diderik, 313
Vaporization, 258
  effect on molecules, 284–8
  see also Vapours
Vapours:
  specific heats of, 36–7, 160–1, 177, 204–5

vapour pressures, 76–7, 181, 249–50, 265, 270, 300
vapour densities, 283–9
sound in, 85
see also Clément and Desormes's 'law'; Vaporization
Vaughan, Benjamin, 58 n.
Velocity of sound:
  correction of Newtonian expression, 68–9, 81–6, 157, 161–5, 168–71, 174–5, 193–5, 253
  determined experimentally, 81–2, 84–5, 165, 170–1
Verdet, Émile, 228 n., 310 n., 316
Vibrational theory of heat:
  defined, 3 n.
  supported by Bacon and Boyle, 14; by Newton, 14, 23 n.; by Macquer, 22, 30; by Fourcroy, 22–3, 30; by Rumford, 4, 23, 99–103, 104, 115, 244, 246; by Davy, 99 n., 104, 115–16, 119–20, 243; by Young, 104, 116–18; by Dulong, 99 n., 244–6, 264; by Carnot, 3 n., 189, 277; by Ampère, 3 n., 277–8
  Lavoisier and Laplace on, 29–30
  mentioned at École Polytechnique, 230; in textbooks, 276–7, 309
  shortcomings of, 280
  not furthered by adiabatic phenomena, 40, 81, 98–9, 103, 107, 119–20
  supported by work on radiation, 3, 116–17, 244, 277–8
  accepted, 308–11
Villefosse, A. M. Héron de: see Héron de Villefosse, Antoine Marie
Viscosity, 312
Volta, Alessandro Giuseppe Antonio Anastasio, Count, 61, 64 n., 66–7, 146, 211–12
Vortex theories, 309–13

Walker, Adam, 19
Warltire, John, 54
Water, composition of, 129, 241
Waterston, John James, 280, 303–4, 305 n.
Watson, Richard, 23 n.
Watt, James, 37, 57–8, 190 n., 265
Webster, Thomas, 95 n.
Wedgwood, Josiah, 58

Wedgwood, Thomas, 100–1
Welter, Jean Joseph, 143, 170–1, 174, 193
Whewell, William, 175 n., 234 n., 238, 291
Wilcke, Johan Carl, 16, 22
Winkelblech, Georg Karl, 319
Wöhler, Friedrich, 256, 291, 319

Wolfe (or Wolf), Nathaniel Matthaeus von, 80 n.
Wollaston, William Hyde, 95–6 n., 266, 290–1
Woolf, Arthur, 178

Young, Thomas, 104, 116–18

PRINTED IN GREAT BRITAIN
AT THE UNIVERSITY PRESS, OXFORD
BY VIVIAN RIDLER
PRINTER TO THE UNIVERSITY